P9-DTI-929

CRABGRASS FRONTIER

CRABGRASS FRONTIER

The Suburbanization of the United States

Kenneth T. Jackson

New York Oxford
OXFORD UNIVERSITY PRESS
1985

Oxford University Press

Oxford New York Toronto
Delhi Bombay Calcutta Madras Karachi
Kuala Lumpur Singapore Hong Kong Tokyo
Nairobi Dar es Salaam Cape Town
Melbourne Auckland

and associated companies in
Beirut Berlin Ibadan Mexico City Nicosia

Copyright © 1985 by Oxford University Press, Inc.

Published by Oxford University Press, Inc.,
200 Madison Avenue,
New York, New York 10016

All rights reserved. No part of this publication may be reproduced, stored in a retrieval system, or transmitted, in any form or by any means, electronic, mechanical, photocopying, recording, or otherwise, without the prior permission of Oxford, University Press.

Library of Congress Cataloging in Publication Data
Jackson, Kenneth T.
Crabgrass frontier.
Bibliography: P.
Includes index.
1. Suburbs—United States—History. 2. Suburban life.
3. Housing—United States—History. 4. United States—Social conditions.
I. Title.
HT384.U5J33 1985 307.7'4'0973 85-4844
ISBN 0-19-503610-7

Printing (last digit): 9 8 7 6 5 4 3 2 1

Printed in the United States of America

For Barbara

and

To the bright, enduring memory of our son,
Kenneth Gordon Jackson II
(1968–1984)

O my son Absalom,
my son, my son Absalom!
would God I had died for thee
2 Samuel 18

Acknowledgments

Over the protracted course of this investigation I have incurred more debts than I could possibly acknowledge here. The National Endowment for the Humanities, the John Simon Guggenheim Memorial Foundation, and the American Council of Learned Societies provided senior fellowships and freedom, while the 1982 Banneker Professorship at The George Washington University enabled me to follow up many leads at the Department of Housing and Urban Development and at the National Archives. Columbia University's Council for Research in the Social Sciences and the Department of History's Dunning Fund generously helped with typing expenses. My research trips around the country were made more pleasant and more useful because of the hospitality and lodging provided by Joyce Dalbey in Houston; by Dale and Douglas Curry in New Orleans; by Sarah and Austin Tothacer in Columbia, S.C.; by Patsy and Lewis Lanter in Atlanta; by Phyllis and Roger Lotchin in Chapel Hill; by Darryl and Michael Ebner and by Carolyn and Jan Benjamin in Chicago; by Juanita and Hiram Caroom in Jacksonville; by Mark Haller in Philadelphia; by Estelle and Bob Crenshaw in Dallas; by Wanda and Alan Goldstein in Dayton; by Anne and Thomas Scheckells, by Thomas M. McNair, and by Patrick W. Murphy in Washington; by Harry L. Davis in Denver; by Paula and Max Schouten, by Sally and John Cunningham, and by Mary and John Heilner in Westchester; by Betty and Robert Hume in San Jose; by Gaither and Byron Smith in San Francisco; by Kenneth M. Batinovich and by Sue and Clelland Downs in Los Angeles; and by Sally and Pete Finch in San Diego. Christine and Val Sharp found me on their doorsteps in Lake Forest, Greenwich, and Ladue on more than a dozen occasions, and each time they opened their homes and refrigerators and proved to be the closest possible friends. In

Memphis Carolyn and John Parish have always treated me as the son they never had, while my mother, Elizabeth Willins Jackson, has been so supportive, kind, and thoughtful over the years that my friends often claim her as their own.

My dependence upon the work of other authors will be obvious to anyone with a passing familiarity with the topic. I have tried to acknowledge specific debts in the notes, but the extent of my borrowing will be evident only to the individuals concerned. For numerous criticisms and suggestions, I am grateful to Alan F. J. Artibise, John F. Bauman, Stanley Buder, Clara Cardia, Michael Conzen, Leonard Dinnerstein, Michael Ebner, Roderick French, David R. Goldfield, Susan S. Hallas, David Halle, Glen E. Holt, Timothy Jacobson, Roger Lotchin, Michael McCarthy, Zane L. Miller, Jon A. Peterson, Stanley K. Schultz, Allen Share, Bayrd Still, Joel A. Tarr, Sam Bass Warner, Jr., Margaret Kurth Weinberg, and Olivier Zunz. Deborah S. Gardner, Camilo J. Vergara, and Carol Willis shared their photographs with me, while George Tremberger took time from his backbreaking schedule to draw maps and charts for this book. At American Heritage Publishing Company, Patrick Bunyan located suitable prints in an enormous collection, as did Janet Parks and Jay Hendrickx at Avery Architectural Library. At the Department of Housing and Urban Development, Frederick J. Eggers, Mary A. Grey, William A. Rolfe, and Joan Gilbert (now of Yale University) helped me through the bureaucratic tangle, while Joseph B. Howerton, Jerry N. Hess, and Charles Gellert showed me where and how to roam in the National Archives. Ben H. Graham, Wallace E. Johnson, and Martin Winter generously took time from their construction businesses to reminisce about the home-building industry.

At Columbia University, where I have taught for the past seventeen years, Stuart Bruchey, Ainslie Embree, Herbert J. Gans, John A. Garraty, Henry F. Graff, Herbert S. Klein, Peter Marcuse, Robert A. McCaughey, Eric L. McKitrick, Robert O. Paxton, Rosalind Rosenberg, Elliott Sclar, James P. Shenton, Alden T. Vaughan, and Gwendolyn Wright have been unselfish scholars and delightful colleagues. William E. Leuchtenburg was the first person I met on Morningside Heights. He has since left New York, but only after encouraging and assisting me in every possible professional and personal way. His careful, thorough scholarship continues to serve as my model. Many former students, especially James Baughman, David Bensman, Eugenie Ladner Birch, William N. Black, Kenneth Cobb, Estelle Freedman, Deborah Gardner, Mark Gelfand, Timothy Gilfoyle, Jacquelyn Dowd Hall, David Hammack, Clifton Hood, Betsy W. Kearns, Thomas Kessner, Veronique Marteau, Eleanora Schoenebaum, David Schuyler, Frank Vos, Ray

Weisman, and Carol Willis may recognize their own insights in the pages which follow. My mentor, Richard C. Wade, coined the phrase "Crabgrass Frontier" and first excited me about urban history two decades ago at the University of Chicago. He knows how important his ideas and his example have been to me ever sincc.

At Oxford University Press, Sheldon Meyer has been a patient and close friend and a valued counselor, while Pamela Nicely has saved me from dozens of careless mistakes. Whatever errors remain are my own.

Several parts of *Crabgrass Frontier* have been previously published elsewhere, and I wish to thank Princeton University Press, the Center for Advanced Study in the Behavioral Sciences, the *Journal of Urban History, Chicago History,* the Columbia Historical Society, Sage Publications, Wadsworth Publishing Company, and Doubleday and Company for permission to reprint portions of earlier essays.

The most important person in this effort has been my wife. I owe this book, as well my two children and most of the other good things in my life, to Barbara's devotion, sacrifice, and inspiration. She enabled me to see things I otherwise never would have noticed, and she helped me to survive in the face of heartbreaking tragedy. For better, and now for worse, she has shared everything with me for twenty-three years, and I hope the dedication conveys some sense of my appreciation and love.

Two weeks before this book was completed our sixteen-year-old son was killed in an automobile accident a few miles from our home. Barbara and I had each previously faced the loss of parents and friends, but no experience in life prepared us for the policeman and the priest at our doorstep on an otherwise ordinary Friday night. Kenneth Gordon Jackson II was too busy with his brother, his friends, his drums, and his sports to be much concerned with his father's book, but he taught me to use the computer on which this manuscript was typed, he set up my first word-processing programs, he formatted and organized my discs, and he invariably helped me when I lost my way. More importantly, he took me away from my desk for thousands of ping-pong battles, dozens of ski trips and little league games, and hundreds of hours of hoops, baseball, and frisbee. Although each of us is unique in all the world, Gordon, as his good friend the Reverend Edd Payne noted, was "more uniquer than most." He was gifted in a hundred wonderful ways, and his laugh, his smile, his quick wit, his love of the outdoors, his affinity for animals of every description, his eagerness for new experiences and challenges, his numerous and devoted friends, and his disregard for conformity and convention set him apart. I will remember my son with joy and love for the rest of my life, and he will bc in my last thoughts.

In the aftermath of Gordon's sudden and unexpected death, Barbara,

Kevan and I were surrounded by friends who cleaned our house, prepared our meals, discouraged our guilt, and shared our grief. None of them read this manuscript, and they know little about this book. But they have had everything to do with our ability, however tortured and uneven, to cope with the emptiness and sadness that remain with us every day. We want to thank all of them, and especially the Reverend Jack Silvey Miller, for helping us to stand again and for loving us when we needed it the most.

Contents

CRABGRASS FRONTIER

Introduction

> Even while you read this whole square miles of identical boxes are spreading like gangrene . . . developments conceived in error, nurtured by greed, corroding everything they touch.
>
> —JOHN KEATS, *The Crack in the Picture Window*

Throughout history, the treatment and arrangement of shelter have revealed more about a particular people than have any other products of the creative arts. Housing is an outward expression of the inner human nature; no society can be fully understood apart from the residences of its members. A nineteenth-century melody declares, "There's no place like home," and even though she had Emerald City at her feet, Dorothy could think of no place she would rather be than at home in Kansas. Our homes are our havens from the world.

This book is about American havens. It suggests that the space around us—the physical organization of neighborhoods, roads, yards, houses, and apartments—sets up living patterns that condition our behavior. As Lewis Mumford has noted, "The building of houses constitutes the major architectural work of any civilization." Obviously, the particular type of man-made setting that results is a function of the interrelationship of technology, cultural norms, population pressure, land values, and social relationships, but even within rigid environmental and technological restraints a variety of physical patterns is possible. Work, religion, and family life may all be conducted within a single space or in separate, specialized structures. Whether that pattern emphasizes togetherness, as in Vienna, or separation, as in contemporary America, is a matter of choice. The 250,000 Dogons of West Africa, for example, are a defiant people who live deep in the interior of Mali and who have thus far resisted centuries of change. They live in rugged and chaotic terrain, and their villages are built straight up the sides of vertical cliffs amid the caves where they bury their dead. So fierce is their reputation that Christian missionaries and proselytizing Muslims have written off the Dogons as hopeless pagans. Yet anthropologists note that their house-compound logically and specifically symbolizes the cosmological principles of creation particular to the Dogon.[1]

In the United States, it is almost a truism to observe that the dominant residential pattern is suburban. The 1980 census revealed that more than 40 percent of the national population, or more than 100 million people, lived in the suburbs, a higher proportion than resided either in rural areas or in central cities. The largest communities have been losing out not only relatively but also absolutely. Of the nation's twenty-five largest cities in 1950, eighteen lost population over the three following decades. Meanwhile, suburbia has become the quintessential physical achievement of the United States; it is perhaps more representative of its culture than big cars, tall buildings, or professional football. Suburbia symbolizes the fullest, most unadulterated embodiment of contemporary culture; it is a manifestation of such fundamental characteristics of American society as conspicuous consumption, a reliance upon the private automobile, upward mobility, the separation of the family into nuclear units, the widening division between work and leisure, and a tendency toward racial and economic exclusiveness.

The term "suburb" is of course vague. The word alone is enough to unleash myths. Only a few people have tried to give it concrete expression. Columnist Erma Bombeck noted a few years ago: "Suburbs are small, controlled communities where for the most part everyone has the same living standards, the same weeds, the same number of garbage cans, the same house plans, and the same level in the septic tanks." Russell Baker has added, only partially in jest, that either America is a shopping center or the one shopping center in existence is moving around the country at the speed of light.[2]

The stereotype is real, embodying uniformity, bicycles, station wagons, and patios. It has been sustained because it conforms to the wishes of people on both ends of the political spectrum. For those on the right, it affirms that there is an "American way of life" to which all citizens can aspire. To the left, the myth of suburbia has been a convenient way of attacking a wide variety of national problems, from excessive conformity to ecological destruction.

Scholars reject the stereotype, but they have not reached any consensus. Indeed, a moment of concentrated reflection will show how stubbornly the concept defies definition. As metropolitan areas have sprawled, suburban ways have evolved increasingly as the ways of people everywhere. Suburbia is both a planning type and a state of mind based on imagery and symbolism. Economists assign suburban status on the basis of functional relationships between the core and the surrounding region; demographers on the basis of residential density or commuting patterns; architects on the basis of building type; and sociologists on the basis of behavior or "way of life." The United States Bureau of the Census defines metropolitan areas as agglomerations with a central city of fifty thousand plus nearby areas with a "significant level" of commuting into the city and a specified amount of urban characteristics. Because the Census Bureau is subject to heavy political pressures, the way it defines "suburbs" and "metropolitan areas" serves more to confuse than to enlighten the serious student. Dictionaries skirt this demographic tangle with descriptions of suburbs as "those residential parts belonging to a town or city that lie immediately outside and adjacent to its walls or boundaries."[3]

Confusing definitions result because so many different types of places are so often labeled suburban. Highland Park in Detroit, Beverly Hills in Los Angeles, and the Park Cities in Dallas are legally independent and yet completely surrounded by large cities. River Oaks in Houston, the Country Club district in Kansas City, and Fieldston in the Bronx are exclusive neighborhoods that are suburban in every way except in law. American city boundaries follow no logical pattern. Newark, New Jersey, stuffed into just twenty-three square miles, is so tiny that a walk of a few blocks from the center of downtown can take one not only to another city, but to another county. Jacksonville, Florida, by contrast, has about the same population as Newark, but it sprawls over an area forty times as large, with the result that much of Jacksonville's 850 square miles seems to consist of uncharted swamp and jungle.[4]

Even if we restrict our attention only to those communities which are self-governing, we quickly observe that American suburbs come in every type, shape, and size: rich and poor, industrial and residential, new and old. Camden, New Jersey, across the Delaware River from Philadelphia, and East St. Louis, across the Mississippi River from the Gateway to the West, are so depressed that their abandonment problems are among the most serious in the nation, while Hillsborough outside San Francisco, Winnetka outside Chicago, and Saddle River outside New York City boast of average home prices approaching half a million dollars. Dozens of peripheral communities are heavily industrial, while others have such rigid zoning restrictions that all apartments are excluded from their quiet precincts. Some suburbs, like Cambridge and New Rochelle, are older than Philadelphia; in contrast, planned communities like Irvine, Reston, and Columbia, are almost new. Ethnic suburbs remain common. Chicago's environs alone include Polish Cicero, Jewish Skokie, black Robbins, and Waspish Lake Forest, as well as places like Oak Park and Evanston that take pride in their heterogeneity.

Despite such extraordinary diversity, one may nevertheless generalize about the American residential experience. Many readers will deny this assumption on regional grounds, arguing that the differences between expanding and prosperous Sunbelt cities like San Diego and Dallas and depressed manufacturing cities like Buffalo and Cleveland are so basic that they constitute different phenomena. American metropolises do vary greatly, largely because of the vastness of a continental nation, the type of transportation that was dominant at the time of their greatest growth, and the structure of the local economy. But similarities among American residential patterns are much more numerous than are differences, especially when age, distance from the city, and socioeconomic class are held constant. To the extent that regional distinctiveness is noticeable, it often runs counter to popular perception. Thus, the typical residential lot in the desert surrounding Los Angeles or Phoenix, where population is sparse and the wide-open spaces prevail, is much smaller than a comparably priced parcel in the suburbs of New York or Boston.

The essential similarities in American suburbanization become clear when we shift to an international perspective. The United States has thus far been unique in four important respects that can be summed up in the following sentence: affluent and middle-class Americans live in suburban areas that are far from their work places, in homes that they own, and in the center of yards that by urban standards elsewhere are enormous. This uniqueness thus involves population density, home-ownership, residential status, and journey-to-work.[5]

The first distinguishing element of metropolitan areas in this nation is their low residential density and the absence of sharp divisions between town and country. In all cultures, the price of land falls with greater and greater distance from city centers. Thus, the amount of space devoted to a single dwelling will always logically be greater on the periphery than at the center. In international terms, however, the structure of American settlement is loose, the decline in density (the density curve) is gradual, and land-use planning is weak. The availability of space and real estate plays a key role in this society's vision of itself. With the broad streets and expansive lawns of American communities, the population is scattered even in the largest conurbations at average densities of fewer than ten persons per acre. As early as 1930, the New York Metropolitan Region, which included the most crowded precincts in the world, spread its 10.9 million residents over 2,514 square miles at the rate of fewer than seven per acre. Other cities have been even more dispersed. Between 1950 and 1970, for example, the urbanized area of Washington grew from 181 to 523 square miles. Such sprawl results from the privatization of American life and of the tendency to live in fully detached homes. Of the 86.4 million dwelling units in the United States in 1980, about two-thirds, or 57.3 million, consisted of a single family living in a single dwelling surrounded by an ornamental yard.[6]

More crowded urban conditions, sharply differentiated from the countryside, are more frequently found in other nations. The outer boundaries of Copenhagen, Moscow, Cologne, and Vienna abruptly terminate with apartment buildings, and a twenty-minute train ride will take one well into the countryside. Similarly, open fields surround the narrow streets and crowded houses of Siena and Florence. Metropolitan Tokyo has swallowed up tens of thousands of tiny farms since World War II, but private building plots rarely exceed one-twentieth of an acre. Unlike Western cities, Shanghai legally includes thousands of acres of productive farmland, but its population is concentrated at the center, where the average density reaches almost one hundred thousand people per square kilometer.

The example of Sweden, which has a standard of living comparable to that of the United States, is particularly revealing. Since 1950 new towns have sprouted around Stockhom, but the highrise, high-density, low-amenity Swedish suburbs, such as Vällingby, nine miles west of the city center, and Farsta, six miles to the south, with their immigrant concentrations and strong dependence upon public transportation, are the physical antithesis of the low-density, automobile-dependent suburbs in the United States.[7]

The second distinguishing residential feature of American culture is a strong penchant for homeownership. This characteristic can best be expressed statistically. About two-thirds of Americans own their dwellings, a proportion which rises to three-fourths of AFL/CIO union members, to 85 percent of all two-person households headed by a 45–64-year-old, and to 95 percent of intact white families in small cities. Overall, the American rate about doubles that of Germany, Switzerland, France, Great Britain, and Norway, and is also many times higher than that of such Socialist nations as Bulgaria and the Soviet Union, where private ownership is technically illegal. Sweden again serves as an instructive example, for in that wealthy nation only about a third of families own either a mortgaged or a debt-free home. This proportion has remained fairly stable since 1945, a period of unprecedented prosperity. Only New Zealand, Australia, and Canada, all with strong frontier traditions, small populations, and a British-induced cultural dislike of cities, share the American experience.[8]

The third and most important distinguishing characteristic of our housing pattern is the socioeconomic distinction between the center and the periphery. In the United States, status and income correlate with suburbs, the area that provides the bedrooms for an overwhelming proportion of those with college educations, of those engaged in professional pursuits, and of those in the upper-income brackets. Despite hopes and claims of a great revival in American cities in recent years, the 1980 census revealed a widening disparity between residents of central cities in metropolitan areas and those of their surrounding suburbs, not only in income but also in employment, housing, living arrangements, and family structure. In 1970, for example, the median household income of the cities was 80 percent of that in the suburbs. By 1980 it had fallen to 74 percent, and by 1983, to 72 percent. Even Boston, widely proclaimed for its downtown renewal and resurgence of middle-class housing, suffered a relative loss of median household income. Because low-income areas, public-housing projects, and minority groups live so close to city centers, economist Richard F. Muth calculated that the median income in American cities tends to rise at about 8 percent per mile as one moves away from the business district, and that it doubles in ten miles.[9]

The situation in other nations provides a striking contrast. In 1976, for example, the wreck of a crowded commuter train near Johannesburg killed thirty-one persons, all but two of whom were black. The racial proportions of that tragedy reflected the fact that in the Union of South Africa the oppressed black population has a long, rush hour journey-to-work, while the inner city is reserved for the gracious homes of the privileged white minority. The South African government forbids the building of houses on the immediate outskirts of major cities. Twenty-five miles north of Pretoria, however, a sea of shanties, made of scraps and wrappings sometimes consolidated with metal bars, covers the land. Officially unrecognized, such settlements in Winterveld accommodate between 500,000 and 700,000 people. Similarly, a few miles from the elegance and comfort of Cape Town, 20,000 black migrants have built illicit shanties from wooden planks, fiberboard, and plastic sheeting in a suburban slum called Crossroads.[10]

Developing countries exhibit similar characteristics. In Cairo the Europeanized and affluent Garden City section lies along the Nile River almost at the center of the metropolis; the major slums are on the southern and northeastern fringes. In Turkey if a person erects four walls and a roof on vacant land overnight he traditionally becomes its owner. As a result, gerry-built night houses, or *gecekondu,* have emerged on the edges of Ankara and Istanbul. In Calcutta and Bombay the only areas with a passable water supply are at the center, where the wealthy live.

The depths of squalor can be found in the thousands of legally defined slum districts, known as *bustees,* which are located throughout the metropolises but mostly around the rim of the cities. In Communist Bulgaria high-level bureaucrats congregate in the middle of Sofia, while the government vainly promises "to eliminate the differences in the way of life between the centre and the periphery and even to make the latter more attractive."[11]

Western European neighborhoods have not lost their cachet just because modern residental subdivisions have been developed. The highest socioeconomic sections of Rome, Barcelona, and Vienna are near the business districts; suburban areas are usually lower-income in character. In Amsterdam affluence characterizes the old center, where rows of restored seventeenth-century town houses line the placid concentric canals. The core has preserved its historical aura and vitality, but the working class has increasingly been forced outward to the suburbs. In Paris, class distinctions tend to be set geographically east or west of the Seine River. The western suburbs—Boulogne, Beuilly, Saint-Cloud, Meridon, Sevres, and Chaville—became fashionable in the nineteenth century and remain solidly middle-class in 1984, but the increasingly Portuguese and Algerian inner suburbs of the north and east, from Saint Denis through Aubervilliers and les Lilas south almost to the Bois de Vincennes, reveal visible urban decay and have long been known as "red suburbs" because of their tendency to vote Communist. At the center of Paris, it remains a privilege—enjoyed by the Baron de Rede, the late Helena Rubinstein, and the late President Georges Pompidou, among others—to live on the Ile de Saint-Louis in the river itself or in neighboring Faubourg Saint-Germain. Blue-collar sections nearby, like Le Marais between the Bastille and the Seine, are organizing defense committees against wealthy gentrifiers.[12]

South American cities also differ radically from their neighbors to the north. In Brazil the exclusion of slum dwellers from the urban cores is so deeply rooted in the culture that the Portuguese word to describe them is *marginais,* and the word used to describe their arrival is *invasaõ*. Pastel-colored squatter settlements—called *favelas* after the name of a flowering tree that grows in profusion on the hillsides—surround Saõ Paulo and Recife. In Rio de Janeiro, the coaches full of tourists heading for the Sepetiba Gulf necessarily pass the shacks at Rocinha, one of three hundred *favelas* scattered around the city. No one knows for sure how many inhabit this shantytown of the Cariocas, but census takers estimate from aerial photographs the population to be 100,000 to 120,000. Similarly, in Buenos Aires, Santiago, Mexico City, and Lima, the most degrading poverty exists on the outskirts, where flush toilets, sewers, running water, and fire and police protection are virtually unknown. In Caracas, the richest city on the continent, the magnificent mountains surrounding the metropolis are dotted with hovels, while the elegant Country Club district is lower in the valley and closer to town.[13]

The fourth and final distinguishing characteristic of the United States residential experience is the length of the average journey-to-work, whether measured in miles or in minutes. According to the 1980 census, the typical American worker traveled 9.2 miles and expended twenty-two minutes each way in reaching his place of employment at an annual cost of more than $1,270 per employee. In larger metropolitan areas the figures were much higher. Precise statistics are unavailable for Europe, Asia, and South America, but one need only think of the widespread practice of going home for lunch, often for a siesta as well, to realize that an easier connection between work and residence is more valued and achieved in other cultures.

This book attempts to account for the divergence of the American experience from that of the rest of the world. How and why did Americans change their assumptions about the good life in the industrial and post-industrial age? Why did the metropolitan areas of the United States decentralize so quickly? What technological and ideological forces created the peculiar shape of the modern metropolis? Have the spatial patterns of American cities—with all they imply about aspirations and ideals—resulted from or caused a set of social values and political policies favoring suburbanization? To what extent has deconcentration involved sacrificing urban facilities in return for maximizing private space? This book then investigates the dynamics of urban land use, the process of city growth through the past, and the ways in which Americans coming together in metropolitan areas have arranged their activities.[14]

This inquiry attempts a broad interpretation and synthesis of American suburbanization. I make no claim to comprehensiveness. Any account that covers all important suburbs is certain to be exhausting before it is exhaustive. But I have sought to integrate intellectual, architectural, urban, and transportational history with public policy analysis, and I have tried to place the American experience within the context of international developments. Because *Crabgrass Frontier* covers a broad geographical and chronological spectrum, it is an essay rather than a monograph and does not attempt precise measurement within a tight conceptual scheme. A good essay, as Oscar Handlin has reminded us, is a product of experience joined to scholarly thought; ideally, it draws together information and illuminates its meaning. Lytton Strachey once complained of the academic tendency to "row out over that great ocean of material, and lower down into it, here and there, a little bucket, which will bring up to the light of day some characteristic specimen from those far depths, to be examined with a careful curiosity.'' My weakness for characteristic specimens is well known, but I have attempted to provide a distillate of ideas about general patterns rather than a series of pedestrian facts about local peculiarities.[15]

Like most broad studies, this one risks overgeneralization. It is difficult to make solid statements differentiating the values and lifestyles of various groups based largely upon residence. The working definition of suburbs in this book has four components: function (non-farm residential), class (middle and upper status), separation (a daily journey-to-work), and density (low relative to older sections). I am aware that a concentration upon low-density, non-farm parcels at the periphery of the built-up portion is arbitrary and imprecise. Low density, for example, means one thing for the nineteenth century, when urban densities normally ranged between 50,000 and 100,000 per square mile and newer areas often had 30,000, and another for the late twentieth century, when many inner cities have been developed at fewer than 15,000 per square mile and many suburban areas often count fewer than 1,000 people in the same physical space. Similarly, some countries have productive agricultural lands which feature higher population densities than the public and unproductive suburbs of the United States.

Several themes recur in this analysis. These include the importance of land developers, of cheap lots, of inexpensive construction methods, of improved transportation technology, of abundant energy, of government subsidies, and of racial stress. Pervasive throughout is the notion that Americans have long preferred a detached dwelling to a row house, rural life to city life, and owning to renting. Following the principle of stratified diffusion, my focus is on the middle and upper classes. Social change usually begins at the top of society. In the United States, affluent families had the flexibility and the financial resources to move to the urban edges first. This fashion for the rich and powerful later became popular with ordinary citizens. Finally, I would argue that history has a fundamental relevance to contemporary public policy, and I would hope that this book indicates that suburbanization has been as much a governmental as a natural process.

1

Suburbs as Slums

> Our property seems to me the most beautiful in the world. It is so close to Babylon that we enjoy all the advantages of the city, and yet when we come home we are away from all the noise and dust.

Written in cuneiform on a clay tablet, this letter to the King of Persia in 539 B.C. represents the first extant expression of the suburban ideal. The desire to combine the best of both farm and city is even older than the letter, however. Today Ur is a desert scrubland with miserable ruins jutting from terrain of sand and mud. It is about 120 miles northwest of the Persian Gulf, in the country we now call Iraq. Four thousand years ago, however, the Sumerian community of Ur in southern Mesopotamia was a place of beauty, graced with towers, palaces, temples, shrines, gardens, and monuments. Between 2,300 B.C. and 2,180 B.C., it experienced great prosperity, and its population of 100,000 spilled beyond the city's gates into what, Sir Leonard Woolley has written, "might be called a suburb of Ur." Similarly, paintings and funerary models of Egyptian cities reveal suburban villas with spacious gardens.[1]

The term *suburb* (or *burgus, suburbium,* or *suburbis*) is of more recent vintage. John Wycliffe used the word *suburbis* in 1380, and Geoffrey Chaucer introduced the term in a dialogue in *The Canterbury Tales* a few years later. By 1500 Fleet Street and the extramural parishes were designated as London suburbs, and by the seventeenth century the adjective *suburban* was being used in England to mean both the place and the resident. According to John Hall's *London: Metropolis and Region,* as early as 1574 the city was beginning to expand beyond its Roman walls to the west along the Thames. Frank S. Smallwood, in *Greater London: The Politics of Metropolitan Reform,* ascribes much of London's early suburbanization to two catastrophes in 1665 and 1666, the plague and the fire. He writes that the first of these was the beginning of a flight to the suburbs that has continued, at an accelerating rate, into modern times."

European suburbs, called *le faubourg* by the French and *die Vorstadt* by the Germans, have a similar lineage. In his remarkable study of medieval Toulouse, John Mundy reports that the twelfth-century use of *burgus* or *suburbium* referred to the housing clusters beyond the Saracen Wall toward the monastery. Outside Paris, the urbane old suburb of Saint-Germain-en-Laye (not to be confused with Saint Germain des Pres on the Left Bank) grew up on the rich flatlands thirteen miles west of the capital. Built around the chateau where Louis XIV, the Sun King, was born in 1638, this tangle of winding streets dotted with 300-year-old stone houses dates back to the fourteenth century. At about the same time, other rich Parisians sought purer air near the greenery of the Parc Monceau and the Bois de Boulogne. Country life offered a welcome solitude, and across Europe the privileged classes periodically vacationed in agricultural areas. And because cities were densely settled, even poorer citizens could walk to rural surroundings less than a mile distant from even the largest cities.[2]

On the North American continent, Boston, Philadelphia, and New York established suburbs well before the Revolutionary War. In the Massachussetts capital, John Staniford advertised a new real-estate development at Barton's Point in 1719 as "laid out in House Lotts with two Streets Cross, that have a very fine prospect upon the River and Charlestown and a great part of Boston." In Philadelphia the first "suburb" was located near Society Hill. There the Shippen brothers owned large tracts of land, and, conscious of its accessibility to the growing city, began to sell off house lots in 1739. A second Philadelphia "suburb" opened in the Northern Liberties in 1741, when Ralph Assheton disposed of his 80-acre estate in small building sites. In 1775 New York suburban residences stretched northward along Greenwich and Bloomingale Roads, while Greenwich Village itself was set apart by two miles of marsh land from the crowded, unhealthy town below Wall Street.[3] In New Orleans, peripheral neighborhoods followed the Parisian tradition and were known variously as Faubourg Ste Marie, Faubourg Marigny, and Faubourg Solet.[4]

Thus, the suburb as a residential place, as the site of scattered dwellings and businesses outside city walls, is as old as civilization and an important part of the ancient, medieval, and early modern urban traditions. However, suburbanization as a process involving the systematic growth of fringe areas at a pace more rapid than that of core cities, as a lifestyle involving a daily commute to jobs in the center, occurred first in the United States and Great Britain, where it can be dated from about 1815. In the next half-century New York, Philadelphia, and Boston exhibited the most extensive changes on their residential peripheries yet witnessed in the world. In London, where the total population was larger but where transportation improvements were less quickly adopted and where the detached house was less readily available, the pace of change was almost as rapid.[5]

The revolutionary nature of the suburbanization process can best be demonstrated by reviewing the five spatial characteristics shared by every major city in the world in 1815. Because the easiest, cheapest, and most common method of getting about was by foot, it is appropriate to refer to such preindustrial agglomerations as "walking cities."

The first important characteristic of the walking city was congestion. When Queen Victoria was born in 1819, London had about 800,000 residents and was the largest city on earth.[6] Yet an individual could easily walk the three miles from Paddington, Kensington, Hammersmith, and Fulham, then on the very edges of the city, to the center in only two hours. In Liverpool, Birmingham, Manchester, and Glasgow, the area of new building was not even two miles from city hall. On the European continent, where medieval fortifications typically posed barriers to outward expansion, as in Vienna, Berlin, Verdun, and Amsterdam, the compact nature of life was equally pronounced. Gross densities normally exceeded 75,000 per square mile and were rarely less than 35,000 per square mile, which is about the level of crowding of New York City in the 1980s.

Although North American cities were newer and smaller than their European counterparts, they exhibited the same degree of intense, inner-city congestion. Lot sizes were small (usually less than twenty feet wide), streets were narrow, and houses were close to the curb. Tiny Elfreth's Alley in Philadelphia, virtually a replica of Restoration London with its little, brick-fronted row houses, survives today as an example of the tight spatial arrangement typical two centuries ago. Meanwhile, large areas only a few miles distant from the Delaware River waterfront were almost completely rural.[7]

The second important characteristic of the walking city was the clear distinction between city and country. In part, this was a legacy from earlier centuries, when the walls of a community were inviolable, indeed almost sacred. As Mircea Eliade has noted, ramparts not only protected against enemies but formed a spiritual boundary of equal significance, for such barriers preserved those within from the evil outside. In fact, it was not until the middle of the nineteenth century, long after the introduction of rifled artillery had rendered municipal fortifications useless, that capitals like Vienna knocked down their thick walls to create grand circular avenues or *Ringstrasses* as the appropriate setting for major public buildings.

Cities in the New World, with the exceptions of Quebec City and Montreal, were rarely surrounded by the massive walls that were typical of European settlements before the Napoleonic Era. Wall Street on Manhattan was named for the temporary and insignificant barricade against the Indians that could not have withstood any determined assault and that was fortunately never tested. Nevertheless, there was no blurring of urban-rural boundaries, and there were no signs announcing the entrance of a traveler into a community. Before the age of industrial capitalism a sharp-edged dot on the map was an accurate symbol for a city. It stood for a site of political and economic power inhabited by a small, specialized part of the total population of any region. There was an obvious visual distinction between the closely built-up residential precincts of a city and rural sections surrounding it, and there were no fast food restaurants, motels, and service stations stretching far along the radial highways.[8]

The third important characteristic of the walking city was its mixture of functions. Except for the waterfront warehousing and red-light activities there were no neighborhoods exclusively given over to commercial, office, or residential functions. Factories were almost non-existent, and production took place in the small shops of artisans. There were no special government or entertainment districts. Public buildings, hotels, churches, warehouses, shops, and homes were interspersed, or often located in the same structure.

The fourth important characteristic of the walking city was the short distance its inhabitants lived from work. In 1815, even in the largest cities, only about one person in fifty traveled as much as one mile to his place of employment. Because the business day was long, and because any distance had to be overcome by horse or foot, there was a significant advantage in living within easy walking distance of the city's stores and businesses. Work and living spaces were often completely integrated, with members of the family, as well as apprentices, literally living above or behind the place of employment.

The final important characteristic of the walking city was the tendency of the most fashionable and respectable addresses to be located close to the center of town. In Europe this affinity for the city's core represented the continuation of a tradition that dated back thousands of years. To be a resident of a big town was to enjoy the best of life, to have a place in man's true home. To live outside the walls, away from palaces and cathedrals, was to live in inferior surroundings. In eighteenth-century Paris, the suburbs were populated largely by persons who were prevented—by taxes collected at the gates or by guild restrictions—from settling in the city proper, or by outcasts of one sort or another who sought to avoid the officialdom of the capital. The very adjectives used to describe them—*faubourien* in French and *vorstädtisch* in German—connoted a working-class environment. In London the sixteenth- and seventeenth-century suburbs were a perennial menace to the maintenance of law and order. In 1580 legislation was adopted to prevent the growth of such districts, but the law was insufficient to stop the proliferation of "base tenements" and disorderly houses on the city's fringe; most likely because it was there that such objectionable businesses as soap making, tanning, and oil boiling were concentrated.[10] As it was impossible to remove all odors from the city, especially because sewers were nonexistent and inhabitants used open peat and wood fires for cooking and heating, the lords tried to locate toward the windward while those who served them were relegated toward the lee.[11]

Although American cities were not rigidly segregated by class before 1815, and although the poor often lived in alleys hard by the more opulent dwellings of the wealthy, there were clear indications that the suburbs were in every way inferior to the core of the city.[12] The word itself had strong perjorative connotations. Ralph Waldo Emerson referred to "suburbs and outskirts of things," while Nathaniel Parker Willis lamented that in comparison with England America had "sunk from the stranger to the suburban or provincial."[13]

William Penn's "Greene Countrie Towne," the largest eighteenth-century American city, serves as an example of this tendency. Norman J. Johnston has plotted the residences of Philadelphia church members of varying status in 1811 and has presented evidence of a clear ranking of residential areas. The wealthy sought dwellings in the heart of the city, not on the edges. Meanwhile, the city's first suburb, Southwark, was populated mainly by artisans—carpenters, shoemakers, tailors—or by persons whose lives were in one way or another connected with the sea (TABLE 1-1).[14] Although not yet the slum it would later become, Southwark counted very few men of wealth or position among its residents. With fewer than four times as many citizens as Southwark in 1790, Philadelphia counted twelve times as many physicians, thirteen times as many merchants and dealers, and twelve times as many lawyers as the suburban community.[15]

Southwark had little status in part because of the long tradition of forcing unwanted business such as slaughterhouses, leather dressers, and houses of prostitution out to the periphery.[16] On May 3, 1799, the *Aurora* thus described the Philadelphia suburbs:

> Persons who are disposed to visit the environs of this city, and more particularly on a warm day after a rain, are saluted with a great variety of fetid and disgusting smells, which are exhaled from the dead carcases of animals, from stagnant waters, and from every species of filth that can be collected from the city, thrown in heaps as if designedly to promote the purposes of death. . . .[17]

TABLE 1-1
Occupational Distribution, District of Southwark
Philadelphia County, Pennsylvania, 1790[a]

White Collar 17%		*Blue Collar* 45%			
Sea Captains	37	Laborers	128	Bees housekeepers	4
Merchants	26	Ship carpenters	56	Cabinet makers	4
Innkeepers	22	Mariners	45	Plaisterers	4
Grocers	20	Shoemakers	39	Painters	4
Shopkeepers	18	House carpenters	32	Porters	4
Schoolteachers	15	Tailors	30	Ship joiners	4
Pilots	14	Blacksmiths	29	Carters	3
Lodgehousekeepers	11	Coopers	26	Caulkers	3
Gentlemen	10	Weavers	17	Mantua makers	3
Gentlewomen	7	Bakers	15	Brewers	3
Clerks	5	Rope makers	15	Wheelwrights	3
Doctors	4	Mates	12	Silversmiths	3
Justices of Peace	4	Joiner and cabinet makers	11	Sailmakers	3
Ministers	4			Sailors	2
Tobacconists	3	Bricklayers	7	Potters	2
Attorneys	2	Ship caulkers	7	Tinmen	2
Constables	2	Butchers	6	Printers	2
Auctioneer	1	Mast makers	5	Barbers	2
Broker	1	Seamstresses	5	Shallop men	2
Beerhouse keeper	1	Boat builders	4	Miscellaneous	25
Customs Officer	1	*Non-Classifiable* 38%		Total	571
Inspector	1				
Nurse	1	Occupation not specified	207		
Sheriff	1	Spinsters and widow	34		
Supervisor	1	Free Blacks	221		
Surgeon barber	1	Slaves	21		
Total	213	Total	483		

[a]The 1267 heads of families seem reasonable if compared with the total population of 5661 for the district in 1790. If anything, my percentages are weighted toward the white-collar side as all questionable occupations were listed under that category.

SOURCE: United States Bureau of the Census, *Heads of Families at the First Census of the United States Taken in the Year 1790. Pennsylvania* (Washington: Government Printing Office, 1908), 208–14.

In 1849, almost exactly fifty years later, this sentiment was echoed by George G. Foster, who noted, "Nine-tenths of those whose rascalities have made Philadelphia so unjustly notorious live in the dens and shanties of the suburbs." He went on to label a suburban prostitution center as "the core of the rottenest and most villainous neighborhood ever peopled by human beings."[18]

This same pattern of decreasing desirability of residence corresponding with increasing distance from the center of the walking city appears in New York. Before the American Revolution, the wealthiest residents of Manhattan lived on the waterfront lanes—especially Dock Street—at the southeastern tip of the island, where they could enjoy proximity to business and the beauty of the upper bay. During the first half-century after independence from England, the most fashionable addresses shifted to the west of Broadway, especially in the vicinity of Columbia College, along Chambers, Warren, and Murray streets. As before, however, the merchant princes lived within walking distance of their emporiums, while the city's poorer denizens had to find habitations along the uptown streets farther north.[19]

Early in the nineteenth century, when New York became the largest city in the western hemisphere and when population pressure was forcing the built-up area northward on Manhattan Island, it was the urban out-casts who initially led the way. On the site of Central Park, which was near the edge of settlement in 1857, Frederick Law Olmsted and Calvert Vaux had to order the eviction of hundreds of ragpickers, junkmen, and drivers who had established squatter settlements there. Similarly, the Yorkville neighborhood on the Upper East Side was labeled at the time as "a district of swamps and thickets and stagnant pools."[20]

The residential options open to black Americans confirm this model of the walking city. Urban slaves were initially required to live in close physical proximity to their owners. This meant that they typically resided in the downtown sections behind the big houses of the white elite. There, behind high walls along the alleys, the slaves were forcibly integrated with the white population.[21]

The growth of the "living out" system, which meant that slaves had some choice of residence, resulted in the movement of blacks to the edges of town. They sought spots as far removed from their masters as possible, which meant retreat beyond municipal boundaries. Thus, the first Americans to flee to the suburbs for racial reasons were black, not white. In Savannah, for example, where the best addresses were the eight or ten squares directly away from the wharf, blacks lived "in the remotest sections of the city, at the extremeties of the Fort, Mamacraw, or Springhill." In New Orleans, the black districts were called "suburb sheds."[22]

The primary exemption to the model of suburbs as slums was the decision of a few very wealthy families connected with urban centers to build homes in the country. Most used such residences only during the summer; a few retreated on weekends as well. This pattern followed not only the English manner but also the tradition of *villeggiatura,* the withdrawal to a country estate which was a central feature of Italian life in the fifteenth and sixteenth centuries after a leisure class had developed in Venice, Rome, and Florence. The land for a circle of three miles around Florence was occupied by opulent manor houses, while Venetian dignitaries of both church and state built similar villas on the Brenta. Hours away from urban centers, such mansions accommodated hunting parties and other forms of lavish entertainment as well as the quiet pleasures of gardening and the contemplative life.[23]

Such country seats on rivers, on lakes, or in other beauty spots within a few hours ride of the city had become de rigueur among the fashionably rich in America even before the Revolution. By 1760 "the country encircling Boston from Danvers to Medford to Middleborough was dotted with gentlemen's seats laid out and built to display the elegance of a rising aristocracy."[24] In the New York area alone, Samuel Ruggles established a weekend home at West Point, Dr. Valentine Mott built a summer residence at what is now 93rd Street and Broadway, Henry Herbert retreated to "Cedars" just out of Newark, Nathaniel Parker Willis put up romantic "Idlewild" on the banks of the Hudson, Washington Irving moved to "Sunnyside," and John Jay built a homestead for his retirement in Katonah.[25] Far up the East River, at the eastern end of present 89th Street, Archibald Gracie, director of the Bank of New York, looked from the windows of his country house across the waters of Hell Gate and over the rolling country of the Bronx,[26] appraising the weather before starting for his office, as mayors 150 years later would do from the same windows. Commuting for Gracie and others, however, was usually not done on a daily basis and remained the exception rather than the rule.[27]

Suburbs, then, were socially and economically inferior to cities when wind, muscle, and water were the prime movers of civilization. This basic cultural and spatial arrangement was essentially the same around the world, and metropolises as different as Edo (Tokyo), London, Melbourne, New York, and Paris were remarkably alike. Even the word suburb suggested inferior manners, narrowness of view, and physical squalor.

2

The Transportation Revolution and the Erosion of the Walking City

Between 1815 and 1875, America's largest cities underwent a dramatic spatial change. The introduction of the steam ferry, the omnibus, the commuter railroad, the horsecar, the elevated railroad, and the cable car gave additional impetus to an exodus that would turn cities "inside out" and inaugurate a new pattern of suburban affluence and center despair. The result was hailed as the inevitable outcome of the desirable segregation of commercial from residential areas and of the disadvantaged from the more comfortable. Frederick Law Olmsted wrote that the city, no less than the private home, had to be divided into various segments that could perform specialized functions: "If a house to be used for many different purposes must have many rooms and passages of various dimensions variously lighted and furnished, not less must such a metropolis be specifically adapted at different points to different needs." Olmsted was writing soon after the Civil War when urban concentrations had generated enough people with enough wealth to provide the market demand for large numbers of private houses near major urban centers.[1]

The Appendix provides a quantitative analysis of the beginnings of this change. Essentially, it demonstrates that enormous growth to metropolitan size was accompanied by rapid population growth on the periphery, by a leveling of the density curve, by an absolute loss of population at the center, and by an increase in the average journey to work, as well as by a rise in the socioeconomic status of suburban residents. This shift was not sudden, but it was no less profound for its gradual character. Indeed, the phenomenon was one of the most important in the history of society, for it represented the most fundamental realignment of urban structure in the 4,500-year past of cities on this planet.[2]

In the most populous metropolitan areas, this shift in residential status between periphery and core began before the Civil War. In Boston, Harrison Gray Otis, a member of an upper-class family, built a considerable fortune by shaping the expansion of the walking city. In 1795 he formed the Mount Vernon Proprietors and several years later purchased eighteen vacant acres on Beacon Street for $18,500. Privy to the decision to locate the new Massachusetts State House on adjacent property, Otis promptly developed his land as housing for Boston's elite. As Walter M. Whitehill has written, "This venture . . . was the largest land transaction ever to have been undertaken in Boston at that time, for it involved a sudden change in the character of an entire region."[3]

Outside of Boston, the most dramatic changes took place in the unannexed villages of Cambridge and Somerville. According to Henry Binford they developed three important suburban characteristics between 1800 and 1850: a set of clear municipal priorities, a preference for residential over commercial expansion, and a stubbornness to remain politically independent from Boston. As the metropolitan population swelled in the 1830s and 1840s, a handful of young, wealthy residents of old Cambridge and Cambridge Port pioneered local commuting. Called "transitional commuters" by Binford, these early suburban residents differed from later commuters because they did not use public transportation and because their original residences were located on the periphery, not in Boston.

In New York City, where extraordinary growth pushed the population of the city and adjacent Brooklyn well past the million mark by 1860, the "genteel" population moved northward from City Hall, especially to the high ground in the middle of the island a few blocks on either side of Fifth Avenue. Building substantial homes along King and Charlton streets or around small parks such as Washington, Gramercy, and St. John's, they brought their elite institutions—the Union Club, the First Presbyterian Church, Grace Church, and Columbia College—with them to the edge of the city.

Change was so rapid in Gotham (Washington Irving's term for New York City) that residents were completely baffled. Phillip Hone, who retired at an early age in 1821 to devote his time to social and public pursuits, including a stint as mayor, complained of being forced out of his comfortable and prestigious home across the street from City Hall by offers for his land which he could not refuse, from people who intended the property for commercial use:

> Old downtown burgomasters, who have fixed to one spot all their lives, will be seen during the next summer in flocks, marching reluctantly north to pitch their tents in places which, in their time, were orchards, corn-

fields, or morasses a pretty smart distance from town, a journey to which was formerly an affair of some moment, and required preparation beforehand, but which constitute at this time the most fashionable quarter of new York.[4]

A house at 29 East 4th Street, now a recognized landmark, illustrates the trend. Built in 1832 as part of a speculator's row on a luxury "uptown" block that was still adjacent to farms, the four-story brick structure was bought three years later by Seabury Tredwell, a prosperous hardware merchant and descendant of Samuel Seabury, the first bishop of the Episcopal Church in New York. The important selling features of the house included an elaborate doorway that combined the best of the Federal and Greek Revival styles, a 4,000-gallon cistern in the rear garden that antedated the Croton Aqueduct water system, interior Greek and Ionic columns separated by polished mahogany sliding doors, and double parlors that ran the full length of the building. Although unusual because various Tredwells lived in it from 1835 until 1933, the home in an agricultural area represented the shift of affluent, bourgeois families from old New York to the developing periphery.

As has usually been the case in the United States, the distribution of population was governed primarily by the desire of property owners and builders to enhance their investments by attracting the wealthy and by excluding the poor. Samuel Ruggles, an 1831 purchaser of the 22-acre Gramercy Farm between 19th Street on the south, 23rd Street on the north, the Bloomingdale Road (now Broadway) on the west, and Second Avenue on the east, was particularly adept at developing high-status real estate. Quickly subdividing the farm into 108 city lots, Ruggles then transformed the equivalent of 42 lots into a private park, 520 by 184 feet. The deed read as follows:

> Samuel Ruggles proposes to devote and appropriate the said 42 lots of land to the formation and establishment of an ornamental private Square or Park, with carriageways and footwalks at the south-western and north-eastern ends thereof for the use, benefit and enjoyment of the owners and occupants of sixty-six surrounding lots of land belonging to the said Samuel B. Ruggles and with a view to enhance the value thereof.

The value of the land was enhanced as the park—admission to which has traditionally been by a special key available only to the chosen—became the center of a wealthy neighborhood and gave an uptown push to the movement of the affluent away from lower Manhattan. By the time of the Mexican War, Gramercy Park had become a bastion of correct society. Dr. Valentine Mott, reportedly "as great a surgeon as Wellington was a soldier," lived at No. 1; Amos Pinchot, noted politician and former governor of Pennsylvania, lived at No. 2. Numbers 3 and 4,

TABLE 2-1
Occupational Distribution, District of Penn (South Penn)
Philadelphia County, Pennsylvania, 1850[a]

New Middle Class[b] 12.5%		*Artisans* 49.3%		*Unskilled* 38.2%	
Storekeepers	4	Carpenters	13	Laborers	33
Merchants	4	Bricklayers	8	Carters	13
Hotelkeepers	3	Shoemakers	8	Miscellaneous	12
Tobacconists	2	Brickmakers	4	*Total*	58
Clerks	2	Cabinetmakers	4	*Percentage*	38.2
Accountant	1	Tailors	4		
Horse Doctor	1	Machinists	3		
Manufacturer	1	Dyers	3		
Salesman	1	Blacksmiths	3		
Total	19	Jewelers	3		
Percentage	12.5	Hatters	2		
		Painters	2		
		Foremen	2		
		Farmers	2		
		Sashmakers	2		
		Weavers	2		
		Misc	10		
		Total	75	*Total Sample*	160
		Percentage	49.3	*Unclassified*	8

[a]Using a systematic random sampling procedure, 10 percent of the total number of households in the district were sampled. If the occupation of more than one family member was given, that of the individual listed first was recorded.

[b]The classifications are those suggested by Sam Bass Warner in *The Private City: Philadelphia in Three Periods of Its Growth* (Philadelphia: University of Pennsylvania Press, 1968), 64–65. Professor Herbert Gutman has advised me regarding the skills of certain of these occupations.

SOURCE: U.S. Bureau of the Census, *Population Schedules of the Seventh Census of the United States, 1850. Pennsylvania.*

considered the most beautiful houses on the park, belonged to Mayor James Harper of the publishing house. Vincent Astor at No. 5, novelist Henrietta B. Haines at No. 10, composer and patron of the arts Samuel L. M. Barlow at No. 11, former Secretary of the Treasury Albert Gallatin at No. 24, and the Steinway piano family at No. 26 were among the early notables.[5]

Residential development in Philadelphia was similar to that of its larger rival. In 1850 the City of Brotherly Love counted about 565,000 inhabitants, its built-up area extended over about ten square miles, and its most important axis of growth was out Ridge Avenue to the northwest. Socioeconomic changes in the Penn District between 1850 and 1860, when that area was directly in the path of advancing settlement (TABLES 2-1 and 2-2) suggest the increasing status of peripheral residence. Arti-

TABLE 2-2
Occupational Distribution, District of Penn (Ward 20)
City of Philadelphia, Pennsylvania, 1860[a]

New Middle Class	26.9%	*Artisans*	48.4%	*Unskilled*	24.7%
Clerks	31	Shoemakers	36	Laborers	69
Merchants	29	Carpenters	33	Car Drivers	7
Grocers	24	Tailors	19	Dressmakers	5
Manufacturers	14	Blacksmiths	15	Conductors	4
Physicians	5	Machinists	10	Paper Hangers	4
Contractors	4	Brickmakers	9	Carters	4
Realtors	4	Butchers	9	Coopers	3
Teachers	4	Moulders	7	Gasworks makers	3
Hotelkeepers	4	Stone masons	7	Tin workers	3
Attorneys	3	Cabinetmakers	7	Domestics	2
Druggists	3	Painters	7	Hatters	2
Clergymen	3	Coachmakers	6	Tanners	2
Chemists	2	Printers	6	Gardeners	2
Gentlemen	2	Jewelers	5	Morocco dressers	2
Tobacconists	2	Plasterers	5	Miscellaneous	16
Millers	2	Saddlers	5	*Total*	128
Traveling Agents	2	Bakers	4	*Percentage*	24.7
One Each	5	Ropemakers	4		
Total	141	Brewers	4		
Percentage	26.9	Stone cutters	4		
		Weavers	4		
		Wheelwrights	4		
		Three Each	12		
		Two Each	18		
		One Each	13		
		Total	255	*Total Sample*	524
		Percentage	48.4	*Unclassified*	33

[a] Both the methodology and the classification system are the same as those employed in TABLE 5.

SOURCE: U.S. Bureau of the Census, *Population Schedules of the Eighth Census of the United States, 1860. Pennsylvania.*

sans and unskilled laborers who walked to work in the adjacent industrial suburbs of Spring Garden and Kensington were the dominant group in both years.[6] But as population soared in formerly empty areas, the proportion of merchants, manufacturers, physicians, and grocers, the "new middle class," in Sam Bass Warner's phrase, approximately doubled.[7] After the Civil War, the number of commuters to the center city continued to increase as businessman opted for "all the beauties of the country, within an easy and cheap communication with the city."[8]

The movement of the affluent toward Beacon Hill in Boston, Gramercy Park and Washington Square in New York, and Germantown near Philadelphia was duplicated in other metropolitan areas. In Cincinnati, travel writer Willard Glazer described the suburbs in 1883 as a "Paradise of grass, gardens, lawns, and tree-shaded roads." In San Francisco by 1860 the city's bankers, merchants, and doctors were moving away from downtown and putting up homes on the heights of Fern (Nob) Hill and Russian Hill. In Chicago most of the high-grade residential areas were still very near the center of the business district at the time of the Civil War, but a tendency for the fashionable to move toward the periphery was clearly apparent by 1873 and was widespread by 1899. In Nashville prominent citizens moved to Edgefield, lying east of the Cumberland River opposite the central business district, after the construction of a suspension bridge in the 1850s. And in Buffalo new elite residential areas were developed at the outer edges soon after 1860.[9]

In smaller cities and towns, suburbs remained predominantly slums until well into the twentieth century. In Calgary, for example, clusters of cheap dwellings and "shack towns" were put up on the narrow flats below the ridge as late as World War I. And in country villages across North America, the best streets were often those toward the center as late as 1970. But by 1875 in the major urban centers, the merchant princes and millionaires were searching for hilltops, shore lands, and farms on which to build substantial estates; crowded cities offered fewer attractions with every passing year.[10]

The First Commuter Suburb

Because land speculators have operated on the edges of cities for thousands of years and because many communities were labeled "suburban" before 1800, there is disagreement about the origins of the modern suburb. Robert A. M. Stern traces it to the booming expansion of London under King George III, when the newly prosperous merchants built small houses in remote villages in emulation of the gentry's country estates. Other possibilities include Clapham and John Nash's Park Village in London, Cambridge outside Boston, Greenwich Village north of New York, Spring Garden and the Northern Liberties near Philadelphia, and New Brighton on Staten Island. None of these communities offered the number of commuters, the easy access to a large city, and the bucolic atmosphere of Brooklyn Heights, which grew up across the harbor from lower Manhattan in the early decades of the nineteenth century.[11]

The Canarsie Indians, calling the high, sandy bank that is now Brook-

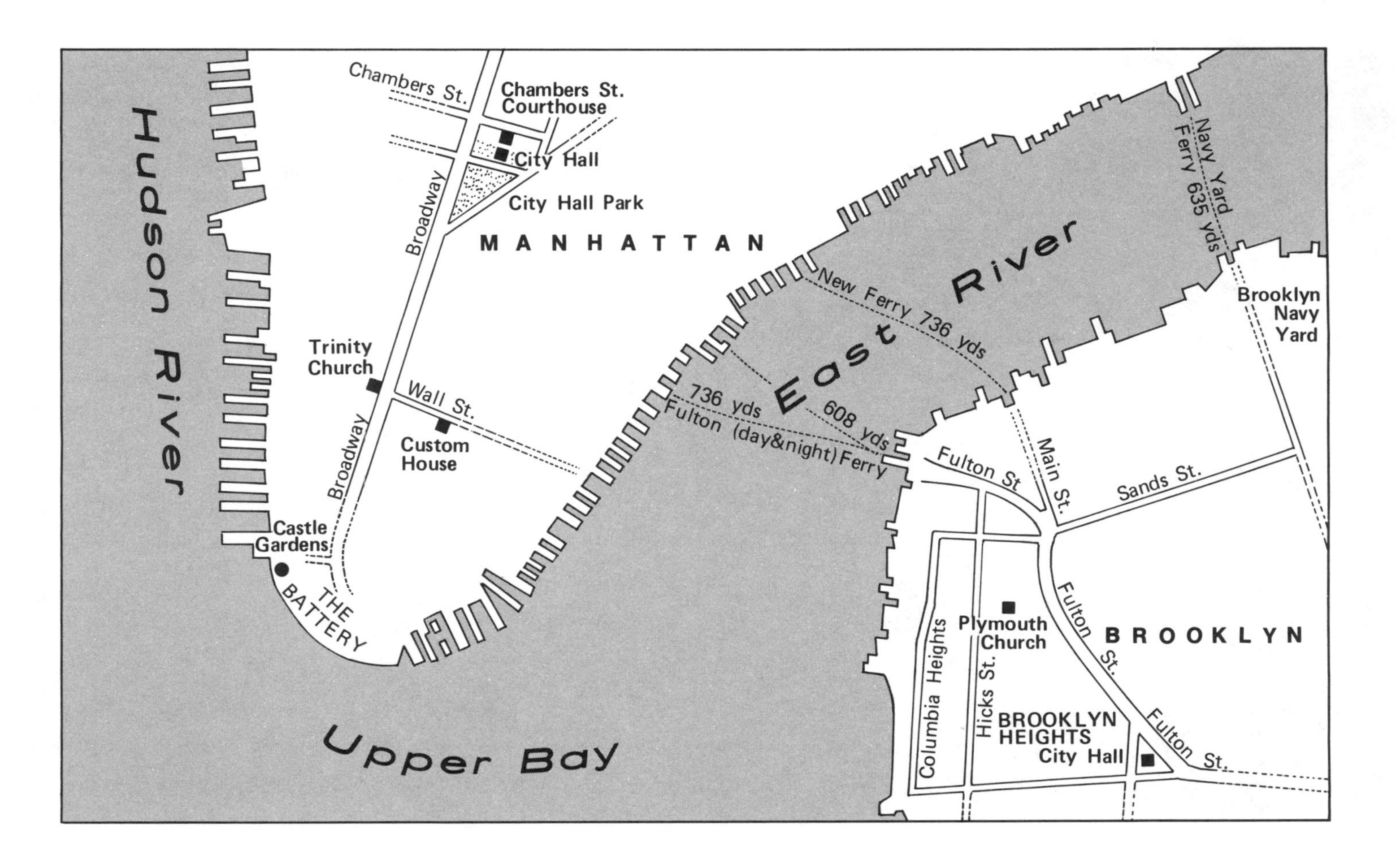
Chambers St.
Chambers St. Courthouse
City Hall
City Hall Park
Broadway
MANHATTAN
Trinity Church
Wall St.
Custom House
Broadway
Castle Gardens
THE BATTERY
Hudson River
Upper Bay
East River
New Ferry 736 yds
736 yds
608 yds
Fulton (day&night) Ferry
Navy Yard Ferry 635 yds
Brooklyn Navy Yard
Fulton St
Main St.
Sands St.
Fulton St.
Plymouth Church
BROOKLYN
Columbia Heights
Hicks St.
BROOKLYN HEIGHTS
City Hall
Fulton St.

lyn Heights "Ihpetonga," lived there for centuries in communal houses that were hundreds of feet long before the Dutch arrived in 1636. But unlike more commercial New Amsterdam, which had been founded across the East River in 1624, Brooklyn remained essentially agricultural until 1800, and its economic ties to the metropolitan entrepôt were slight. As late as 1810, it was occupied mostly by farms, and its population was less than five thousand (TABLE 2-3).[12]

TABLE 2-3
Population of New York City, Brooklyn, and Flatlands, by Decades, 1790–1890

Year	*New York City*	*Brooklyn*	*Flatlands*
1790	33,131	1,603	423
1800	60,489	2,378	493
1810	96,373	4,402	517
1820	123,706	7,175	512
1830	202,589	15,394	596
1840	312,710	36,233	810
1850	515,547	96,838	1,155
1860	813,669	266,661	1,652
1870	942,292	396,099	2,286
1880	1,206,299	566,663	3,127
1890	2,515,301	806,343	4,075

[a]The purpose of this table is to demonstrate the effect of geography upon community growth. Flatlands and Brooklyn were both nineteenth-century towns within Kings County. Brooklyn, however, was immediately adjacent to lower Manhattan, while Flatlands was on the eastern edge of the county, about eight miles overland from New York.

SOURCE: Ira Rosenwaike, *Population History of New York City* (Syracuse, 1972), especially Tables 4 and 15.

In the next four decades, however, the town of Brooklyn was transformed. Regular steam ferry service to New York City (then consisting only of Manhattan) began in 1814, and one year later the *Brooklyn Star* predicted that the town "must necessarily become a favorite residence for gentlemen of taste and fortune, for merchants and shopkeepers of every description, for artists, artisans, merchants, laborers, and persons of every trade in society." The accuracy of this prophecy soon became apparent. With its tree-shaded streets, pleasant homes, access to Manhattan, and general middle-class ambiance, Brooklyn attracted those who sought respite from the extraordinary bustle and congestion of Gotham. Walt Whitman, whose office at the Brooklyn *Eagle* overlooked the Fulton Ferry slip, frequently commented on the suburb's phenomenal growth.

"There," he said, "men of moderate means may find homes at a moderate rent, whereas in New York City there is no median between a palatial mansion and a dilapidated hovel." Writing often of the river, the fog and the ferries, Whitman captured the commuting character of the place he called "Brooklyn the Beautiful:"[13]

> In the morning there is one incessant stream of people—employed in New York on business—tending toward the ferry. This rush commences soon after six o'clock. . . . It is highly edifying to see the phrenzy exhibited by certain portions of the younger gentlemen, a few rods from the landing, when the bell strikes . . . they rush forward as if for dear life, and woe to the fat woman or unwieldy person of any kind, who stands in their way.

Additional ferry lines soon expanded the commuting possibilities. In 1836, the South Ferry began regular runs from Whitehall Street to Atlantic Street in Brooklyn, and in 1846 the Hamilton Ferry began connecting Hamilton Avenue with the battery in Manhattan. By 1854 the Union Ferry Company had consolidated a dozen competing lines and was providing 1,250 crossings daily at a standard one-way fare of two cents.[14] Williamsburg, for example, was served by six steam ferries leaving Peck Slip every ten minutes and Grand Street every five minutes during the working day. Its population shot from 3,000 in 1835 to 40,000 in 1852. By 1860 the various East River ferries were carrying 32,845,000 passengers per year (or about 100,000 per working day), and by 1870 the patronage had increased to 50 million per year. Indeed, the folk wisdom of the day held that when there was fog in the harbor, half the business population of Manhattan would be late for work. Actually, only 40 percent of Brooklyn wage earners worked in Manhattan in 1860, but they tended to be above average in wealth and position. As the ferry districts became distinct from the more rural areas of the rest of Kings County, so the political form began to change. In 1834 Brooklyn leaders won city status from the state legislature over the opposition of Manhattan representatives. By that time suburban landowners and speculators were anxious to subdivide their farms into city lots, and they perceived that a stronger government would provide the streets and services that would aid the rapid development of the periphery. But high taxes, hopefully, would not be necessary, as real-estate investors were quick to emphasize. In 1854, when James Cole and Sons advertised 450 lots near Carroll Park, the firm noted optimistically:

> The distance of these lots from the Merchants Exchange, New York, is the same as to Tenth Street in that city, but requiring only half the time to pass, on account of the rapid transit over the long ferries between the two

> cities . . . ; the consolidation of Brooklyn and Williamsburgh will have a tendency to reduce and permanently fix the tax on property which is at present far less than the taxes of New York, while the actual cost of real estate in Brooklyn does not exceed one-fifth of what it is in corresponding parts of that city.[15]

Whether the attraction was easy access, pleasant surroundings, cheap land, or low taxes, the suburb was growing at a faster rate than the city by 1800, and in almost every decade until the Civil War its population approximately doubled. One wag noted that Brooklyn "sold nature wholesale" to real-estate developers, for sale to homeowners at retail. In 1853, at single sales, 360 lots were sold in Bushwick, 150 in Fort Hamilton, 600 in Brooklyn proper, and 950 in Newton. Even on the farthest outskirts of the built-up region, the influx of middle-class families soon became apparent. In 1850, for example, the little village of Bedford (now part of Bedford-Stuyvesant, the largest black neighborhood in the Northeast), was made up of a variety of occupations, reflecting its essentially rural orientation. By 1880, however, when it had become part of the expanding metropolis, very few laborers remained, and the farmers had disappeared.[16] Newspaper advertisements offered a home in the suburbs, no farther from the heart of Manhattan than many tenements, for 10 to 40 percent down and payments spread over three to five years. Developers were not only building elegant structures for the wealthy featuring "clean sea breezes and a glorious view of greater New York and its harbor," but inexpensive dwellings for the middle classes.[17]

Far from seeing the growth of Brooklyn as an advantage, New York newspapers, politicians, and land developers by 1850 brooded over the intense competition and expressed concern over "the desertion of the city by its men of wealth." The Association for the Improvement of the Condition of the Poor moaned that "many of the rich and prosperous are removing from the city, while the poor are pressing in," and as early as 1840 Alderman Daniel F. Tieman was opposing the improvement of the ferry service (an operation controlled and regulated by the City of New York by legislative fiat to the continuing chagrin of Brooklyn) on the assumption that even more people might thereby be induced to forsake Gotham for the more bucolic atmosphere across the river. Instead, Tieman and his supporters suggested a fare rise of more than 700 percent (from 3 to 25 cents).[18] The *New York Tribune* summed up the indictment against suburban competition on January 21, 1847:

> Property is continually tending from our city to escape the oppressiveness of our taxation; many who have made fortunes here carrying them away to be expended and enjoyed, while thousands who continue to do

> business here, reside, and are taxed elsewhere on the same amount. Thus, while every suburb of New York is rapidly growing, and villages twenty and thirty miles distant are sustained by incomes earned here and expended there, our City has no equivalent rapidity of growth, and unimproved property here is often unsalable at a nominal price.

As a result of the continuing exodus and spillover from New York, Brooklyn was gradually transformed from a suburb into a major city in its own right, the fourth largest in the country in the latter part of the nineteenth century. By 1890 it counted more than 261,000 foreign-born residents, equal to its entire population only thirty years earlier. With this increase in size, the decay, noise, and fast lifestyle so many had fled Manhattan to avoid followed them across the river. Even though Brooklyn's initial budding was due to a quiet environment easily accessible to the central business district of the world's busiest seaport, its later growth was the result of the development of its own commerce and industry. By 1890 Brooklyn possessed large hat factories, chemical works, foundries and iron mills, candy companies, and coffee and syrup mills. Only Chicago had a larger dressed-meat operation, and no place on earth had larger sugar-refining and grain-depot operations.[19]

Hezekiah Beers Pierrepont

Brooklyn's meteoric rise as the premier suburb of the first half of the nineteenth century was mirrored by the career of Hezekiah Beers Pierrepont.[20] Born in 1768 and the grandson of one of the founders of Yale College, he struck out on his own at an early age because he disliked Latin and Greek. Working first at the Customs House in New York City and later as a financier, Pierrepont quickly demonstrated an ability to make money. In 1793 he formed a partnership with his cousin, moved to Paris, and began importing provisions into France, where scarcity prevailed as a result of the recent revolution. Successful despite the Reign of Terror (Pierrepont witnessed the execution of Robespierre on July 28, 1794), he subsequently expanded his operations to India and China. His foreign ventures turned sour in 1797, however, when his heavily laden cargo vessel, the *Confederacy,* was lost to pirates. All efforts at recovery failed, and the 29-year-old entrepreneur was bankrupt.[21]

Returning to New York City in 1800, Pierrepont took a more common route to fortune. In 1802 he married Anna Marie Constable, the daughter of merchant William Constable, the largest single landowner in New York State. Receiving half a million acres in Oswego, Jefferson, Lewis, St. Lawrence, and Franklin counties as a wedding present,

Pierrepont abandoned foreign commerce as too risky and decided to service the more predictable demand for alcohol. For this purpose, he purchased a gin distillery at the foot of Joralemon Street in Brooklyn. Soon thereafter, in 1804 he bought the nearby Benson farm and its spacious residence—Four Chimneys—for his bride.[22]

The brewery did not prosper, so Pierrepont soon turned his energies to land speculation and community boosterism. More than any other individual, he recognized the future prospects of his little village, reasoning correctly that great profits would accrue to those who acted with dispatch. He expanded his local real-estate holdings (the Robert De Bevoise farm and part of the Joris Remsen farm were among his acquisitions) to a total of sixty acres, including eight hundred feet of frontage on the New York harbor, and he became active in local politics.[23] Pierrepont was a member of the Committee of Fifteen which obtained a village charter for Brooklyn in 1816, and he served thereafter as a trustee. Similarly, he made his influence felt in the opening and widening of roads that would effect his holdings. Unhappy with the first street plan presented to the trustees, Pierrepont hired his own surveyor, developed an alternative proposal more attractive to his financial interests, and pushed it through to adoption. He then had his property marked off into 25-by-100-foot lots. In thus combining the roles of land speculator and local politician, Pierrepont was establishing a tradition that would typify residential settlements of all varieties in the United States.

A close friend and financial supporter of Robert Fulton, who developed the first successful steamboat in the hemisphere in 1807, Pierrepont initially held his lots off the market and bided his time until the improvement of East River transportation would enable him to sell them at handsome prices. In 1814 Fulton began the world's first steam ferry service between Beekman Slip in Manhattan (renamed Fulton Street in 1816 in honor of the inventor who died in 1815) and Fulton Street in Brooklyn. Each of the ferries was large enough to accommodate two hundred passengers and numerous horses and wagons and fast enough to make the crossing in eight minutes. Pierrepont meanwhile was formulating his plan for a high-class residential area for affluent businessmen who could purchase several contiguous lots and combine them into one prestigious property suitable for a large mansion. In 1820 the "Heights" still boasted only seven houses, but an 1823 advertisement for "Lots on Brooklyn Heights" was a harbinger of change:

> Situated directly opposite the southeast part of the city, and being the nearest country retreat, and easiest of access from the center of business that now remains unoccupied; the distance not exceeding an average fifteen to twenty-five minute walk, including the passage of the river; the ground elevated and perfectly healthy at all seasons; as a place of resi-

> dence all the advantages of the country with most of the conveniences of the city.
>
> Gentlemen whose business or profession require daily attendance into the city, cannot better, or with less expense, secure the health and comfort of their families than by uniting in such an association.[24]

As we have seen, the appeal of Brooklyn Heights was as Pierrepont expected. In 1841 of the eighty-four heads of households residing on land purchased from this early developer, thirty-nine worked in "the city," twenty-six of them as merchants. Included among the new suburbanites were William S. Packer, a wealthy, retired fur merchant, and Abiel Abbott Low, a shipping magnate successful in the China trade.[25]

Hezekiah Pierrepont died in 1838, before he realized his final dream of constructing a public promenade along the edge of the waterfront.[26] But his heirs continued the division and sale of the estate lands, sometimes in large parcels intended for further subdivision. And Pierrepont himself did live long enough to see the impeccable dwellings, some of them conventional urban row houses, of Montague, Remsen, and Pierrepont streets, which were as peaceful as New York City was bewildering and noisy, contribute to an image of Brooklyn that was attractive to the middle class.

Just as Hezekiah Pierrepont symbolized the first generation of Brooklyn suburban growth, Edwin Clark Litchfield typified the second. With a fortune based on railroad promotion in the Middle West, Litchfield began in 1852 a series of extensive land purchases in Park Slope. The tract he ultimately assembled included the high ground that is now the central portion of Park Slope as well as a broad area of meadow and marsh down near the Gowanus Canal. Litchfield's grand design called for residential development of the upper section and commercial and industrial use of his properties near the canal, but his first step was to construct an elaborate Italianate residence—Grace Hill—to set an aristocratic tone for the neighborhood. The project was spectacularly successful; the area became among the most desirable in the metropolitan region, and Litchfield's home became the Brooklyn headquarters of the New York City Department of Parks and Recreation. Part of his yard was incorporated into Prospect Park, the most noble of Frederick Law Olmsted's many creations.

Other Ferry Suburbs

Brooklyn was the first and most important of the modern "ferry suburbs," but water commutation was also significant elsewhere. Before the

Holland (1927) and Lincoln (1937) tunnels were completed, there were a score of ferry lines across the Hudson River to New Jersey. In 1821 Colonel John Stevens, inventor of the screw propeller and the owner of the entire island of "Hobuck," established regular steam service to Hoboken, which soon became a well-known pleasure spot.[27] John Jacob Astor built a summer house there in 1829, and soon thereafter the Weehawken Palisades became the site of elaborate estates of wealthy New Yorkers. Eventually, thousands of ordinary pleasure-seekers crossed the Hudson every weekend in the summer months to enjoy its delights. By the early 1850s, ferries shuttled between Jersey City and Manhattan every fifteen minutes, and by 1870 daily commuting to Gotham had become an established feature of middle-class life in Hudson County, with the Jersey City ferry alone bringing more than three thousand passengers per day into the city.[28]

No other American harbor generated nearly as much water-borne commutation as that of New York, but daily ferry riding was also a common pre-Civil War experience between Philadelphia and Camden, Newport and Cincinnati, and Pittsburgh and Allegheny City. In San Francisco wealthy merchants began building residences on the opposite side of the bay in Oakland and Alameda in the 1850s, while in Boston the two-cent and five-minute all-water journey to Noodle's Island and East Boston was daily attracting more than a thousand riders.[29]

The Origin of Public Transportation Systems—The Omnibus

The application of steam power to water travel had a substantial impact on only a limited number of cities. But many large communities were confronted with growth that was so substantial that it outmoded pedestrian movement as a workable basis for organizing urban space. Buildings became larger and streets more crowded, so that congestion could be relieved only by making more territory accessible to city residents. The obvious solution was the establishment of ground transportation networks in urban areas. Prior to 1825 no city anywhere possessed a mass-transit system—which may be defined as operation along a fixed route, according to an established schedule, for a single fare.[30] Horse-drawn carriages for hire—sometimes known as taxis or hackneys—represented the public mode for short trips, while stagecoaches served a similar function for more distant journeys. The first transit system came about almost by accident in 1826, when a retired French Army officer purchased some public baths several miles outside Nantes in western France. To improve accessibility to his business, Baudry initiated a short stage line between the center of Nantes and the resort. The coaches were popular almost immediately, but Baudry noticed that most of the riders were getting off at intermediate pick-up points rather than continuing on to the baths. Capitalizing on his discovery, Baudry promptly expanded the number and the routes of his short-haul stages, which he dubbed "omnibuses" after the name over the door of a hatmaker named Omnes. The experiment was successful, and on January 30, 1828, the first bus in Paris traveled from the Madeleine to the Bastille. By 1832 Baudry's rudimentary transit system had been copied in Bordeaux, Lyons, and London. Essentially, the omnibus combined the functions of the hackney and the stagecoach.[31]

Abraham Brower introduced omnibus service to North America with operation along Broadway in New York in 1829; enterprising opportunists then took the idea to Philadelphia in 1831, to Boston in 1835, and to Baltimore in 1844. The typical pattern was for a city government to grant a private company—usually a small businessman already in the livery or freight business—an exclusive franchise to operate coaches along an existing street. In return, the company agreed to maintain certain minimum standards of service. Over the years, the cost of a single ride averaged about ten cents, although many firms sought to develop commuter service to new residential areas by offering annual season tickets that permitted an unlimited number of rides at an average expense of four cents per day.[32]

By midcentury, the omnibuses had become a big business and had added noticeably to the traffic congestion that had already become a standard feature of urban life. New York City alone counted eighty licensed coaches in 1833; 108 in 1837; 255 in 1846; 425 in 1850; and 683 in 1853, when twenty-two separate firms were competing for the trade, and when average waiting time on some corners was less than two minutes. At one intersection on lower Broadway coaches reputedly passed on an average of every fifteen seconds. So important did the service become to Gotham's economy that when a fire consumed most of the equipment of the busiest West Side line, downtown merchants and city officials stumbled over themselves in helping to get the omnibuses back on the streets. And when the little boxes on wheels appeared in Baltimore in 1844, the *Sun* enthusiastically predicted that they would enable "persons to reside at a distance from their places of business in more healthy locations without loss of time and fatigue in walking."[33]

Although the omnibus represented an obvious improvement over walking, the unpadded benches, poor ventilation, and rude, bad-tempered drivers did cause fatigue and exposed the limitations of ground level transportation. The heavy, twelve-passenger vehicles were severely hampered by the condition of city streets, which were paved with uneven cobblestones at best and deeply rutted at worst. Riding in such circumstances was an emotional and physical trial even for the hearty. As the *New York Herald* complained in 1864: "Modern martyrdom may be succinctly described as riding in a New York omnibus."[34] As if such ordeals were not enough, omnibus travel speed was invariably less than five miles per hour and often no quicker than the pace of a brisk pedestrian. Not surprisingly, even in New York City, which had the most extensive omnibus network, only about 25,000 persons, or about one resident in thirty, used this form of public transportation on a daily basis in 1850. The significance of this primitive system, however, was that it encouraged an influential minority of urban citizens to develop what Glen Holt has called the "riding habit."[35]

Steam Railroads

By the middle of the nineteenth century, steam engines with the ability to do the work of a hundred horses were coming into common use. In particular, the steam railroad had a profound influence in reshaping the American city. The railroad was of course an English invention, having been first developed in 1814 by George Stephenson, an unlettered engine-wright in a coal mine. It was his idea that coals could be the power source to transport other coals to the waterside. Exactly half a century later, in *Our Mutual Friend,* Dickens described oozing London's edge, "where Kent and Surrey meet, and there the railways still bestride the market gardens that will soon die under them." In North America, the first important line to begin laying rails was the Baltimore and Ohio Railroad in 1829.[36]

The first railroads were designed for long distance rather than local travel. But they sought revenues wherever they could find them and very early on built stations whenever their lines passed through rural villages on the outskirts of the larger cities. In the nation's largest city rudimentary commuter travel by steam railroad began in 1832 and by 1837 the New York and Harlem Railroad offered regular service to 125th Street.[37] The achievement led to the construction of a new hotel and to the prophecy of the *New York Herald* that "this and other improvements will make Harlem a fashionable rival to Hoboken, New Brighton and other summer resorts."[38] Additions to this same line led to central Westchester County by 1844 and to the New York *Tribune* prediction, "The line of this road will be nearly one continuous village as far as White Plains by 1860." Meanwhile, the New York and New Haven Railroad along Long

Island Sound reached New Haven in 1843, and the Harlem River line toward Albany reached Peekskill in 1849. Along these tracks population grew by more than 50 percent in the first decade after initial construction, as real-estate developments sprang up in Rye, Tarrytown, and New Rochelle.[39] Between 1850 and 1860, the population of Westchester County as a whole grew by 75 percent, and that portion nearest the city more than doubled. As early as 1855, English observer W. E. Baxter noted that suburban villas were "springing up like mushrooms on spots which five years ago were part of the dense and tangled forest; and the value of property everywhere, but especially along the various lines of railroad, has increased in a ratio almost incredible. Small fortunes have been made by owners of real estate at Yonkers, and other places on the Hudson River."[40]

Because annual commutation rates (forty-five dollars to Bronxville in 1853) were too high for most wage earners, the railroads helped to foster the image of Westchester as a leafy enclave for the well-to-do. But even before the Civil War, the southernmost stations at Fordham, Morrisania, Tremont, and Mount Vernon were becoming centers of middle-class residence. Morrisania in particular was completely transformed between 1850 and 1865. A small village on the Boston Post Road in the south central part of what is now the Bronx, Morrisania was the scene of large-scale building activity after the opening of a railroad station in the mid-1840s. A particularly important 400-acre subdivision in the 1860s was called Old Morrisania and located near the Gouverneur Morris mansion.

Just as the northbound railroads opened up Westchester County (including what would later become the Bronx), largely replacing the steamboats that had run to Yonkers and Peekskill, so did the Long Island Railroad and the New York and Flushing Railroad enable former Manhattanites to commute from the east. The original purpose of the Long Island Railroad was to link New York City and Boston. Because of "the hills of Connecticut and the many wide and deep New England rivers," a direct mainland route was not feasible. Railroad promoters believed that the solution was to be found on Long Island, which "followed the general shoreline of Connecticut and which had neither rivers nor hills." The initial route, which extended "six miles or more north of the south shore, through the unimproved and barren pine-plain area," terminated at Greenport, where passengers boarded a steamboat to Stonington, Connecticut, and from there a mainland train to Boston. The first run was made on July 27, 1844, with great success. Meanwhile, almost as an afterthought, the LIRR had established commuter service along the route by 1860.[41] The New York and Flushing Railroad, obviously more interested in short-haul traffic, was unabashedly the instrument of real-estate speculators. By bringing villages in what is now the borough of Queens within one hour of Manhattan in the late 1850s, the railroad led to an influx into Newton, Maspeth, and Flushing.[42]

Railroad commuting was well established in many other cities before the Civil War. On the other side of the Hudson River from Manhattan, Jersey City was the hub of a combined ferry/railroad route which enabled commuters to traverse the sixteen miles from South Orange to New York City in less than an hour. And during the 1850s, the rail route between Newark and Jersey City was one of the busiest in the world. By 1859 in Philadelphia, more than forty trains were making commuter stops in the northwestern suburb of Germantown, while in Chicago the northern town of Evanston was growing rapidly because of the frequent service of the Chicago and Milwaukee Railroad.[43]

Boston, however, had a larger proportion of suburban riders than any other community. Along the main line of the Boston and Worcester passengers could travel to Brookline by 1834, and within a decade seven different Hub-based companies were pioneering with such devices as commuter and family fares and free weekend round trips for Bostonians who might be tempted to purchase land in outlying areas. The new price schedules occasionally reduced the price of railroad commutation to that of an omnibus ride within the city of Boston. The tactic worked as railroads and realtors propagandized together in behalf of suburban living.[44] "Somerville, Medford, and Woburn," read an advertisement for the Boston and Lowell, "present many delightful and healthy locations for a residence, not only for the gentleman of leisure, but the man of business in the city, as the cars pass through these towns often during the day and evening, affording excellent facilities for the communication with Boston." Suburban developers reciprocated by including railroad timetables in the advertisements and reminding potential purchasers that every lot was "within a few minutes walk of the station."[45]

Commuter fares, which were often sold on an annual basis and which ran about sixty-two dollars to Lynn and fifty dollars to Newton in the early 1840s, became particularly important as competition among the railroads increased and prices fell. When some annual suburban round-trip fares fell to as little as thirty dollars per year, the steam railroad became financially competitive with the omnibus.[46] As a result, there were fifty-nine commuter trains coming into Boston every day from fifteen miles or less (another forty-five came from longer distances) as early as 1849. Places like Dedham, Milton, Quincy, Dorchester, Brighton, Newton, Medford, Melrose, Malden, Winchester, Somerville, and West Cambridge were being labeled by the *Boston Evening Transcript* (April

4, 1855) as railrod suburbs,[47] and Ralph Waldo Emerson characterized "the readers and thinkers of 1854 as the men on the morning train into the city." Henry David Thoreau was even swept up in the enthusiasm, noting in his journal that "five times a day I can be whirled to Boston within an hour."

The enthusiasm of the railroad barons and real-estate developers was not always shared by the general population. Ministers denounced the companies for running trains on Sundays, and farmers regarded trains as unnatural enemies, which shattered their solitude, frightened cows out of giving milk, and blackened the countryside with cinders and soot. Like their contemporary Henry David Thoreau, they objected to whistles and clattering wheels; unlike Thoreau, they occasionally tore up tracks, burned down stations, and caused wrecks by pulling spikes out of the roadbed.[48]

The disruptive steam railroads encountered even more serious difficulties in built-up areas. In New York City in 1839, a locomotive boiler exploded at Fourteenth Street, killing the engineer and injuring twenty passengers. As a result of this and other accidents, fear of the big engines became widespread, and on more than one occasion mobs even tore up tracks on the Bowery. City fathers in both New York and Brooklyn soon banned locomotives from the populous parts of their town (below 14th Street in Manhattan) and required that horses pull the trains into open countryside when steam engines could take over. Similar restrictions in Philadelphia forced the railroads to terminate west of the Schuylkill River.[49]

One way to quell such fears was to miniaturize the engines and to disguise the locomotives to look like passenger cars rather than fearsome monsters. The result was the introduction of so-called "dummy" engines, which were carefully constructed so as to appear harmless. In the long run, however, such efforts at concealment proved unnecessary. As accidents became less frequent, the steam locomotive became an accepted part of the urban landscape. More importantly, the locomotive came to be regarded as a positive force for good, as a means of enabling families to escape the congestion of the crowded city. This "moral influence" of the railroad led supporters of mechanical power to support actively railroad extension. When in 1849 some citizens opposed the extension of the Harlem Railroad south to Chambers Street, the New York *Tribune* countered:

> We hope yet to see every part of our City penetrated by railroads, so that nearly every citizen may take a car within two blocks of his store or shop, and be swiftly carried out to his residence amid green fields and waving

forests for a trifle, and the unhealthy packing of ten thousand human beings into three or four blocks of buildings will be gradually overcome.[50]

The Horse Railway

The very rapid adoption of the steam railroad in the United States after 1830 led to the suspicion that placing the omnibus on iron rails and iron wheels would vastly improve service. Initially developed by John Mason on regular railroad tracks between Prince and Fourteenth streets in lower Manhattan in 1832, the horse-drawn streetcar, popularly known as the horsecar, promised to combine the low cost, flexibility, and safety of animal power with the efficiency, smoothness, and all-weather capability of a rail right-of-way. The horsecar was, in short, a blend of the virtues of omnibus and train.

The great expansion of horse-drawn railways came after 1852, when Alphonse Loubat developed a type of grooved rail that lay flush with the pavement. This was an essential improvement because the earliest horsecars had used rails which protruded six inches or more above street level, seriously interfering with coach and wagon traffic and preventing adoption of the innovation. By 1855 the horsecar was forcing the omnibus off the major thoroughfares and onto the secondary routes in New York, and by 1860 the same process was taking placc in Baltimore, Philadelphia, Pittsburgh, Chicago, Cincinnati, Montreal, and Boston. The first tracks for street railways in Europe were laid at Birkenhead in 1860 and London in 1861.[51]

The great advantage of the horsecar obviously lay in its use of rails, which made possible a much smoother ride at a speed (6–8 mph) almost twice as fast as the omnibus, an important consideration if one wanted to live at a distance from work. Moreover, the reduced friction enabled a single horse to pull a thirty- to forty-passenger vehicle that had more inside room, an easier exit, and more effective brakes than the typical omnibus. All these advantages lowered operating costs, ultimately reducing the average fare for a single ride from fifteen cents on the omnibus to ten cents on the horsecar. The only person whose ride was not noticeably improved was the driver, who sat unprotected from the weather on an open platform. It was thought that if the platform were enclosed, the driver's attention and alertness might be compromised.[52]

The usual pattern was for tracks to be laid radiating outward from the center of the city in a linear fashion. The horsecar tracks followed the main roads and generally were developed toward the emerging wealthy neighborhoods on the periphery. Indeed, the very availability of quick, efficient mass transportation enhanced property attractiveness, a tendency not lost on land speculators. Subdividers nestled their plots close to the tracks and advertised their proximity to the new convenience. In Boston, where developer George Brastow of Somerville was the leading figure in forming the Middlesex Horse Railroad Company, the horsecar lines tended to extend real-estate development in the same direction as the earlier omnibus. In St. Louis mass transportation extended the built-up area to the northeast, while the equally attractive southwest grew more slowly because of the paucity of horsecar routes. And in Oakland, California, real estate tycoon E. C. Sessions organized the Oakland Fruit Vale Railway Company in 1875 to service his newly subdivided Highland Park and Diamond districts. A second Oakland land baron, Walter Blair, founded the Broadway and Piedmont Railroad in 1876, with results that were noted by the *Oakland Daily Evening Tribune* on December 12, 1876:

> An example of the value of a street railroad as an investment to improve or open up real estate can be seen in the Broadway and Piedmont Railroad which was completed last spring. This was a private enterprise projected by Walter Blair and Samuel and Montgomery Howe solely to give access to their landed property and to enhance the value of the property along the line. The increase in the value of property along the line as well as the high volume of sales which has accompanied the new line has proved the wisdom of the enterprise.[53]

The way in which the new transit system altered the habits of the well-to-do almost immediately after its introduction is revealed in an 1859 diary entry of Sidney George Fisher, the scion of a wealthy Philadelphia family:

> These passenger cars, as they are called, but which are street railroads with horse power, are a great convenience. Tho little more than a year old, they have almost displaced the heavy, jolting, slow and uncomfortable omnibus and are destined soon to banish it and hacks also entirely. They are roomy, their motion smooth and easy, they are clean, well cushioned and handsome, low to the ground so that it is convenient to get in or out and are driven at a rapid pace. They offer great facilities for traversing the city, now grown so large that the distances are very considerable from place to place. They traverse the city in its length and breadth and save time and expense. Today I took the 6th St. line at the turnpike and went to Shippen St., then walked to Elizabeth Fisher's wharf. Then, took the Pine St. line and went up to Mrs. Hone's in Pine above 19th St. Then took the Spruce St. line and came down to 10th St. and went from thence to Fisher's house. Coming home in the afternoon, I took the 5th St. line at Walnut St., close to Mr. Ingersoll's door and came out to the Germantown turnpike. Their remarkable success proves how much they were needed.[54]

The privileged Fisher was hardly typical of working people, who had neither the leisure nor the income for such frequent crosstown trips. But the "riding habit" quickly caught on with the middle class, so much so that crowding became an early and persistent problem. A typical experience was related in a New York newspaper: "People are packed into them like sardines in a box, with perspiration for oil. The seats being more than filled, the passengers are placed in rows down the middle, where they hang on by the straps, like smoked hams in a corner grocery."[55]

Nowhere was the horsecar impact more immediate or more profound than in the nation's largest city. In 1853, their first full year of operation, the New York horse railways carried about 7 million passengers. By 1856 they had laid twenty-three miles of track on little Manhattan Island, and both the Second and the Third Avenue lines had reached 60th Street. By 1860 track mileage had risen to 142 and ridership had reached 36 million, or about 100,000 per day.[56] By that time, the boundaries of a 45-minute commute reached the southern edge of Central Park, then under construction, and in periods of light traffic went beyond 80th Street. Although in 1856 the *Tribune* noted, "Out of town, which a few years ago meant above Canal Street, now means across the river or bay, far down by the seashore, or in the fast receeding forests of adjoining counties," the practical impact of the horsecar was to provide a tremendous real-estate boost to upper Manhattan. By offering a comfortable means of land transportation, at least in comparison with available alternatives, the streetcar temporarily diverted the migration of taxpayers via ferry to Brooklyn and to New Jersey and insured the high-density residential coverage that would distinguish New York from other large cities.[57]

Most importantly, the horsecars contributed to the development of the world's first integrated transportation systems. In New York, Philadelphia, Boston, and Chicago, they connected with the omnibuses which provided crosstown service, with the steam railroads which provided long-distance commuter service, and with the ferries which constantly trundled back and forth across the adjacent waterways. Many of the horsecar patrons were commuters on the railroads or ferries who used the horsecars for shorter trips from the terminals to their place of work. By the mid-1880s, there were 415 street railway companies in the United States operating over 6,000 miles of track and carrying 188 million passengers per year, or about 12 rides for every man, woman, and child who lived in a city of at least twenty-five hundred persons. The impact of such a ridership was in many respects forseen by Fisher in 1859:

> A beneficial effect of this will be to enable everyone to have a suburban or villa or country home, to spread the city over a vast space, with all the advantages of compactness and the advantages, moreover, of pure air, gardens, and rural pleasures. Before long, town life, life in close streets and alleys, will be confined to a few occupations, and cities will be mere collections of shops, warehouses, factories, and places of business.[58]

Although the horsecar did not extend the building radius of the city nearly as much as the electric streetcar and the automobile would, Fisher's prediction was particularly appropriate for the United States, for it was there that the horsecar was most extensively and intensively adopted. As a contemporary summed up the impact: "It is hardly too much to say that the modern horse-car is among the most indispensable conditions of metropolitan growth. In these days of *fashionable effeminacy* and *flabby feebleness,* which never walks when it can possibly ride, the horsecar virtually fixes the ultimate limits of suburban growth."[59]

Horsecar railways were built much more slowly in Europe, usually ten to fifteen years later than in North America. In 1869 only a few lines were operating on the continent, and as late as 1875 the total ridership of Paris, London, Vienna, and Berlin combined was much less than that for New York City alone. In Tokyo, the largest city in the Orient, the horsecar was not even introduced until 1882.[60]

Comparisons with Queen Victoria's powerful and technologically advanced Great Britain are instructive. In Leeds the omnibus was the dominant mode of transit until the 1890s, a generation after those clumsy wagons had been mostly withdrawn from the streets of American cities. Similarly, the horsecar itself was to remain in active English service until World War I, long after the electric trolley had replaced the ubiquitous horse in United States streetcar systems. This rapid American exploitation of new means of transport was a reflection of a national temperament which emphasized mobility and change. In Great Britain, by contrast, no less a personage than the Duke of Wellington, the victor over Napoleon at Waterloo, thought it a mistake to build railroads because they would "only encourage the common people to move about needlessly."[61]

Transportation Innovation and Suburban Growth

Transportation change is not a sufficient explanation for the initial development of the suburban trend. If that were the case then every great city that shared the wealth of the Industrial Revolution and the technology of the horsecar and the railroad would have exhibited a similar residential pattern. To be sure, as Adna Ferrin Weber noted in his exhaustive *The Growth of Cities in the Nineteenth Century,* metropolitan growth on every continent was most rapid on the fringes. In London a new environment was taking shape in Barnes, Hampstead, Putney, Hammersmith, and St. John's Wood by the end of the eighteenth century, and by the 1840s the vastness of these suburbs was filling visitors with wonder. Utilizing both private carriages and public omnibuses, the aristocracy was a transforming the small villages to the west into fashionable places of residence. In 1849 Thomas Babington Macaulay reported that the "chiefs of the mercantile interests" had already removed their families to "suburban country seats surrounded by shrubberies and flower gardens. . . . Lombard Street and Threadneedle Street are merely places where men toil and accumulate. They go elsewhere to enjoy and to expend." In fact, the great mansions and landscaped parks of "Imperial Kensington" were so influential that those citizens of London interested in social status were finding it necessary to remove themselves to the "court suburb" by midcentury. As in the United States, the upper classes were the first, and the working classes the last to move into the commuting suburbs. Even today London remains a city of low population density—a collection of villages anxious to avoid the type of concentration typical of the rest of Europe.[62]

This pattern was hardly inevitable. The privileged groups in large American and British cities could have retained their convenient domiciles in the core and left the shabby periphery to the poor. This is what happened in Europe, Asia, and South America, where new neighborhoods were typically densely settled and inhabited by the less fortunate, rather than the well-to-do. And even these poorer denizens crowded as close to the city as they could get. One mile beyond Paris in 1850 placed one in a wilderness of gypsum quarries and bond hills; at a similar distance from the walls of Rome the traveler was quite literally in an immense expanse of emptiness. In the 1890s, Weber calculated that the population density of fifteen American cities averaged twenty-two people per acre as compared with 157.6 for thirteen German cities, and he speculated about the phenomenon:[63]

> It has sometimes been urged that this is largely the result of the development of the electric street railway in America, but the causal connection is not apparent. . . . It should rather be said that the American penchant for dwelling in cottage homes instead of business blocks after the fashion of Europe is the cause, and the trolley car the effect.

For the underlying causes of the increasingly stratified and segregated social geography of great American cities, as well as their relatively low density as compared to Europe, we must look not just to transportation technology and the powerful mechanical forces unleashed by the Industrial Revolution but to the development of new cultural values.

Home, Sweet Home: The House and the Yard

Probably the advantages of civilization can be found illustrated and demonstrated under no other circumstances so completely as in some suburban neighborhoods where each family abode stands fifty or a hundred feet or more apart from all others, and at some distance from the public road.

—FREDERICK LAW OLMSTED

A separate house surrounded by a yard is the ideal kind of home.

—MARY LOCKWOOD MATTHEWS, *Elementary Home Economics* (1931)

In 1840 suburbs had not yet developed into a recognizable entity, distinct from either the city or the farm. Peripheral towns were merely lesser versions of small cities. Outlying residents looked upon urban centers as agents of progress and culture. It was in the cities that the latest innovations developed: Philadelphia with a marvelous public waterworks in 1799, Boston with free public education in 1818, New York with public transportation in 1829. The eastern cities imported the elegant style of the Georgian London town house, they provided gas lamps and public health systems, and in every way they offered urban services superior to those of any suburb.[1]

William Dean Howells (1837–1920), America's foremost man of letters after the Civil War, experienced firsthand the relative advantages of city, small town, and suburb. An outlander from Martins Ferry and Hamilton, Ohio, Howells located in Boston as a young adult and rose with spectacular speed to become editor of the *Atlantic Monthly*. Moving successively to suburban Cambridge, to the Back Bay near the center of Boston, back to Cambridge, to suburban Belmont, back to an apartment hotel in Cambridge, to the old Beacon Hill neighborhood, and finally and permanently in the 1890s to New York City, Howells knew better than most the problems caused by the lack of urban services in the suburbs. Writing in 1871 of residence just a few miles from Boston, he noted, "We had not before this thought it was a grave disadvantage that our street was unlighted. Our street was not drained nor graded; no municipal cart ever came to carry away our ashes; there was not a water-butt within half a mile to save us from fire, nor more than a thousandth part of a policeman to protect us from theft."[2]

Peripheral towns patterned themselves after urban models and sought to project an image of dynamic growth; with the right combination of luck, grit, and leadership, any one of them could grow into a really big city. The example of Brooklyn, vigorously competing with mighty New York, was an inspiration. Although known by the sobriquet "City of Churches," the upstart community was not simply an "overgrown village" or a "bedroom" for Gotham. Brooklyn early developed the institutions that enabled it to become a leading metropolis in its own right—colleges, art museums, opera companies, music academies, libraries, and fire, police, and sanitation systems.

Even the nomenclature of outlying communities suggested connections with a metropolis or aspirations to urban greatness. Thus the nineteenth century produced in a single region a South Chicago, North Chicago, South Chicago Heights, and Chicago Heights. Meanwhile, a few miles distant from Detroit, founders of a new community took the name of Birmingham, after a smoky English industrial metropolis, even though in the next century it would become not a center of manufacture but a leafy residential retreat for wealthy executives. This predeliction for urbanism led some boosters to incorporate their dreams into town names—as in Oklahoma City, Carson City, and Kansas City—in the hope that the wish might father the fact.

By 1890, however, only half a century later, the suburban image was quite distinct from that of large cities. No longer mini-metropolises, peripheral communities, like Brookline outside of Boston, followed a different path. Moreover, the expectations about residential space shared by most Americans today had become firmly implanted in middle-class culture. This shift had many dimensions and sprang from many causes, but the suburban ideal of a detached dwelling in a semirural setting was related to an emerging distinction between *Gemeinschaft,* the primary, face-to-face relationships of home and family, and *Gesellschaft,* the impersonal and sometimes hostile outside society. In 1840 only New York and Philadelphia had as many as 125,000 residents, and the factory system was in its infancy. The typical urban worker toiled in an establishment employing fewer than a dozen persons. By 1890, when the Bureau of the Census announced that the Western frontier no longer existed, the

United States had become the world's leading industrial nation. In that year the country was already one-third urban and the population of the Northeast was well over one-half urban (defined by the census as communities of 2,500 or more persons). New York was closing on London as the world's largest city, while Chicago and Philadelphia each contained about one million inhabitants. Minneapolis, Denver, Seattle, San Francisco, and Atlanta, which hardly existed in 1840, had become major regional metropolises. Perhaps more important was the rise of heavily layered government bureaucracies and of factories employing hundreds and sometimes thousands of workers. As more people crowded together in public spaces, families sought to protect home life by building private spaces. Conviviality and group interaction, despite the massive growth of fraternal societies in the late nineteenth century, gave way to new ways of thinking about the family, the house, and the yard, and, ultimately, to new ways of building cities.

Family and Home

In both Christian and Jewish culture, the family has always occupied an exalted station. It represents the chosen instrument of God for the reproduction of the species, the nurturing of the young, and the propagation of moral principles. But as the French social historian Philippe Aries has noted, the family as a tightly knit group of parents and children is a development only of the last two hundred years. Prior to the eighteenth century, the community was more important in determining an individual's fate than was his family. In pre-Napoleonic Europe, about 75 percent of the populace lived in squalid hovels, which were shared with unrelated individuals and with farm animals. Another 15 percent lived and worked in the castles and manor houses of the rich and powerful, where any notion of the nuclear family (father, mother, and children in isolation) was impossible. In cities the population was arrayed around production rather than biological units. Each household was a business—a bakery, hotel, livery stable, countinghouse—and apprentices, journeymen, servants, and retainers lived there along with assorted spouses and children. Much of life was inescapably public; privacy hardly existed at all. In every case, the image of the home as the ideal domestic arrangement was missing. Even the word *home* referred to the town or region rather than to a particular dwelling.[3]

In the eighteenth century, however, the zone of private life began to expand, and the family came to be a personal bastion against society, a place of refuge, free from outside control. Aries notes how the arrangement of the house and the development of individual rooms reflected this desire to keep the world at bay and made it possible, in theory at least, for people to eat, sleep, and relax in different spaces. The new social and psychological concept of privacy meant that both families and individuals increased their demand for personal rooms. In the United States, especially in the suburbs, intricate floor plans soon allowed for distinct zones for different activities, with formal social spaces and private sleeping areas.[4]

Although this attitudinal and behavioral shift characterized much of European and Oriental culture, the emerging values of domesticity, privacy, and isolation reached fullest development in the United States, especially in the middle third of the nineteenth century. In part, this was a function of American wealth. In Japan the family, and especially the household, has been the central socioeconomic unit since the fifteenth century, a notion that fits with the Buddhist ideal of suppressing individual desires if they are not in conformity with the best interests of the house. Social and economic conditions in Japan, however, imposed such severe restrictions on residential space that dwellings there were (and continue to be) dwarfishly small in comparison with the West. Houses there are regarded as little more than shells required to keep out the rain, for the focus is the business of living going on within the structure.[5]

Aside from America's greater wealth, an important cultural dimension to the shift should be noted. In countless sermons and articles, ministers glorified the family even more than their predecessors had done, and they cited its importance as a safeguard against the moral slide of society as a whole into sinfulness and greed. They made extravagant claims about the virtues of domestic life, insisting that the individual could find a degree of fulfillment, serenity, and satisfaction in the house that was possible nowhere else.[6] As the Reverend William G. Eliot, Jr., told a female audience in 1853: "The foundation of our free institutions is in our love, as a people, for our homes. The strength of our country is found, not in the declaration that all men are free and equal, but in the quiet influence of the fireside, the bonds which unite together in the family circle. The corner-stone of our republic is the hearth-stone."[7]

Such injunctions took place as industrial and commercial capitalism changed the rhythm of daily life. Between 1820 and 1850, work and men left the home. The growth of manufacturing meant that married couples became more isolated from each other during the working day, with the husband employed away from home, and the wife responsible for everything connected with the residence. The family became isolated and feminized, and this "woman's sphere" came to be regarded as superior to the nondomestic institutions of the world. Young ladies especially were encouraged to nurse extravagant hopes for their personal environment and for the tendering of husband and children. For example, Horace Bushnell's *Christian Nurture,* first published in 1847, described how the home and family life could foster "virtuous habits" and thereby help assure the blessed eternal peace of "home comforts" in heaven.[8]

Whether women regarded the family as a training ground for the real world or as an utter retreat from the compromises and unpleasantries of competitive life, they were told that the home ought to be perfect and could be made so. Through the religious training and moral behavior of its inhabitants and the careful design of the physical structure, a simple abode could actually be a heaven on earth. "Home, Sweet Home," a song written by John Howard Payne in 1823, became the most widely sung lyrics of the day, as Americans identified with the restless wanderer yearning for his childhood home.

Although most celebrations of the private dwelling were written by men, if any one person presided over the new "cult of domesticity," it was Sarah Josepha Hale, editor of *Godey's Lady's Book,* a Philadelphia-based periodical intended for middle-class readership. Her verse in praise of the home found its way into many publications and was typical of a broad effort to institutionalize the female as homemaker and queen of the house. Hale's vision, and that of almost everyone else, assumed that man's was the coarser sex; women were softer, more moral and pure. The only respectable occupation for adult females (unless they were governesses) was that of wife and mother. Dependence was not only part of woman's supposed nature, but also of English and American law. Married women had scant legal identity apart from their husbands, whose control over their wives' bodies, property, and children was all but absolute.

Like verse and prose, pictures and prints with domestic themes were published in millions of copies and in considerable variety. At midcentury, the new technology of reproducing pictures encouraged the craft businessmen Currier and Ives to establish a firm producing lithographs for magazines and books. Among the most popular of the early Currier and Ives series was one of four prints on the "seasons of life," which clearly associated happiness and success with home settings and the family.

Although most writers were too sentimental and mawkish to talk about such matters as mortgage financing and structural engineering, at the core of their thought were new notions about the actual and symbolic value of the house as a physical entity. Yale theologian Timothy Dwight was especially blunt:

> The habitation has not a little influence on the mode of living, and the mode of living sensibly affects the taste, manners, and even the morals,

> of the inhabitants. If a poor man builds a poor house, without any design or hope of possessing better, he will . . .conform his aims and expectations to the style of his house. His dress, his food, his manners, his taste, his sentiments, his education of his children, and their character as well as his own, will all be seriously affected by this ugly circumstance.[9]

The single-family dwelling became the paragon of middle-class housing, the most visible symbol of having arrived at a fixed place in society, the goal to which every decent family aspired. It was an investment that many people hoped would provide a ticket to higher status and wealth. "A man is not a whole and complete man," Walt Whitman wrote, "unless he owns a house and the ground it stands on." Or, as *The American Builder* commented in 1869: "It is strange how contentedly men can go on year after year, living like Arabs a tent life, paying exhoribitant rents, with no care or concern for a permanent house." The purchase of one's home became more than a proxy for success; it also conferred moral rectitude. As Russell Conwell would later note in his famed lecture, "Acres of Diamonds," which he repeated thousands of times to audiences across the country:

> My friend, you take and drive me—if you furnish the auto—out into the suburbs of Philadelphia, and introduce me to the people who own their homes around this great city, those beautiful homes with gardens and flowers, those magnificent homes so lovely in their art, and I will introduce you to the very best people in character as well as in enterprise in our city, and you know I will. A man is not really a true man until he owns his own home, and they that own their homes are economical and careful, by owning the home.[10]

On the simplest and most basic level, the notion of life in a private house represented stability, a kind of anchor in the heavy seas of urban life. The American population, however, was very transitory. The United States was not only a nation of immigrants, but a nation of migrants. Alexis de Tocqueville observed in 1835, "An American will build a house and sell it before the roof is on," and recently urban historians have demonstrated that in fact residence at the same address for ten years was highly unusual in the nineteenth century. The best long-term data on mobility concerns Muncie, Indiana, site of the classic *Middletown* studies. During the five years between 1893 and 1898, some 35 percent of Muncie families moved; between 1920 and 1924, the proportion rose to 57 percent; during a five-year period in the 1970s, it dropped to 27 percent. Compared to other advanced societies the figures seem to be substantial.[11]

Despite such mobility, permanent residence was considered desirable, and, then as now, homeownership was regarded as a counterweight to the rootlessness of an urbanizing population. The individual house was often no more than one in a series of houses, yet it assumed to itself the values once accorded only the ancestral house, establishing itself as the temporary representation of the ideal permanent home. Although a family might buy the structure planning to inhabit it for only a few years, the Cape Cod, Colonial Revival, and other traditional historical stylings politely ignored their transience and provided an architectural symbolism that spoke of stability and permanence.

Business and political leaders were particularly anxious for citizens to own homes, based on the hope, as Friedrich Engels had feared, that mortgages would have the effect of "chaining the workers by this property to the factory in which they work." A big employer like the Pennsylvania Railroad reportedly was unafraid of strikes because its employees "live in Philadelphia and own their homes, and therefore, cannot afford to strike." Or, as the first president of the Provident Institution for Savings in Boston remarked, "Give him hope, give him the chance of providing for his family, of laying up a store for his old age, of commanding some cheap comfort or luxury, upon which he sets his heart; and he will voluntarily and cheerfully submit to privations and hardship."[12]

Marxists and feminists saw this threat because they did not share the vision of tranquil, sexually stratified domesticity in isolated households. In Europe Charles Fourier agreed with Engels that the family was based on the domestic enslavement of women, while in the United States, Charlotte Perkins Gilman, Melusina Fay Peirce, Victoria Woodhull, and a group of "material feminists" proposed a complete transformation of homes and cities to end sexual exploitation. Their formula for a "grand domestic revolution" included kitchenless houses and multi-family dwellings. The idea was that some women would cook all the food or do all the laundry, and that regular salaries would attend such duties. On both sides of the Atlantic Ocean, communitarian socialists conducted hundreds of experiments with alternative lifestyles, and many of the most active spokesmen specifically denounced the ideal of the female as the full-time homemaker and the man as absent bread-winner. As a Fourierist journal remarked in 1844, the semirural cottage "is wasteful in economy, is untrue to the human heart, and is not the design of God, and therefore it must disappear." As Fourier wished, in many areas of the world and among working class and minority populations in the United States, larger groupings would often be more important than the nuclear unit for reproduction, child-raising, and the economic functioning of the individual.[13]

The isolated household became the American middle-class ideal, however, and it even came to represent the individual himself. As Clare Cooper has noted, just as the body is the most obvious manifestation and encloser of a person, so also is the home itself a representation of the individual. Although it is only a box and often the unindividualized result of mass production and design, it is a very particular box and is almost a tangible expression of self. Men and women find in their homes the greatest opportunity to express their personal taste. Gaston Bachelard has gone further and suggested that much as the house and non-house are the basic divisions of geographical space, so the self and the nonself represent the basic divisions of psychic space. Not surprisingly, Anglo-Saxon law and tradition regard a man's home as his castle and permit him to slay anyone who breaks and enters his private abode. The violation of the house is almost as serious as the violation of the self.[14]

Real Estate

It is no accident that land is called real estate. For many centuries the ownership of land has been not just the main but often the only sure basis of power. In most primitive societies, where people belonged to the land rather than the reverse, private property was unknown. For example, in the simple agricultural communities of early modern Europe, the village made an annual and temporary allotment of land to families for cultivation. The farm remained the permanent possession of the village itself. This communal form of landholding continued in Russia until the Revolution and is common today in India and among groups that live primarily by hunting, fishing, or herding.

In most western European societies, civilization brought with it the idea of private property, and emerging divisions between rich and poor were reflected in an unequal distribution of real estate. From roughly 200 B.C. to 200 A.D. in Italy and western Europe, and for centuries before that in Greece, money had little value until it was invested in land. Indeed, the one-third of Roman citizens who did not own land were not even eligible for the military draft.[15]

When the titles and ranks of European nobility became but a reflection of the size and location of ancestral estates, real property became even more prized. In eighteenth-century France, the drive to own land was stronger than any other in the ancien regime. It reached into every level of society, from the Parisian shopkeeper or artisan who dreamed of a vegetable garden, to the duke who wanted yet another forest for the hunt. Of course, wealth could be acquired by other means—by trade, by fighting, by favors to the crown—but wealth had to be protected by power, and power was in land. As Mark Girouard has written: "From the Middle Ages until the nineteenth century anyone who had made money by any means and was ambitious for himself and his family, automatically invested in a country estate."[16]

The tendency was particularly strong in England, where the Dukes of Westminster offer an extreme example of the pattern. In 1677 Sir Thomas Grosvenor married Mary Davies, who inherited the 500-acre Manor of Ebury, an area of marsh and meadow just outside London. As the city spread westward and the manor was populated by the prosperous and aristocratic, the holdings became elegant Mayfair and Belgravia and busy and crowded Pimlico. The dukedom was created in 1874, and as was typical of British nobility, the family invested in such country estates as Eccleston in Cheshire and Enniskillen in Northern Ireland. With much of central London in their portfolio, the Grosvenors became the richest family in the United Kingdom—in 1985 their land is said to be worth more than $1 billion and includes the United States Embassy on Grosvenor Square and two renowned hotels, Claridge's and the Grosvenor House.

The English love for the land and antipathy for the city were reflected in New York City and Albany. Both communities were originally settled by the Dutch, whose desire to "live and die in a spatially compact domestic and occupational locale" reflected their homeland pattern. In both cities, this was expressed by a preference for town-centered activities and the subdivision of space into thin lots. After the English took over in 1664, Albany was turned into a garrison town with little cosmopolitan life. In their dislike of urban life, the British established large country estates outside the city and centered their lives on the land, as had their forebears. New York City escaped such a fate because its population was the most heterogeneous in the colonies.[17]

The idea that land ownership was a mark of status, as well as a kind of sublime insurance against ill fortune, was brought to the New World as part of the cultural baggage of the European settlers. They established a society on the basis of the private ownership of property, and every attempt to organize settlements along other lines ultimately failed. The principle of fee-simple tenure enabled families to buy, sell, rent, and bequeath land with great ease and a minimum of interference by government. It became, in Sam Bass Warner's phrase, "the freest land system anywhere in the world."

Whether well-born or an indentured servant, practically everyone set himself quickly to the task of organizing the landscapc into private parcels and somehow procuring a share of the division. The American dream was in large part land. A few families, such as the Van Cortlandts, the Morrises, the Schuylers, and the Livingstons, received truly impressive grants from the Crown, and their senior members were literally known as 'lords of the manor'' when the colonies were part of the empire. A number of these great estates survive in shadowy form two centuries after the American Revolution, especially on the east bank of the Hudson River, where pillared porticos, gothic gables, and castellated turrets remain as visible signs of the eclectic tastes of the rich and powerful. Elsewhere in the country, large holdings became a prerequisite for social acceptance, and something like grace—a reward for faith in God's land—attached to real estate success. Writing in 1940 of his plantation boyhood in Mississippi, William Alexander Percy recalled, ''Training in a profession, though ornamental, was unnecessary for a gentleman, but of course you couldn't be one at all unless you owned land.''[18]

The original Americans—called Indians by the Europeans—did not join the rush. Unfamiliar with the concept of permanent land ownership, they believed instead that the soil, like the wind, the rain, and the sun, could be used, but not possessed. The typical Indian assumption was that each human was as much a passer-by on the land as the wild creatures were passers-by to him, and that no person had any more right to a particular habitat than any other person or any other living thing. This notion of land as a social resource was swept aside, however, partly because Indian ways were held in contempt, partly because of the vastness of the national domain, and partly because the harmful effects of individual ownership were not apparent for generations.

Although the European immigrants appropriated Indian lands without many moral or financial niceties, the colonists did agree with the original Americans that meadows and fields should serve a useful rather than an ornamental purpose. In rural areas, this meant that the value of a parcel was a function of the amount of corn or potatoes it would yield or the number and size of farm animals it would nourish. In cities, value was determined by the number and importance of the shops and houses that could be fitted on the lot. In both scenarios, land was thought of primarily as economically productive.[19]

The Yard

Between 1825 and 1875, middle-class Americans adopted a less utilitarian expectation about residential space. They no longer needed herbs and vegetables from gardens, and, thanks to the mowing machine, a smooth lawn replaced the rough meadow cut by scythe or sheep. The suburban dream demanded an enlargement of open areas. In particular, the ideal house came to be viewed as resting in the middle of a manicured lawn or a picturesque garden. First, rural cemeteries, later parks, and then suburban cottages were advocated for the benefit of "aesthetic and moral nature," as well as physical health. During the 1850s, editorials in the New York periodical, *The Crayon,* written by the painter Asher B. Durand (1796–1886), stressed the need to link "the life contemplative with the life practical."[20]

The revolutionary change in attitude that this represented can best be appreciated by recalling that for the first four thousand years of urban history, congestion had meant security, with the very walls of the city representing safety from invading hordes or rampaging bandits. European-style protective fortifications were never essential for defending the New World, but in colonial America the Puritans did believe that eternal salvation could best be won in a cohesive, tight community, and they regarded the wilderness as the dark and terrifying home of Satan's minions. The earliest settlers almost had to be forced out into the forested interior, and at the first sign of danger they raced immediately to the nearest settlement.

As a result of this hostile view of nature and because public transit was either lacking entirely or uncomfortable and slow, wealthy Americans preferred attached or row houses. This way of building had a long tradition elsewhere. It reached its most intense development along the canals of Amsterdam and the streets of other Dutch cities, but the method was as old as Rome and was common in medieval towns. The row house was adopted in America from the beginning of white settlement; Jamestown built its first half-timbered row of "faire framed houses" in 1610. The famed Philadelphia row house made its first appearance in 1691 and was duplicated with local versions of the same form in other cities, especially after 1794, when Charles Bulfinch built a majestic row in Boston. By 1800 it had become the basic form of residential building in New York, Boston, Philadelphia, Baltimore, Providence, and other large communities on the eastern seaboard. A variation in Montreal in the 1850s led to "a terrace landscape," or the uniting of a homogeneous group of attached houses behind a single monumental facade. On the edges of each big American city, the built landscape consisted of block after block of small houses of one or two stories closely set along narrow streets, although acres of open land lay vacant adjacent to the settlements. A large home on a tiny lot in a densely-settled neighborhood was considered a perfectly appropriate residence for a high-status family prior to 1875.[21]

In all American cities, as elsewhere in the world, front and side yards were almost nonexistent, while the small rear yards were apt to be covered with back-alley dwellings, structures that were notorious not only in the larger metropolises, but even in smaller manufacturing centers like Watertown, Bridgeport, and Paterson. Rear areas were usually less than twenty-five feet deep, and the little space that was not built upon was typically rancid, disreputable, and overrun by rodents. Because regular garbage collection was rare before the Civil War, most families threw their refuse out the doors to scavenging dogs and pigs. Except for regular visits to the privy vault or outhouse, most people avoided the back yard entirely; a social occasion there would have been unthinkable.[22]

House siting, like architecture, followed no clear pattern before 1860. Not only was there no conventional way to determine where on a given lot a house should be placed, but also there was no rule as to how the land between houses, if any, was to be divided; how much open space should be devoted for the grounds, or how important or large the front yard should be relative to the rear yard. The tendency was to have no front yard at all. A glance today at Newport, Rhode Island, or Charleston, South Carolina, or brief strolls through old residential neighborhoods like Benefit Street in Providence (1756–1758); Chestnut Street in Salem (1790–1800), North Water Street in Edgartown (1810–1840), Society Hill in Philadelphia (1780–1830), or Washington Square (1825–1845) or Gramercy Park (1831–1855) in New York City will indicate how houses nestled up to the street, with a prominently placed front door that invited entrance.[23]

Quite self-consciously, small towns copied the compact arrangements of larger cities. In Bedford, New York; Essex, Connecticut; New Market, Virginia; Emmettsburg, Maryland; and dozens of similar communities, the old homes stand directly on the street. The setting for one of Bayard Taylor's nineteenth-century novels is the fictitious town of Tiberius in the 1850s. Connected to the outside world by a new branch of the New York Central Railroad, Tiberius took special pride in its business district, where "houses were jammed together as compactly as possible, and huge brick blocks, with cornices and window caps of cast-iron, staring up pompously between one-story buildings of wood, saying to the country people on market: 'Behold, a city!' "[24]

The attraction of "jammed together" houses receded with each passing year after 1840, however, and by 1870 detached housing had clearly emerged as the suburban style, different from the manor house and the farmhouse, both of which involved a lifestyle economically connected to the land, and different from the city row house, which occupied too little real estate. The preferred site became a semirural homestead, and the most conspicuous theme of architectural pattern books was private space. Drawings typically depicted an isoalted structure surrounded by a substantial garden. Occasionally, a double house appeared, but it provided separate entrances and thick party walls, and designers emphasized that it was only a transitional structure for families on the way up. The row house did not completely disappear; as late as 1920, some 71 percent of the District of Columbia's population lived in row houses (versus 14 percent in detached houses and 15 percent in apartments), while in New York and Philadelphia developers remained with the attached style until after World War II. Without question, however, grass had become an essential aspect of the suburban dream.[25]

The idea that a large, weed-free lawn was a necessity did not emerge full-blown in any particular decade. To paraphrase Richard Hofstadter, there are no absolute discontinuities in history. In Europe a few eccentrics, such as Rousseau, had always romanticized the countryside and argued that the ills of life were derived from the arid rituals of an overrefined civilization. But most restless and intelligent souls had opted for the city, which offered freedom, independence, and variety. In 1802 even William Wordsworth could look out over London from Westminster Bridge and write: "Earth has not anything to show more fair." In the United States, especially before 1830, when paving was rare and nature was threatening, the city street represented progress and the control of man over his environment.

By 1840, as humankind was removed from the real troubles of nature, an idealized view of the outdoors was emerging. Historians have often focused attention on the new appreciation for grandeur and natural beauty that was fostered by the European romantic movement by the Napoleonic era and by American artists before the Civil War. What the blue mountains of Cadore had been to Titian in the sixteenth century, the New York Catskills were to the Hudson River School in the first half of the nineteenth century. Thomas Cole, and somewhat later John Frederick Kensett, reveled in the landscape—the breathtaking views, the virgin hemlocks, the black locust trees, the stupendous river—and in so doing offered a more lyrical view of nature than had previously been typical of American artists. Similarly, popular writers like Washington Irving, James Fenimore Cooper (*The Last of the Mohicans* is a good example), and William Cullen Bryant celebrated the hills and lakes, the valleys and rivers, of their still-semiwild continent.

Epidemic disease was another powerful impetus for making one's escape from the crowded city. In Europe, from the thirteenth century onward, the dread of plague emptied inner precincts at every rumor of pestilence. In the United States, periodic outbreaks of smallpox, yellow fever, and cholera took a heavy toll in every community, particularly in the warm summer months. Sometimes it seemed as if the very survival of cities might be at stake. As a Philadelphia citizens group pleaded in 1793: "If the fever shall become an annual visitant, our cities must be abandoned, commerce will desert our coasts, and we, the citizens of this great metropolis, shall all of us, suffer much distress, and a great proportion of us be reduced to absolute ruin." As might be expected, scarcely a single suburban advertisement in the middle decades of the nineteenth century failed to contain the boast that residence among open spaces was more healthy than life in cities.[26]

The lyrical view of nature, supplemented by the dread of epidemics, was transferred to residential experience with the introduction of the villa and the bungalow. Defined by John Claudius Loudon's 1839 *Encyclopaedia of Architecture* as "a country residence, with land attached, a portion of which, surrounding the house, is laid out as a pleasure ground . . . with a view to recreation and enjoyment, more than profit," the villa represented a very different concept of a house—a place with a yard. In 1833 Alexander Jackson Davis published *Rural Residences,* a volume replete with villa designs, and thereafter the notion of a decorative lawn became more popular. The word *bungalow* (from *bangla* meaning of, or belonging to Bengal) originally referred to any Bengali house in India, and it later came to mean a house sitting freely in a garden. Because Calcutta was located very literally in a swamp, Europeans living there adopted the practice of placing residential structures on top of piles of dirt created by digging out the space around the proposed house. In America the term ultimately came to mean a one-story dwelling with a distinctive, very low, wide pitch to the roof that was particularly popular between 1900 and 1930.[27]

By 1870 separateness had become essential to the identity of the suburban house. The yard was expected to be large and private and designed for both active and passive recreation, in direct antithesis to the dense lifestyle from which many families had recently moved. The new ideal was no longer to be part of a close community, but to have a self-contained unit, a private wonderland walled off from the rest of the world. Although visually open to the street, the lawn was a barrier—a kind of verdant moat separating the household from the threats and temptations of the city. It served as a means of transition from the public street to the very private house, as a kind of space that, by the very fact of its having no clearly defined function, mediated between the activities of the outside and the activities of the inside. The sweeping lawn helped civilize the wild vista beyond and provided a carpet for new outdoor activities such as croquet (a lawn game imported from England in the 1860s), tennis, and social gatherings. More importantly, lawns provided a presumably ideal place to nurture children. As Emerson noted in his "Journal" in 1865: "There is no police so effective as a good hill and wide pasture in the neighborhood of a village, where the boys can run and play and dispose of their superfluous strength and spirits." Thus, if a man wanted to enjoy a cigar outside or join his family for a spirited game, he could do so on his own property.[28]

In New York and other large urban areas, the 25-foot-wide lot remained standard, but even there the small front yards seemed a generous improvement over the old in-town row houses which abutted the sidewalk or street. On the outskirts of the same cities, legal covenants written into property deeds from the 1880s onward required that structures be set back from the street by a minimum number of feet. Real-estate advertisements emphasized these requirements as well as those requiring that houses cost at least a given amount as a guarantee that the neighborhood would remain desirable. The change in the visual appearance of the community has been best expressed by Lewis Mumford: "Rows of buildings no longer served as continuous walls, bounding streets that formed a closed corridor: the building, divorced from its close association with the street, was embosomed in the landscape and deliberately absorbed by it."[29]

Unlike Europeans, jealous of their possessions and of their privacy, Americans did not build walls around their houses. The new suburban yard in the United States followed a naturalistic or romantic approach. It was inspired by the English, with antecedents in the Orient, and seemed well suited to the spaciousness of New World suburbs. The style sought to use the existing terrain, with gently curving paths, irregular groupings of trees and shrubs, and rustic pavilions.

This open American pattern contrasted with the more formal style of French, Italian, Asian, and even English cultures, where the lawn, usually tiny, was used for the display of individual plants along with garden ornaments—statues, vases, or fountains. There the house was enclosed by a wall and the resulting courtyard was hidden from view. Only in the Southwest, and especially California, where dense hedges, white walls, or high fences completely privatize the lawn, does the romantic ideal break down. The basic thrust is similar: to separate the family by real estate from intruders into private space.

The idealization of the home as a kind of Edenic retreat, a place of repose where the family could focus inward upon itself, led naturally to an emphasis on the garden and lawn. Grass of course was not new. Men have hunted over the grasslands, grazed livestock on them, and farmed them for thousands of years. Archeologists have found traces of cereal grasses—corn, wheat, barley, oats, rice—around the remains of Stone

Age dwellings. Genghis Khan sent his hard-riding horsemen across Asia and into Europe not only to conquer and rule but also to gain new grazing lands for his nomadic people. And the range wars of the American West set cattlemen against farmers over the question of whether frontier grasslands should be considered public or private property.

As a functionless carpet of green in front of a house, however, the grass lawn is a more recent development. The English lawn had an aristocratic birth inside medieval castle walls, where meadow grasses were kept short by beating and trampling them underfoot. The green was a place for lords and ladies to dance and dally, and no one seemed to mind when wild flowers crept in. Eventually, this small plot with formal boundaries was replaced by the sweeping lawns of what came to be known as the English garden. (In England one still refers to the grounds about a house as a garden, not as a yard). Full of copses and winding streams, wandering paths and thickets, the English garden was long on sweeping vistas of turf. One of the best known of the landscape gardeners, Batty Langley, claimed in *New Principles of Gardening* (1728) that the formal gardens of Hampton Court, the palace on the Thames that was Henry VIII's favorite, would look much better if they were stripped of "those trifling plants of Yew, Holly, etc. . . . and made plain with grass."

Although the elaborate lawn would be attainable only by the wealthy in England, in the United States carefully tended grass became the mark of suburban respectability. In 1870 Frank J. Scott published *The Art of Beautifying the Home Grounds* and Jacob Weidenmann issued *Beautifying Country Homes: A Handbook of Landscape Gardening,* the first American books devoted entirely to "the methods by which every land-owner may improve and beautify his suburban home effectively and with economy." Explicit in such books was the notion that the only reason for living in the city was to make enough money to retire to the country. The well-manicured yard became an object of great pride and enabled its owner to convey to passers-by an impression of wealth and social standing—what Thorstein Veblen would later label "conspicuous consumption." Such a large parcel of land was not a practical resource in the service of a livelihood, but a luxury in the service of gracious living. As Weidenmann noted in his very first sentence: "The location of the house . . . should be sufficiently back from the public road to afford ample room for an unbroken ornamental lawn."[30]

The propitious development of a reliable and inexpensive lawn-mower made grass cutting, in theory, a good method of weekend relaxation. "Country gentlemen will find in using my machine an amusing, useful and healthful exercise," claimed Edwin Budding of Gloucestershire when he applied to the British Patent Office in 1830 for a patent on the world's first lawn-mower. The heavy early models had to be drawn by two men or by a horse, but by the 1860s inventors had developed machines light enough to be pushed by a woman or boy. Scott was particularly ecstatic about the supposed benefits of such exercise: "Whoever spends the early hours of one summer day, while the dew spangles in the grass, in pushing these grass cutters over a velvety lawn, breathing the fresh sweetness of the morning air and the perfume of the new mown hay, will never rest contented in the city." Although generations of adolescents would learn to curse such a vision, the yard had assumed a parklike function by the last quarter of the century.[31]

New ideas about the house and the yard did not enter the nation's consciousnes through the efforts of any person or group of individuals. Dozens of people, including the park planner Frederick Law Olmsted, the social reformer Charles Loring Brace, and the Transcendentalist thinker Ralph Waldo Emerson, helped create a new suburban vision of community. But three authors whose productive lives spanned the years between 1840 and 1875—Catharine Beecher, Andrew Jackson Downing, and Calvert Vaux—were the most important voices in shaping new American attitudes toward housing and residential space.[32]

Catharine Beecher

The eldest child of Lyman Beecher, a famed Calvinist preacher in the Puritan tradition, Catharine Beecher was born in 1800 into a family in which the missionary fires burned brightly. In addition to her father, her seven brothers were all ministers, including Henry Ward Beecher, the leading Protestant clergyman in the United States between 1850 and 1887. From his pulpit in Brooklyn's fashionable Plymouth Church (Congregationalist), he preached to an audience of thousands every Sunday. His reputation was so great and his oratory so spellbinding that an alleged adulterous affair with a female parishioner and a sensational trial scarcely reduced his influence. Catharine's sister, Harriet Beecher Stowe, wrote the inflammatory novel, *Uncle Tom's Cabin,* that helped lead to the Civil War. Another sister, Isabella, was one of the leading feminists of her generation. And Catharine herself, who never had her own home and family and was rarely on friendly terms with her closest kin, became the nineteenth century's leading theorist of the virtues and requirements of domesticity.[33]

With the death of Mrs. Beecher when Catharine was sixteen, Harriet was five, and Henry was an infant, Catharine assumed the primary female responsibility in the huge Beecher household. In 1823 her fiancé, a professor at Yale, died in a shipwreck, and the following year she took charge of a Hartford girl's school. Nine years later, Catharine accompanied her famous father and her siblings to the "western wilderness" of Cincinnati, where she almost immediately founded the Western Female Institute. Although the school closed its doors four years later, Catharine remained in the Queen City until her death in 1878.

Throughout her long life, Beecher believed fervently in the moral superiority of women over men, a position outlined in the first of her twenty-five books, the privately printed *Elements of Mental and Moral Philosophy, Founded Upon Experience, Reason, and the Bible* (1831). Catharine was not a feminist, however. She opposed the women's rights movement as soon as it emerged as a national organization, insisting that woman's relation to man should be one of dependence and subservience. "Heaven has appointed to one sex the superior and to the other the subordinate station," Beecher intoned. Unlike Angelina Grimké and other militants who sought immediate female self-realization, Beecher believed that women could best achieve their goals by being so unassuming and gentle that men would yield to them.

Beecher's national influence began with her *Treatise on Domestic Economy, For the Use of Young Ladies at Home and at School,* which first appeared in 1841. An immediate popular success, it was frequently adopted as a textbook and was reprinted dozens of times over the next thirty years. Because the "cult of true womanhood" linked the home with piety and purity, Beecher sought to connect architectural and landscape design with her domestic ideal. Covering such topics as the care of infants and the proper procedure for every household activity, room by room and day by day, the *Treatise* was the first American book to offer plans for the practical dwelling; the recommendation was for a substantial one- or two-story cottage with such amenities as parlors, dining-rooms, sleeping areas, and indoor privies. Although her designs were technically conventional—the houses were boxes with a central core of fireplaces—the book provided a vision of a healthy, happy, well-fed, and pious family living harmoniously in a well-built, well-furnished, well-kept house. Beecher followed inventions closely and by the end of her life suggested that advanced technology in cooking, heating, and lighting be applied to residential building.[34]

Beecher did not specifically refer to suburbia, but she assumed that family life could best thrive in a semirural setting. She believed that "implanted in the heart of every true man, is the desire for a home of his own." Devoting five chapters of the *Treatise* to yards and gardens, she argued in favor of the physical and social separation of the population into the female-dominated sphere of home life, preferably suburban, and the male-dominated sphere of the business world, usually urban.[35]

In subsequent articles in *Harper's New Monthly Magazine,* in several books on housekeeping and family care that Catharine Beecher published between 1869 and 1873, and especially in a compendious volume she co-authored with sister Harriet Beecher on *The American Woman's Home, or the Principles of Domestic Science* (1869), Catharine Beecher acknowledged that her ideas were "chiefly applicable to the wants and habits of those living either in the country or in suburban vicinities as give space or ground for healthful outdoor occupation in the family service." Throughout all her writings, she expressed her affinity for the picturesque cottage and used her immense influence to popularize the desirability of a bucolic and quiet family life. Even though she never had a husband, children, or a large place of her own, Catharine Beecher imparted common sense and practical counsel to members of her sex throughout a broad and strange land.[36]

Andrew Jackson Downing

In 1841, the same year that Beecher's *Treatise on Domestic Economy* appeared, Andrew Jackson Downing published the first American book to deal with the art of landscape gardening in both a scientific and philosophical way. He explained his purpose as follows: "Hundreds of individuals who wish to ornament their places, are at a lost to proceed, from the want of some leading principle." Borrowing many of his ideas from the English author John Claudius Loudon, whose *Encyclopaedia of Cottage, Farm, and Villa Architecture* appeared in 1839, Downing became the most literate and articulate architectural critic of his generation and the most influential single individual in translating the rural ideal into a suburban ideal.[37]

Born in Newburgh, New York, in 1815, Downing inherited a nursery from his father and soon moved into the more creative and lucrative field of landscape design. His *Treatise,* which was to go through eight editions and sixteen printings before 1879, appeared when he was only twenty-six years of age, but he had already achieved quite a reputation in the Hudson Valley as a horticulturist. His books were notable for making architecture and landscape gardening entertaining and enjoyable for a general audience.

Like Catharine Beecher, Downing raised his voice in support of the private home. "We believe," he wrote in 1850, "above all things under heaven, in the power and virtue of the individual home." To this Hudson River Valley native, the house was less a symbol of status than of character. Well-designed and conceived, it would foster republicanism rather than ostentation. "A house built only with a view to animal wants," he wrote, "will express sensuality instead of hospitality . . . gaudy and garish apartments will express pride and vanity." Downing was a strong proponent of the English cottage, an idea born in the 1820s and brought to America in the 1830s. It was a house form particularly appropriate for the growing middle class, and it was in the United States that it became a national type.[38]

In spite of his humble origins, the failure of his nursery in 1847, and the bankruptcy from which he was saved only by borrowing money from his friends, Downing was a snob and an aloof aesthete. He divided domestic architecture into three classes: the cottage, a servantless small dwelling; the farmhouse, a larger but equally utilitarian building; and the villa, a substantial structure "requiring the care of at least three or more servants." The villa was his ideal. Only the affluent and the comfortable could afford the spacious grounds and open spaces that he felt would allow one to live in absolute harmony with nature. Modeling his villas on the ideal of the English country gentleman, he recommended grounds of hundreds of acres and regarded plots of fewer than five acres as unduly crowded and apt to introduce city ways to the serenity of the countryside.[39]

Although Downing's cottage plans were in fact more elaborate and expensive than he assumed they would be, and although he did not recommend the inexpensive balloon frame construction method,[40] his lively articles and books popularized simple and functional buildings, offering a suburban ideal to which even working men could aspire. In 1845 he became the editor of a journal of rural art and rural taste known as *The Horticulturist.* Thereafter, his essays on parks, homes, and landscape architecture reached a large and predominantly middle-class audience. His counsel was year-round country living. "In the United States," he wrote in 1848, "nature and domestic life are better than society and the manners of towns. Hence all sensible men gladly escape, earlier or later, and partially or wholly, from the turmoil of cities."[41]

Downing seldom referred directly to suburbia, but he accepted the romantic concept of the inferiority of urban residence and deplored the practice of building cities "as though there was a fearful scarcity of space." For the man who absolutely had to wring a livelihood from "the nervous hand of commerce" in the city, Downing noted approvingly that the new steam railroads "cannot wholly escape doing some duty for the

Beautiful, as well as the useful,'' by opening land for suburban development. The iron horse had ''half-annihilated'' old notions of time and space by allowing the city worker to go home to a country cottage. Commuting, therefore, would enable him to ''breathe freely'' and to keep alive his love of nature. Advising Americans to make their homes in the countryside, Downing argued:

> The love of country is inseparably connected with love of home. Whatever, therefore, leads men to assemble the comforts and elegancies of life around his habitations, tends to increase local attachments, and render domestic life more delightful; thus not only augmenting his own enjoyment but strengthening his patriotism, and making him a better citizen.[42]

Tract developments would not suffice for this purpose. As David Schuyler has noted, Downing was particularly scornful of the country builder who ''covers the ground with narrow cells, and advertises to sell or rent them as charming rural residences.'' He specifically condemned the new suburban town of Dearman, about twenty miles north of Manhattan on the Hudson River (near Tarrytown). The heavily wooded and hilly site was potentially glorious and offered frequent steamboat, stagecoach, and railroad connections with New York City. With imaginative use of new landscape-gardening concepts such as curvilinear roads and irregular lots shaped according to thc contours of the terrain, Dearman might have been a model suburb. But the developers built ''mere rows of houses upon streets crossing each other at right angles and bordered with shade trees.''[43]

To counter subdivisions designed only to maximize lucre, Downing offered a suburban plan that would bring out the best of both the manmade and the natural environments. Focusing on a large, commonly owned park at the proposed village center, Downing's ideal suburb would feature single-family cottages on lots with street frontages of at least one hundred feet, or about four times the width of the average plot in nearby New York. Broad avenues leading away from the park would assure adequate circulation, while individual gardens would enable each family to achieve a small measure of independence. Thus, the community-oriented park and the privately oriented houses would foster the union between human culture and nature.[44]

Towards the end of his short life, Downing's *The Architecture of Country Houses* (1850), which was reprinted nine times fore 1886, offered one of the earliest discussions of the linkage between home and citizenship. He argued that the nuclear family was ''the best social form'' and that ''the *individual home* has a great social value for a people.''

"Place the house in the countryside," he said, and there, "in the little world of the family home . . . truthfulness, beauty and order" would have dominion.[45]

Downing never saw most of his ideas reach fruition. On July 28, 1852, while still a young man, he set out with his wife for Newport to superintend the construction of a villa. Bidding friends farewell at the Newburgh waterfront, they boarded a Hudson River side-wheeler, the *Henry Clay,* for the journey downriver to New York City. As was common at the time, the captain engaged in a race with another steamer to determine which was the better and faster ship. Just a few miles above Manhattan (at 254th Street), the overheated vessel burst into flames. A strong swimmer, Downing could have made his way to shore, but the gentle and well-mannered man perished trying to rescue fellow passengers, at least eighty of whom died.[46]

Messages of condolence came from the President and from across the ocean. *The New York Daily Tribune* called him "a man of genius and high culture," while *The New England Farmer* noted that "the death of no man in the nation could be a greater loss." But his friend and colleague Calvert Vaux said it best:

> Andrew Jackson Downing was not only one of the most energetic and unprejudiced artists that have yet appeared in America, but his views and aspirations were so liberal and pure that his artistic perceptions were chiefly valued by him as handmaids to his higher and diviner views of life and beauty. . . .
>
> He was fortunately not a man of promise only, but of rich performance; and although cut off in the very prime of a hearty, active, ever-expanding life, he had already lived and labored to such good purpose that he can scarcely be said to have left his work unfinished. . . .[47]

Calvert Vaux

Taking up where Downing left off, Calvert Vaux became a major influence in the development of an American suburban architecture. Born in England in 1824, Vaux trained in the profession in London. In 1850 he met Downing in Paris and so impressed the noted horticulturist that he returned with him to Newburgh, where the two set up an office. Within twenty years, he had won a reputation as one of the nation's most prominent landscape architects, and he spent much of his career defending and propagating Downing's ideas. Although he was only an average designer, Vaux was personable and persuasive, and his devotion to what he considered good taste was constant. He persuaded Frederick Law Olmsted to undertake with him in 1856 the plan for Central Park, and he later collaborated on such national treasures as Prospect Park in Brooklyn, South Park in Chicago, and both the Metropolitan and the Natural History Museums in New York.[48]

Vaux deplored the conventional American homes of his time, with their basement dining-rooms and stuffy sleeping quarters. *Villas and Cottages* (1857), contained thirty-nine designs built in the Hudson Valley in the early 1850s; most were relatively low-priced rural and suburban cottages. Each example was complemented with detailed floor plans and a commentary stressing the importance of landscape, the materials used in construction, and the arrangement of interior spaces. One-third of the book was given over to a lengthy introduction that provided an illuminating insight into the thought of a perceptive observer of the national temper. Comparing his adopted land with Europe, Vaux noted:

> One important evidence of a genuine longing for the beautiful may be at once pointed out. Almost every American has an equally unaffected, though not, of course, an equally appreciative, love for "the country." This love appears intuitive, and the possibility of ease and a country place or suburban cottage, large or small, is a vision that gives zest to the labors of industrious thousands. This one simple fact is of marked importance; it shows that there is an innate homage to the natural in contradistinction to the artificial—a preference for the works of God to the works of man . . .[49]

Although his commissions were often from the well-to-do, Vaux rejected ostentation and pretense. He regarded Italianate and Gothic dwellings as unliveable because of their excessive halls, passages, corridors, vestibules, and staircases, and he told his readers to look to their personal needs rather than to foreign styles for the best guide to comfort and beauty. Addicted to the natural landscape, he was quick to condemn the banality of the gridiron street plan:

> One especial disadvantage that rural art labors under in America is, that the plans of country towns and villages are so formal and unpicturesque. They generally consist of square blocks of houses, each facing the other with conventional regularity; and this arrangement is so discordant with the varied outlines characterizing American scenery, that Dame Nature refuses, at the outset, to have anything to do with them, and they never seem afterward to get any better acquainted with her. Except, perhaps, in a very large city, there is no advantage gained by this intense monotony of arrangement, and it is much to be regretted that in many new villages that are being erected the same dull, uninteresting method is still predominant.[50]

The Anti-urban Tradition in American Thought

Beecher, Downing, and Vaux were part of an Anglo-American culture that had never placed a high value on city life. Even before the Industrial Revolution transformed many English cities into gloomy slums, London inspired oppressive horror among such major authors as Daniel Defoe, Henry Fielding, Alexander Pope, and William Wordsworth. The very thought of re-creating Old World conditions filled Thomas Jefferson with dread. During an eighteenth-century epidemic of yellow fever, he derived consolation from the thought that it might discourage the growth of future urban centers. "I view large cities," Jefferson wrote in a famous passage, "as pestilential to the morals, the health, and the liberties of man. True, they nourish some of the elegant arts, but the useful ones can thrive elsewhere, and less perfection in the others, with more health, virtue, and freedom, would be my choice." Henry David Thoreau was equally abrupt. "A man's health," he wrote in 1862, "requires as many acres of meadow to his prospect as his farm does loads of muck."

In no country was there a single intellectual or popular attitude toward cities. Paris has long been considered among the most desirable of capitals, but when Baudelaire contemplated it late in the nineteenth century, he saw "Hospital, brothel, purgatory, hell, prison,/ Where every monstrosity blossoms like a flower." In England, Archibald Alison was the sheriff of Lanarkshire when he excoriated cities on moral as well as demographic grounds in the 1840s: "It is there that vice has spread her temptations, and pleasure her seductions, and folly her allurements: that guilt is encouraged by the hope of impunity, and idleness fostered by the frequency of example." But no one was more cynical than John Ruskin, who denounced "that great foul city of London there—rattling, growling, smoking, stinking,—a ghastly heap of fermenting brickwork, pouring out poison at every pore."[51]

In the United States many talented writers testified to the magnetic quality of the American metropolis, and they celebrated the economic growth and material progress that urbanization helped make possible. Pulp fiction, such as Horatio Alger's *Ragged Dick* (1868), depicted the city as the locus of nearly unlimited opportunity, while more talented writers, such as Walt Whitman, valued New York for the stimulation that could be derived from it. The rich and well-born were especially taken with urban pleasures. Sidney George Fisher, the Philadelphia diarist, lamented that a "man of my education cannot live among farmers in the country. The moment you leave the neighborhood of a city you are in the midst of barbarism, except in a very few spots in America."[52]

On balance, however, the American metropolis was more a symbol of problems and of evil than of hope, love, or generosity. William Dean Howells, Henry George, Edward Bellamy, and Jacob Riis shocked their nineteenth-century readers with city tales of "hopeless-faced women deformed by hardship" and of "the festering mass of human wretchedness," while American politicians gloried in the frontier tradition and told their audiences that tillers of the soil represented the nation's best hope for the future. Such sentiments reached fullest expression at the 1896 Democratic National Convention in Chicago, where the booming voice of a young Populist congressman, William Jennings Bryan, thrilled a massive throng: "Burn down your cities and save our farms, and your cities will spring up again as if by magic. But destroy our farms, and grass will grow in the streets of every city in the land."

The traditional American distrust of population concentrations was heightened in the nineteenth century, when every decennial census revealed that a larger proportion of the citizenry was rejecting agrarian life for the better opportunities of crowded settlements. Especially troublesome was the notion that size itself seemed to confound every temporary solution to periodic crises. As gains were made in public health, fire prevention, water supply, and sanitation, more severe emergencies rose to take their place.

Noise and air pollution, for example, appalled travelers. Unlike the major cities of Europe, South America, and Asia, America's metropolises were centers of manufacture. Industry in the steam era, when railroads offered the best method for shipping, tended to concentrate as close to the distribution points as possible. Smokestacks belched soot into the air of every city, and nearby sections soon turned to slums. No one with options wanted to live in close proximity to important rail lines or to heavy industry. Henry James's *The Bostonians* describes such an urban scene as "a few chimneys and steeples, straight, sordid tubes of factories and engine shops . . . loose fences, vacant lots, mounds of refuse, yards bestrewn with iron pipes, telephone poles and bare wooden backs of places."

As cities became larger, noisier, and more fearsome, the specter of danger replaced the earlier notion of the city as refuge. Nineteenth-century health officials in every advanced nation noted that average life expectancy was much higher in rural than in urban areas and that, cut off from a fresh infusion of laborers from without, cities would soon be without inhabitants. For the sick, developers promised that the suburban environment would produce a cure, as in this 1871 Louisville advertisement: "[South Park] is just the place for dozens of families of the city with tendency to consumption, since a home in this pine forest would prolong their life many years.'' Another speculator boasted, ''The atmosphere at WARWICK VILLA is delightful, cool, bracing and envigorating: NO MALARIA, coal soot, smoke, dust or factories.''[53]

Almost worse than pestilence was immoral behavior, which shifted from an earlier association with the frontier and the wild West to a clear urban emphasis. Irresolute, unsupervised, and alcoholic men and women too often gave in to wicked temptations. Law enforcement proved so difficult to provide that reformer Josiah Strong proclaimed that ''the first city was built by the first murderer, and crime and vice and wretchedness have festered in it ever since.''

Suburbia, pure and unfettered and bathed by sunlight and fresh air, offered the exciting prospect that disorder, prostitution, and mayhem could be kept at a distance, far away in the festering metropolis. In 1860, for example, Robert Campbell and Edward Willis purchased eighty acres in what is now the South Bronx (then called South Morrisania) for $204,000. They subdivided the property into one thousand 20-by-100 foot lots, leveled the hills, and filled the hollows. The development went on sale in 1862, but it was the infamous New York City Draft Riots in July 1863 that gave Campbell's venture its biggest boost. Emphasizing safety and accessibility, advertisements for the building parcels suggested, ''Those who wish to secure a quiet home sufficiently remote from the city to be out of its turbulence and yet within a convenient business distance had better seek out North New York.''[54]

The changing ethnic composition of the urban population also increased middle-class antipathy to the older neighborhoods, as Poles, Italians, Russians, and assorted eastern and southern Europeans, most of them Jews or Catholics, poured into the industrialized areas after 1880. Although only one-third of all Americans lived in cities in 1890, two-thirds of all immigrants did. By 1910 about 80 percent of all new arrivals at Ellis Island were remaining in cities, as were 72 percent of all of those ''foreign born.'' Toward the end of the nineteenth century, mayors in New York, Chicago, and Boston were being elected by immigrant votes, and the possibility was raised that urban officials might be unwilling to use the police against labor radicals, most of whom came from Europe. To this fear were added specific programs to tax property so as to create public improvements and jobs to benefit working-class voters. The observation of Lord Bryce that municipal government was ''the one conspicuous failure of the United States'' was often quoted. The import of such projections was not lost on middle-class families, who often took the opportunity that low price and good transportation afforded to move beyond city jurisdictions.

By romanticizing the benefits of private space and by combining the imagery of the New England village with the notion of Thomas Jefferson's gentleman farmer, individuals like Catharine Beecher, Andrew Jackson Downing, and Calvert Vaux created a new image of the city as an urban-rural continuum and spawned a remarkable generation of landscape architects like Charles Eliot, Robert Morris Copeland, and H. W. S. Cleveland, who proposed fundamental changes in the form of the metropolis. By the 1870s the word *suburb* no longer implied inferiority or derision. Maxwell, for example, called Cincinnati's suburbs the city's "crowning glory," and Frank J. Scott published a pretentious volume on *Suburban Home Grounds*. The new suburbs were the precise opposite of the kind of dense human settlement that had characterized the planet for millenia. Formerly, open spaces like the Piazza San Marco in Venice had been scattered behind and between buildings. The open style of the American suburb, in contrast, scattered a few houses in the midst of open spaces.[55]

So great was the process of suburbanization that in 1855 *The Crayon* complained of the absence of walls surrounding American cities: "There is something in a wall which divided the city from the country, and while it shuts man into the former, by a kind of stimulant to contrariness drives him out into the latter. Here city grows into country; we never know when we leave one or enter the other." Lewis Mumford, the self-proclaimed authority on such matters, has described the ideal:

> To be your own unique self; to build your unique house, mid a unique landscape; to live in this Domain of Arnheim a self-centered life, in which private fantasy and caprice would have license to express themselves openly, in short, to withdraw like a monk and live like a prince—this was the purpose of the original creators of the suburb. They proposed in effect to create an asylum, in which they could, as individuals, overcome the chronic defects of civilization while still commanding at will the privileges and benefits of urban society.[56]

Although there were many critics of the isolated household, after the Civil War the detached house and the sizeable yard became the symbols of a very distinct type of community—the embodiment of the suburban ideal. The solid and spacious houses that lined the tree-arched avenues and fronted the winding lanes of dozens of suburbs exuded success and security. They seemed immune to the dislocations of an industrializing society and cut off from the toil and turbulence of emerging immigrant ghettoes. Pitched roofs, tended lawns, shuttered windows, and separate rooms all spoke of communities that valued the tradition of the family, to the pride of ownership, and the fondness for the rural life. By the 1890s country life periodicals that had nothing to do with farming were devoting their issues to a "simple life" of large, free-standing houses amidst ample acreage and appropriate foliage.

Such residences were attainable only by the middle and upper classes. For most Americans life consisted of unrelenting labor either on farms or in factories, and slight relaxation in decrepit lodgings. But the image had a growing attraction in a society in which urbanization's underside—the slums, the epidemics, the crime, the anomie—was so obvious and persistent a problem. The suburban ideal offered the promise of an environment visibly responsive to personal effort, an environment that would combine the best of both city and rural life and that would provide a permanent home for a restless people. There was irony in this retreat from commercialism and industry because, amid the dense foliage, somewhere below the streets, pipes and wires brought the latest domestic conveniences to every respectable home. And, as Gwendolyn Wright has observed, technology—the steam railway, the streetcar, the water system—made suburbanization possible.[57]

Romantic Suburbs

> Home is the place where, when you have to go there, they have to let you in.
>
> —ROBERT FROST

At midcentury, even as Andrew Jackson Downing, Catharine Beecher, and Calvert Vaux wrote of the desirability of a semirural lifestyle, there was as yet no precedent for developing a suburb as a completely planned and separate unit. John Nash's Regent's Park in London was picturesque in its landscape design, but its residences were densely packed and close to the street, and the community never had a separate existence. In the United States Downing, Beecher, and Vaux had in fact written more of home than of community, while Hezekiah Pierrepont, Edwin Litchfield, and Samuel Ruggles, like most land developers, had essentially been speculators rather than planners. By the 1850s, however, with an exploding urban population and new transit modes that made commuting feasible, the stage was set for the planning of the suburb as a unit, as a romantic community in harmony with nature.

The Gridiron System and the Winding Lane

The most distinctive nineteenth-century planning development was related to the physical design of the street. Throughout history all human settlements have set aside some rights-of-way for general use. If they had not, the urbanized area would have been a solid mass of buildings and private property and would have suffocated from a lack of movement. Although streets have served variously as play areas for children, market space for merchants, and foraging grounds for hogs, their purpose has always been to provide public circulation spaces in towns, and they have traditionally been just wide enough to accommodate two passing carts.[1]

Most ancient cities had no recognizable street pattern. Their narrow passageways twisted and turned at random, and when viewed from above or on a map, they resembled arteries in the human body. Hippodamus, the most famous personality in Greek town planning, introduced straight and parallel streets into Peiraeus about 450 B.C., and thereafter the "gridiron system," as it later became known, was a common feature of the city states. After the decline of Greece, this rectangular arrangement fell into disfavor for fifteen hundred years, but by the sixteenth century it had regained its status. After the model of Sir Christopher Wren's London, Philadelphia adopted its checkerboard style in 1682, Savannah in 1733, and New York City in 1811. When Pierre Charles L'Enfant put forth his plan for the nation's capital in 1791, the arrangement was a series of radial streets superimposed on a gridiron. Thus, in the District of Columbia, the curving roads of Anacostia are a signal that the neighborhood was not part of the Washington master plan and that it was considered separate from the more affluent sections.[2]

The straight, right-angled system simplified the problems of surveying, minimized legal disputes over lot boundaries, maximized the number of houses that fronted on a given thoroughfare, and stamped American cities with a standardized lot, often twenty-five feet wide and one hundred feet deep. In other words, as John Randel candidly admitted upon submitting the Manhattan grid plan of 1811, it facilitated "buying, selling, and improving real estate."[3]

The psychological significance of the clean, efficient, utilitarian grid went even deeper. The pervasive right-angled plot, which enabled such efficient speculative subdivision, and so limited the utility and beauty of the city, personfied the antinaturalism that influenced nineteenth-century urban form. Rectangular streets testified to man's capacity to overcome the hostility of the land and to civilize a continent. "Curved lines, you know," observed Daniel Drake, "symbolize the country, straight lines the city." Early planners associated the grid system with success and refused to make any deviation, even when the configuration of the terrain suggested it. The result led Lord James Bryce, after visiting numerous American cities, to complain that grid monotony "haunts one like a nightmare." Almost a century later, Lewis Mumford noted sadly: "The rectangular parceling of ground promoted speculation in land-units and the ready exchange of real property; it had no relation whatever to the essential purposes for which a city exists."[4]

The grid was so popular in the half-century after the adoption of the United States Constitution that it was even applied to the wilderness. Through the instrumentality of the Northwest Ordinance in 1787, Thomas

Jefferson and his fellow Enlightenment rationalists administered America's trans-Appalachian expansion in conformance with a philosophy of universal neatness. Similarly, although Western cities were settled by different peoples at different times, their straight streets crossing each other at right angles suggested that urbanity was prized, that the gridiron system had practical advantages, and that the goal was to be as large as possible. It also gave at least the illusion of orderliness and prosperity that settlers associated with the big cities of the East. Across the Ohio Valley and into the Great Plains, one can follow the establishment of the ubiquitous grid upon the landscape. Not only were the streets laid out in perfect symmetry, but also in Oklahoma City (1890), Salt Lake City (1870), and Dodge City (1872) they were carefully numbered, the better to suggest future prosperity and metropolitan status. The carving of the nation into a giant gridiron culminated in the Homestead Act of 1862, which divided each square mile into quarter sections of 160 acres, all of them bordered by straight lines.

Criticism of the grid focused on the overcrowded tenement conditions that were seen as an inevitable result of rectangular blocks in cities. Poorly lighted, inconvenient, unattractive, and conducive to disease, such streets were thought to scar permanently their unfortunate residents. According to Frederick Law Olmsted, even the most elegant homes suffered from the grid, and he regarded the fashionable New York brownstone as "really a confession that it is impossible to build a convenient and tasteful residence in New York, adapted to the civilized requirements of a single family, except at a cost which even rich men generally find prohibitory." Landscape architect H. W. S. Cleveland was equally succinct in 1873 when he suggested that the indiscriminate stuffing of cities into gridiron boxes was "as absurd as would be the assertion that the convenience and comfort of every family would be best served by living in a square house, with square rooms, of a uniform size."[5]

One solution to such objections that became common on the developing periphery late in the nineteenth century was patterned after the grand, tree-lined boulevards of Baron Georges Haussman in Paris. Such prototypical thoroughfares as Euclid Avenue in Cleveland, Elm Street in New Haven, Beacon Street in Brookline, Drexel Boulevard in Chicago, Summit Avenue in St. Paul, and Eastern Parkway in Brooklyn featured single-family houses with large, carefully tended lawns. Extraordinarily wide, these elaborate roads were seen as extensions of the developing park system, intended to provide a pleasant pathway from one open space to another. Some were planned as arboretums exhibiting many varieties of trees and shrubs in the center strips. Because of the emphasis on spaciousness and on greenery, the typical block size along these elegant rights-of-way was about twice the size of the 800-by-200-foot block that had been standard in the pre-Civil War city.

Meanwhile, the use of uniform setback lines and the preference for centering a house to equalize both side yards created a homogeneous statement that enabled residents to eradicate many vestiges of the heterogeneity that characterized the cities they had fled. For example, in 1843, deeds for the lots in the Linden Place subdivision in Brookline, Massachusetts, included the provision that houses be erected at least thirty feet from the street and "that the only buildings to be erected or placed upon said parcels shall be dwelling houses." As the century progressed, deeds forbade sales to "any negro or native of Ireland."[6] Even the apartment house, which was just beginning to gain acceptance in the 1870s as an appropriate home for the affluent, came to be adorned with grand arched entrances, sculptured fountains, circular carriage drives, and lush greenery.[7]

More important than the grand avenue in creating a new image for high-status suburban residence was the winding lane. Just as the grid was ideal for the row house, the undulating pattern was best suited to the suburban cottage being popularized by Beecher, Downing, and Vaux. First introduced in suburban areas in the 1850s as a design feature, a gentle turn was indicative of the pastoral and bucolic pace of the home rather than the busy and efficient system of the office or factory. Like the natural landscape, the curvilinear road was intended to be picturesque, because as practically every suburban developer would ultimately learn, the image of the bending road—not a short cut, not a thoroughfare, not a commerical strip, not a numbered street—was part and parcel of the suburban ideal. It offered the aesthetic order of unified design rather than the mechanistic order imposed by grid subdivision. By 1873 Cleveland found it hardly conceivable "that any sane man will attempt seriously to defend the rectangular system when applied to a tract comprising much inequality of surface." Only the "selfish greed of real estate proprietors," he argued, prevented the disappearance of the grid.[8]

Alexander Jackson Davis and Llewellyn Park

In the decade before the American Civil War, the world's first picturesque suburb was developed in the eastern foothills of New Jersey's Orange Mountains. Heavily wooded, with rolling hills and clear streams, it afforded a spectacular view of Manhattan and was only thirteen rail miles from New York City over the new Delware, Lackawanna, and

Western Railroad. In 1852 Llewellyn S Haskell, a prosperous drug merchant, began purchasing property in West Orange. He added land every year until by 1856 he and eight partners owned four hundred acres, mostly on the south slope of a mountain. A member of a religious cult known as the Perfectionists, who believed that by correct living they might attain the perfect existence on earth, Haskell had a passion for natural beauty that dated from his youth in New Gloucester, Maine. As a businessman in Manhattan, he was enthusiastic about landscape planning, and he gave determined and influential support to Central Park. His specific aim in Llewellyn Park was to create a picturesque community, "a retreat for a man to exercise his own rights and privileges."[9]

Haskell propitiously selected Alexander Jackson Davis (1803–1892) to prepare the site plan. The son of the editor of a Protestant review in New York, Davis was the most prolific architect of his generation. A true romantic who studied at a New York art school organized by the painter John Trumbull, Davis had strong feelings about the role of imagination in the creative process. He was inspired by the majestic Hudson River and by the rolling Berkshire Mountains and was a close friend of Thomas Cole and William Cullen Bryant.

In 1837 Davis published *Rural Residences,* which was intended as a means for the "improvement of American country architecture." Although privately printed, *Rural Residences* broke new ground as the first "house pattern book." Each of its illustrations was accompanied by a summary of materials and construction methods and by an estimate of cost. By 1852 Davis had designed dozens of baronial homes for the wealthy—the Italianate villa of Edwin Clark Litchfield in Brooklyn and the Gothic Revival "Lyndhurst" on the Hudson River, among them. He also built a few Swiss chalet-style dwellings for estate workers, but few of these Helvetian oddities have been preserved.[10]

Davis received the commission for Llewellyn Park at about the same time that he learned of the death of his close friend, Andrew Jackson Downing. Davis and Haskell had a fruitful collaboration; indeed, the precise demarcation between their achievements is impossible to ascertain. Haskell, the more active of the two, clearly set the parameters of the community plan and insisted that the theme of natural beauty be paramount in every decision. However, he recognized the genius of his associate, writing in 1854: "We thank Mr. Davis, the Michaelangelo of his time, for what he has done for us. No other man could have combined nature and art."[11]

Designed "with special reference to the wants of citizens doing business in the city of New York, and yet wishing accessible, retired, and healthful homes in the country," Llewellyn Park introduced two features—the curvilinear road and the natural open space at the center—that were unprecedented in modern residential experience.[12] Both took full advantage of the landscape. Contrasting sharply with the gridiron layout then popular in most urban areas, the seven miles of gracefully winding lanes followed the natural contour of the land and were appropriately named Tulip, Mountain, and Passive. This undulating street pattern was an essential part of a picturesque and romantic environment. Although Thomas Jefferson introduced winding paths to the United States in his design for the gardens at Monticello, Llewellyn Park was the first to incorporate purposefully undulating roads for an entire community. The goal was to preserve the rural character of the grounds, instead of allowing them, as one contemporary critic said, to assume "the rectangular forms of a village, which are a repetition of city lots on an inferior scale."[13]

The second innovative design feature of Llewellyn Park was the fifty-acre "Ramble." Intended as a completely natural open area, without any formal layout except pedestrian walkways which curved through the woods and connected with the cliff walk along the ridge of the mountain, the Ramble followed a tumbling stream. Haskell retained the underbrush and trees as they were found. Care of the Ramble was entrusted to a Committee of Management, elected by the landowners, and title to the property itself was permanently placed in the hands of three self-perpetuating trustees.

Davis and Haskell sought to reinforce the pleasant environment in a variety of other ways as well. The founders stipulated in the original covenant that no factory, shop, slaughterhouse, or other place of industry could ever invade their peaceful refuge. Lot sizes averaged more than three acres, and fences were prohibited because they interfered with the natural scenery. Owners were free to landscape their lawns according to individual preference, but every effort was made to harmonize each site with the natural fall and character of the land. Even the rustic bridges were designed so that they enhanced rather than detracted from the original setting.

Llewellyn Park received quick and enthusiastic approval, and it was labeled "the most sensible real estate development in American history." But residence in the parklike surroundings was possible only for the well-to-do, and it has retained its snobbish ambience ever since. Alexander Jackson Davis moved there after his retirement, and inventor Thomas Alva Edison made it his home for most of his active life. Most heads of households were successful businessmen and professionals who could afford an expensive residence and the time and cost of railroad commutation to Manhattan. They sought the quiet, secluded neighborhood that Haskell and Davis originally envisaged, and they endorsed the century-old gatehouse and prominent signs warning: "Private Entrance. Do Not Enter." Llewellyn Park, then as now, provided an environment free from the nuisances of congestion, industry, and poverty which were characteristic of city life. Most importantly, it introduced to landscape architecture the notion that a beautiful natural setting could be created for a group rather than an individual family; it thus began the tradition of carefully planned suburbs that was to be a central concern of American and British architects in the twentieth century.[14]

Frederick Law Olmsted and Riverside

Like Alexander Jackson Davis and Llewellyn S. Haskell, Frederick Law Olmsted was also a protégé of Andrew Jackson Downing. In the tradition of many nineteenth-century men of accomplishment, Olmsted had experience in many occupations during his long life (1822–1903). Apprenticed as a civil engineer, he worked as an experimental farmer on Staten Island in the 1840s, wrote an influential study of slavery in the South in the 1850s, and administered the United States Sanitary Service during the Civil War. Best known as the designer, architect-in-chief, and superintendent of New York City's Central Park, he created in Central Park a world-renowned open space that influenced the establishment of similar oases in every major city in America.[15]

After his well-publicized success with Central Park, Olmsted became the nation's best-known landscape architect in the generation after the Civil War. Unlike Downing, Olmsted saw the suburb not as an escape from the city, but as a delicate synthesis of town and wilderness. He hoped that peripheral growth would provide an advance upon dense settlement and become a "sensible and permanent movement." In fact, he defined the "strictly suburban" community as one of "detached dwellings with sylvan surroundings yet supplied with a considerable share of urban conveniences." He also made the interesting prediction that "no great town can long exist without great suburbs."[16]

Olmsted and his partner, Calvert Vaux, laid out sixteen suburbs, among them Brookline and Chestnut Hill in Massachusetts, Sudbrook and Roland Park in Maryland, and Yonkers and Tarrytown Heights in New York. His first and most influential residential creation was Riverside. Begun as just one of dozens of Chicago suburbs that took advantage of the expanding commuter railway system in the decade following the Civil War, Riverside was nine miles west of State Street and the first suburban station on the Chicago, Burlington, and Quincy Railroad (the Burlington

Route). The 1600-acre site was particularly attractive; Olmsted and Vaux pronounced it "the only available ground near Chicago which does not present disadvantages of an almost hopeless character."[17]

With the exception of Llewellyn Park, speculators in most suburbs laid out streets and lots in the familiar grid pattern and set aside only small parcels of land for parks, churches, and other public purposes. Riverside was different. When Emery E. Childs and a group of Eastern investors established the Riverside Improvement Company in 1868, it seemed possible that the undeveloped site on the Des Plaines River might be based on a philosophy that transcended mere money-making.[18]

Childs gave Olmsted and Vaux virtually complete freedom. They meticulously planned the water supply, drainage, lighting, schools, and recreational facilities and set aside seven hundred acres for public use. Parks were an essential part of the overall design; the most prominent being a 160-acre reserve along a three-mile stretch of the river. But a series of smaller parks possessing "the character of informal village-greens, commons and playgrounds" created other unfenced areas for family recreation. A special dam across the river backed up water for pleasure-boating.

"Rural attractiveness" required more than public spaces, however. In keeping with precepts of landscape theory first promulgated by Downing twenty-five years earlier, Riverside offered generous lots (usually 100 by 225 feet) and an environment that combined "the conveniences peculiar to the finest modern towns with the domestic advantages of a most charming country." Curved roadways were adopted to "suggest and imply leisure, contemplativeness, and happy tranquility;" the grid, according to Olmsted, was "too stiff and formal for such adornment and rusticity as should be combined in a model suburb." To convey a feeling of spontaneity, Olmsted and Vaux planted trees at irregular intervals. To give a sense of openness, they insisted that houses be set back thirty feet from the street. To suggest prosperity and elegance, they required that homeowners maintain immaculate gardens.[19]

Riverside's architecture matched its planning. William LeBaron Jenney, later to become famous as one of the fathers of the skyscraper, exercised a veto power over all construction plans and personally designed the distinctive water tower and the three-story Riverside Hotel. He built his own home there in a modified Swiss style, which, he claimed, matched the overall plan: well-organized, yet informal and rustic.

The most unusual feature of the Riverside design was a limited-access highway to Chicago. Believing that the separation of work and home life was necessary in the modern world, Olmsted thought that commuting should be a pleasant experience. He proposed a road so fine that it could compete with the railroad. No stores or industries would be permitted to front on the drive and disturb the natural vista. Central lanes would be reserved for carriages and horseback riders, while the outside lanes would handle freight and provide access to nearby houses. On pleasant days such a drive would afford the harried businessman opportunity for "taking air and exercise amid delightful vistas on his way to work." Unfortunately, the tight control over public land which Olmsted and Vaux exercised within Riverside could not be extended beyond its boundaries, and the modern turnpike remained only a dream.[20]

Riverside itself was not an immediate financial success. Investment capital was scarce because the great Chicago Fire of 1871 necessitated that all available funds go toward the rebuilding of the city itself. Land values in the model suburb fell over the first five years of its existence, and the Riverside Improvement Company went bankrupt in the Panic of 1873. However, Olmsted and Vaux managed to achieve a substantial portion of their goal. Riverside was acclaimed as the most complete realization of Olmsted's conception of a proper residential district, and its picturesque surroundings ultimately attracted "the more intelligent and more fortunate classes" for whom it was intended. Although its tree-lined, curvilinear streets have been long since absorbed into Chicago, Riverside remains a monument to what David Schuyler has called "the nineteenth-century search for an urban compromise." The town helped set the pattern for future attempts to preserve natural topography in innovative urban design. Olmsted himself retained his conviction that well-planned suburbs were "the most attractive, the most refined and the most soundly wholesome forms of domestic life, and the best application of the arts of civilization to which mankind has yet attained."[21]

Alexander T. Stewart and Garden City

The most ambitiously planned suburb of the nineteenth century, as well as the most conspicuous failure, was the brainchild of Alexander Tunney Stewart (1803–1876), a Scottish immigrant whose elaborate dry-goods emporium, built in 1846 at the corner of Broadway and Chambers streets in New York City, is usually regarded as the world's first department store. In 1869 his long-time architect, John Kellum (1807–1871), learned of the imminent sale of 7,170 (later increased to 8,670) acres of the common lands of the Town of Hempstead, a rural Long Island community about twenty miles east of the city. At the time, Stewart had an annual income of about $2 million, a Fifth Avenue mansion, and a notable art collection, and he was searching for a new challenge.[22]

Because the sale lands were publicly owned, and because the townspeople were especially anxious to avoid working-class development, bidding for the tract was competitive, and the winner was to be chosen by public referendum. Stewart entered a bid of $385,000 (about $55 per acre) for the property and, in a letter to the *Hempstead Sentinel,* announced his intention to build "attractive buildings and residences" for a population "desirable in every respect as neighbors, taxpayers, and citizens," rather than "tenement houses and public charities of a like nature." The Town of Hempstead accepted the Scotsman's offer, in part because of his stated high-class intentions and in part because his chief rival for the tract was suspected of planning to build a cemetery or a penitentiary. But the impression nevertheless spread, fueled by an August 7, 1869, article in *Harper's Weekly,* that Stewart's purpose was the erection "of homes for the working classes of New York and Brooklyn." As George L. Hubbell, the manager of Garden City in its second stage of development (1897–1918), noted; "His sole purpose was to make a good investment and to develop and sell the property as would any other developer."[23]

The site of Garden City was a natural prairie about thirteen miles long by 2½ miles wide, with gentle slopes here and there and an abundant water supply. The Long Island Railroad passed near the development, and Stewart constructed a spur to the main line at his own expense to make commuting more convenient. A shrewd investor, Stewart understood that the price of land was a function of its accessibility to places of employment. He also knew that the high cost of rail commutation (about one dollar per day to New York City in the 1870s) would preclude the possibility of Garden City becoming a haven for the working classes.

The physical plan of Garden City, which was largely the work of Kellum, differed from that of Llewellyn Park and Riverside in that it incorporated the familiar gridiron street system. Unlike the Manhattan plan of 1811 and numerous urban layouts in the West, however, the Garden City grid was not arbitrarily imposed on the land. Kellum broke up the potential montony in several ways. First, he inserted diagonal avenues which conformed to the natural drainage of the land. Second, he allowed the measurements of the individual blocks to vary with the topography. Finally, he planned several parks of 50 to 150 acres each to interrupt the regularity of the streets.

Perhaps the most important single feature of the Garden City plan was the unusual size of the streets and the individual lots. On the average, the blocks were from 1,000 to 1,500 feet long, and they were separated from each other by an average residential width of 500 feet. The enclosed area was approximately five times the size of a typical New York City or Brooklyn block. Even the quietest Garden City streets were at least eighty feet wide (with fifty feet for the road bed and thirty feet for sidewalks and trees) versus an average of sixty feet for the much busier east-west arteries of Manhattan. Similarly, the typical building lot in Garden City was about 1½ acres, extraordinarily large for a suburb and about twenty-five times as big as its New York counterpart. Such dimensions, as in Llewellyn Park and Riverside, were calculated for the needs of very prosperous families. After providing space for a house set back seventy-five feet from the street and after building a stable and possibly a garden shack, there remained room for the obligatory garden and lawn.

The name *Garden City* may have derived from the nickname of Chicago before the 1871 fire, but more probably it was simply descriptive of the planned greenery which was expected to enhance the appeal of the community. The idea of turning unused land into a blooming area was a metaphor repeated in many press articles: "Hempstead Plains, hitherto a desert, will be made to blossom as the rose"; and "Hempstead Plains . . . will, within a few years, be improved from a waste to a garden." In keeping with the theme, thousands of trees were planted along the newly surveyed streets in 1871, while lakes, shrubbery, and serpentine walkways were to make the parks, and especially the largest one, worthy suburban successors to Central Park in New York City.

The most important departure from previous practice, and from common sense as well, was Stewart's decision not to sell property, but to rent the houses at annual leases of $250 to $1,000. The idea was to prevent Garden City from "degenerating in the future to the standard of scores of country villages." To that end, tenants were to be screened by an estate manager, who would check their financial, family, religious, and social status. "The one and only thing demanded" of potential residents were "references of an unquestionable character."

The Garden City leasing scheme recalled European systems of social control which had never been incorporated into American traditions. It was similar to the procedure in small industrial towns—such as Lowell—where mill hands rented small quarters from the all-powerful company. Stewart, however, was attempting to control the lives of affluent businessmen, not the powerless workers of a mill village. As the *New York World* noted:

> He is attempting a daring experiment, nothing less than a community . . . (with) all the appliances of municipal life, without a single other person having an interest in a foot of the whole domain. He proposes to be landlord, mayor, and alderman, in fact the whole municipality. All inhab-

itants pay him rent and purchase goods in his stores. This may succeed, but it would be a miracle should it do so.[24]

There was no miracle. Although Stewart intended to develop initially only five hundred acres in the western part of his tract, the death of Kellum in 1871 and of Stewart himself in 1876 removed from the scene the individuals most able to make the community flourish. The original plan made no provisions for schools or churches, a critical omission in a settlement of several hundred houses. And the isolation from communal institutions was aggravated by the isolation of homes from each other. As has been written: "Even urban dwellers longing for space could hardly have appreciated the lonely distances between homes on the hot, dusty plain." As early as 1873, the nearby *Flushing Journal* noted sadly:

> It strikes the observer as strange to see the many fine houses—certainly a score as good as anything in Flushing—already built and unoccupied, forty smaller houses erecting, some of them enclosed, a fine brick depot, a large hotel and no inhabitants—no one to stop at the hotel, no one to live in the houses, and the graded streets not deserted because never used by the busy throng.[25]

By the 1890s, after two decades, only thirty of Garden City's sixty houses were rented. At that time a Rochester newspaper reported: "Garden City is one of the most curious phases of suburban development, if an obvious failure can be called a development." Not until land was sold for private ownership did Garden City begin attracting many residents to fill its empty lots. Thus, the most enduring lesson of A. T. Stewart's planned surburb was negative; it demonstrated that affluent families would not support a rental market for expensive detached homes.[26]

Lewellyn Park, Riverside, and Garden City are only a few examples of planned suburban growth in the United States before 1875. There were other communities that started out as semiutopian ventures, and some achieved limited success. Perhaps the earliest American attempt came in 1849 with the organization of the Industrial Home Owners Society Number One. Led by John Stevens, the group was made up of New York City residents who were frustrated by high rents and overcrowding, and who sought a working-class community where people of moderate means could afford individual cottages with agreeable neighbors and a short commute. In 1850 Stevens purchased 367 acres from five farmers in Westchester County, divided the land into quarter-acre parcels on a regular grid pattern, and held a lottery to distribute the plots to the thousand-plus members of the association. By 1854 the new residents had constructed more than three hundred homes, planted shade trees along the new streets, built a commercial section around the railroad station, and incorporated their community as Mount Vernon. The village bustled with energy and commerce, and famed newspaper editor Horace Greeley supported the experiment as an alternative for those who were unwilling to "Go West."[27] But Industrial Home Owners Society Number One floundered when hard times in the 1850s forced individual members to sell their holdings to outsiders. There was no clause in the manifesto to keep the settlement a closed one, and the newcomers were more affluent than the original workers. By 1860 there was little to distinguish Mount Vernon from dozens of growing villages around New York. By 1880 it was itself a small city.[28]

The second abortive attempt at early suburban planning came in 1851, when William M. Shinn laid out Evergreen Hamlet near Millvale, Pennsylvania. Unfortunately, the community was across the Allegheny River from Pittsburgh and beyond commuting range. Only six of a planned sixteen houses were built, and the association was dissolved in 1865.[29]

More unusual and more successful was Vineland, New Jersey, founded in 1861 by Charles K. Landis as a town where cooperative land use could promote private comfort. As in Llewellyn Park, Riverside, and Garden City, Landis bought enough land to assure complete control. He oversaw Vineland's design and restricted the use of private parcels, and by 1865 he had created a thriving community of several hundred settlers. Rather than relying on a single group whose disfavor could cripple the whole enterprise, he attracted an economic mixture of families, to whom he sold rather than rented. And rather than leaving the schools and churches for the villagers to build, Landis included careful provisions for educational, religious, and commercial institutions. Vineland succeeded because it worked within existing demands for housing and land; Garden City failed because it tried to change the traditional pattern of homeownership.[30]

Llewellyn Park, Riverside, and Garden City were better publicized than other planned communities, however; and the removal of Americans to the metropolitan fringes was heavily influenced by their example, even though Garden City and Riverside were only impressive failures. It was later in the nineteenth century that other planned developments—most conspicuously Roland Park in Baltimore, Redleaf Park near Philadelphia, and Pinehurst in North Carolina—also captured the imagination of architects and developers and generally achieved more financial success than their predecessors. However, what Llewellyn Park, Riverside, and Garden City did was set the sociological and architectural pattern for hundreds of communities that developed in the twentieth century. Before these original three communities were built, the choice had been between dense development or a rural environment. By eschewing the gridiron system and by cooperating with nature instead of stamping out every trace of the original topography, Haskell, Davis, Olmsted, Vaux, Stewart, and Kellum evolved a new form of urban settlement. The communities built by these men were in the romantic tradition popularized by Downing, Beecher, and Vaux, as well as by Thoreau and Thomas Cole. Llewellyn Park, Riverside, and Garden City differed from the rural cottages of Downing and Beecher, where it was the individual who moved out to the countryside alone and landscaped his property while the surrounding area was to be enjoyed romantically in its wild splendor. But collectively and as individual examples, they pointed out specific ways in which commercial land development could attract families—or as in the case of Garden City not attract—away from the cities by creating a complete environment that fulfilled expectations of a tranquil life, close to nature, with urban comforts. They followed the Anglo-American ideal of the English garden—cozy and irregular, wayward and inspirational—rather than the severe and mathematical French garden, where geometries were kept in perfect shape, gravel was raked several times a day, and hedges took on the form of cone and cylinder, obelisk and cannonball.

Llewellyn Park, Riverside, and Garden City were all blatantly elitist, with their large plots, generous open spaces, and expensive homes. Indeed, they demonstrated two important truths: that quality single-family homes in a planned environment could not be built at a profit for the working classes, and that those who could afford the luxury of a substantial home on a large plot would not be satisfied with anything less than full ownership.

5

The Main Line: Elite Suburbs and Commuter Railroads

In 1861 when Edmund Ruffin of South Carolina proudly fired a cannon at Fort Sumter in Charleston's harbor, thus beginning the Civil War, the United States was an essentially agricultural land of fewer than 34 million inhabitants. Only two of its cities had as many as half a million citizens, and most of its people had never seen a railroad track, or a three-story building, or a crowd of a thousand persons.

In 1913, when Henry Ford introduced the moving assembly line for his Model T, the American nation had become the world's leading industrial power, a colossus of flame and steel whose annual output exceeded even that of the Kaiser's Germany. New York was about to become the world's largest city, surpassing London, and almost half the United States population had become urban.

The rapid growth of the economy—coupled with the quickening pace of the Industrial Revolution—provided the basis for the organization of business on a scale undreamed of in the antebellum period. Jay Gould, Cornelius Vanderbilt, Leland Stanford, Collis P. Huntington, and Mark Hopkins in railroads; John D. Rockefeller in oil; Andrew Carnegie in steel; J. P. Morgan in banking; Henry O. Havemeyer in sugar refining; Solomon Guggenheim in mining; Gustavus Swift, Philip Armour, and Michael Cudahy in meat packing; and James B. Duke in tobacco were simply the most famous of the captains of industry, more recently labeled robber barons, who harnessed the human and material resources of an abundant land and superintended the growth of giant corporations like Standard Oil of New Jersey and the United States Steel combine.

These men at the top of the business pyramid were handsomely rewarded for their efforts. Elaborate balls and parties, huge mansions replete with dozens of liveried servants, and grand tours of Europe symbolized the "era of excess." But such men were acutely aware of the tenuous nature of their achievement and of the rapid intellectual, ethnic, social, and political changes that were undermining previous beliefs and values. The new industrialists knew that they were scorned by the old Yankee families in Boston, the Knickerbocker establishment in New York, and the first families of Philadelphia, and they sought something substantial to fall back on. In order to justify the risks, the long hours at the office, the sacrifices for family and posterity, and in order to gain a larger measure of social acceptance, the robber barons sought security in a country estate, an impressive physical edifice that would represent more stability than any urban residence. In England some five hundred country houses were built or remodeled between 1835 and 1889. In the United States the brewing, shipping, railroad, iron, and banking millionaires followed this British tradition of the country gentleman or the French pattern of an aristocratic chateau in the Loire Valley. The American nouveaux riches embraced the notion of conspicuous consumption in the form of ornamental real estate and decided that the most fashionable way to display great wealth was to invest in a rural estate of appropriately grand dimensions. In the short space of about twenty years between 1885 and 1905, they built neo-Gothic, neo-Renaissance, and Georgian structures as expansive and expensive as any in Jane Austin's England—one gentry estate after another.[1] As Barr Ferree noted in 1904:

> Country houses we have always had, and large ones too; but the great country house as it is now understood is a new type of dwelling, a sumptuous house built at large expense, often palatial in its dimensions, furnished in the richest manner, and placed on an estate, perhaps large enough to admit of independent farming operations, and in most cases with a garden which is an integral part of the architectural scheme.[2]

A few of these manor houses were intended solely for summer use and were put up in remote locations like Newport, Rhode Island; Bar Harbor, Maine; Loudoun County, Virginia; and Saratoga Springs, New York. Some were completely isolated, such as the Vanderbilt's mammoth "Biltmore," in Asheville, North Carolina. In Delaware, "Chateau country" took on new meaning as members of the du Pont family and the highest executives in the company gobbled up land for other estates. The most common location, however, was within reasonable commuting distance of a major metropolis.

Although every important city spawned a few huge estates, the north shore of Long Island—from Great Neck to Lloyd Harbor—best epitomized the desire of wealthy Americans to take up country residence. Lured by the island's natural beauty, by its bays and coves and rolling hills, and especially by its proximity to the nerve center of American industry and finance, socially prominent families began spreading over the farmlands in the 1870s. Men like John S. Phipps, whose father was a founder of the United States Steel Corporation, and Marshall Field, Chicago's great merchant prince, built lavish mansions with square pavilions, balustraded arcades, and hundreds of acres of lawns, ponds, gardens, and polo fields. The spacious interiors of such structures contained European objets d'art, gilded mirrors, crystal, and old brocades. A typical estate was Greentree, the 500-acre property of John Hay Whitney, master of one of this country's great fortunes. E. J. Kahn, Whitney's biographer, said that Whitney's adopted daughter Kate, who was Franklin D. Roosevelt's grandchild, once took her children through the White House. "After inspecting it," Mr. Kahn wrote, "they pronounced it nice enough, but hardly on a par with Greentree." Even in 1985, when most of the larger properties have been turned over to public use or subdivided for town houses, the area still evokes the opulence of debutante balls, stately yachts, and senile dowagers.[3]

The Residential Options of the Upper Middle Class

Although the owners of the Georgian manor houses on Long Island and elsewhere were hardly typical—at no time did the owners of such domains represent more than a fraction of 1 percent of the population—they did set a well-publicized example of stylish suburban living that the merely comfortable attempted to follow. The number of families sufficiently well-to-do to own their own homes and to have at least one domestic servant expanded enormously in the half-century after the Civil War, whether as senior upper-level executives in large companies; as plant managers; as owners of small businesses which shared in the general economic boom; as lawyers, physicians, and other professionals serving the complicated requirements of an expanding urban population. Estimates of the number of such families vary and are necessarily imprecise in any event, but it is likely that about 10 percent of the residents of the major American cities reached this level of comfort in the 1880s.

This privileged urban group had essentially three residential options in the latter three decades of the nineteenth century—to remain in a private dwelling within the city, to move to an elegant apartment house, or to relocate on the growing edges. The first option was especially difficult because of rapid physical and demographic changes in large cities. The older residential areas were often close to the central business district, and property values were rising rapidly in anticipation of a change in land use. Speculators bought such properties in the hope that the houses could be torn down to make room for commercial development, or perhaps for manufacturing or warehousing activities. While waiting for such a transition, they subdivided the single-family homes for as many as eight families, renting the result to poorer groups that needed cheap lodging close to workplaces. Thus, at the same time that their homes were becoming less attractive for personal residential uses, the older families found that they could sell at a profit. With the exception of rare neighborhoods like Nob Hill in San Francisco, Gramercy Park in Manhattan, and Beacon Hill in Boston, most of the elite sections lost their cachet, and their residents were replaced by newcomers.[4]

The second option, that of apartment dwelling, only became viable in the 1870s. Although elegant Parisians had been living on "shelves" since the days of the ancien régime, respectable American families considered apartment living as sexually racy until after the Civil War, for the very logical reason that busybodies could not monitor the comings and goings of visitors when entranceways did not front on the street. Immigrants and laborers were aware of these conventions also. Many crowded together in the same structure, but any family of even modest social aspirations insisted on a private dwelling, however humble. In fact, the word "tenement" described *any* dwelling housing three or more families.[5]

Change came in 1870, when the first upper-income apartment building in the United States was built by architect Richard Morris Hunt for Rutherford Stuyvesant, one of the most respected men in New York. Offering residents the security and convenience of a full-time concierge, the Stuyvesant Apartments on Eighteenth Street overcame the risqué image of the French flat and attracted couples of impeccable social standing. Previously, it had been thought that the presence of several unrelated families on the same floor would encourage promiscuity.[6]

The success of the Stuyvesant encouraged other builders, and by the 1880s, when multi-unit buildings were appearing by the hundreds in New York City and by the score in Philadelphia, Boston, Chicago, and Baltimore, quality apartment buildings had become a source of civic pride and urban boosterism. In 1879 the most lavish residential structure yet seen in the United States was put up on the corner of West 79th Street and Central Park West. Known as the "Dakota" because it was so far from the center of things that one wag said it might as well be in the Dakotas, it was richly ornamented and featured separate servant's entrances, service stairways, and service elevators. Meanwhile, single-family housing starts in New York City plummeted from thirteen hundred in

1886 to forty in 1904, a shift hastened in the late 1890s when it became possible to draw electric current for elevators from street conduits.

"Flat fever" was most pronounced in New York City, but even in Gotham the third option available to the potential home-buyer—suburban residence—was popular. The idea of the house as an expression of a middle-class dream, promoted by a kind of soft-sell, image-oriented salesmanship, was, as we have seen, not new in the Gilded Age. Andrew Jackson Downing, Catharine Beecher, and kindred spirits had been promoting such notions since the early 1840s. In the 1880s, however, these ideas began to assume the aspects of a movement. Affluent towns grew phenomenally in the last quarter of the nineteenth century, and by 1900 places that hardly anyone had even heard of twenty-five years earlier—like Fernwood, Darby, Overbrook, Ardmore, Haverford, and Bryn Mawr in the Philadelphia area alone—had become synonymous with stylish living. As was said about a prominent Massachusetts suburb: "In 1845, Brookline was a rural town near Boston, with some suburban and urban features; by 1885, though much underdeveloped land remained, Brookline had become an integral part of a greatly enlarged city."[7]

An important part of this suburban growth was related to the unprecedented expansion of railroad service in the United States between 1865 and 1900. In the latter year the American nation had more miles of track than the rest of the world combined. Because the railroads were typically organized by urban businessmen, the routes radiated out from every big city. The horsecars of course offered more frequent and less expensive service to outlying towns, but the railroads were twice as fast and their accommodations much more comfortable. Building upon a cadre of pre-Civil War commuters, railroad executives rapidly expanded the short runs from the downtown stations to the nearby towns and countryside. Some of the stops had existed as villages for centuries in relative isolation from their neighbors, but the steam railroad dissolved the barriers of time and distance to bring them into a more metropolitan orbit.

The term "Main Line" derives from the Philadelphia experience. In the 1860s and 1870s, the Pennsylvania Railroad decided to straighten the meandering track along the primary route to Pittsburgh. Rather than fight the farmers along the way, the railroad simply bought them out. After shifting the right-of-way, the railroad then went into the real-estate business, selling land to large developers and to individual purchasers. It also improved commuter services to such western suburbs as Swarthmore, Villanova, Radnor, and Stratford. Some areas with improved schedules, such as Chestnut Hill bordering Fairmount Park and Germantown, were within the boundaries of the city itself and were as dignified and pretentious as any outside the municipality. But most were outside the corporate limits, and the most fashionable were along the 'Main Line.' In the 1930s Philip Barry's play, "The Philadelphia Story," revolved around the life of the privileged on the Main Line (it was made into a motion picture in 1940 with Katharine Hepburn, Cary Grant, and James Stewart), and by the 1960s, Nathaniel Burt could conclude that the gentry, in a proportion of ten to one, lived in the suburbs and preferred to remain there.[8]

Commuter railroads were established all across the nation. In Cincinnati in the 1870s, the Marietta Railroad began to advertise the allure of schools, churches, and residential restrictions to businessmen who could afford large, comfortable houses in parklike communities and who might be induced to foresake the increasingly immigrant city. To the south, the Memphis and Selma (later absorbed by the Frisco) and the Memphis and Charleston (later the Southern) Railroads both had a thriving commuter business to the Bluff City in the 1880s. In San Francisco, steam-operated systems connected with the cable cars to take passengers to such nearby destinations as Seal Rocks and Cliff House. Everywhere, the early commuter railways acted like magnets, drawing successful businessmen from the quick-paced cities to little towns along the tracks.

The Railroad Suburbs of Chicago

In proportion to its size, Chicago grew more rapidly than any other large city in the world between 1850 and 1900. Constantly written about and visited and known as the "city of broad shoulders" because of its heavy industries and sprawling stockyards, the Illinois metropolis had become the nation's railroad hub by 1870. Trunk lines radiated out in every direction—the Santa Fe, the Chicago and Northwestern, the Chicago and Milwaukee, the Rock island; the Burlington; the Illinois Central—to name only a few. To Rudyard Kipling, visiting in the 1890s, Chicago was a "real city," unlike other places in the American West. It was also monumentally corrupt, and its packinghouse and factory neighborhoods were teeming masses of poor immigrants. "Having seen it," Kipling said of the Windy City, "I earnestly desire never to see it again."

Not surprisingly, Chicago had many commuter villages strung out like beads along the major rail lines. As early as 1855 the *Chicago Daily Democratic Press* claimed: "The New York and Philadelphia suburbs will now meet a rival, and the glories of the Hudson River, with her sylvan shade and suburban palaces, will not henceforth shine beyond all parallel." Evanston, Wilmette, Winnetka and Highland Park were on the Chicago and Northwestern Railroad; Aurora and Hinsdale were on the Burlington Route; Kenwood and Hyde Park were on the Illinois Central; and Morgan Park, Englewood, and Blue Island were on the Rock Island railroad. As early as 1873, Everett Chamberlin noted

> Chicago, for its size, is more given to suburbs than any other city in the world. In fact, it is doubtful if any city of any size, can boast an equal number of suburban appendages. The number of suburbs of all sorts contiguous to Chicago is nearly a hundred, and they aggregate a population of 50,000 or more, represented by 5,000 or 6,000 heads of families, all of whom do business in the city, and form a large per cent of the passenger list of the 100 or more trains that enter and leave the city daily.[9]

Fifteen years later, Chicago's railroad commuters had multiplied tenfold to a new total of seventy thousand and the population of the suburbs exceeded three hundred thousand.

The Chicago railroad suburb most noted for its social pretensions, its careful planning, and its exquisite topography was Lake Forest. Located on Lake Michigan about twenty-eight miles north of the Loop, Lake Forest was laid out in 1857 by Almerain Hotchkiss, a skillful St. Louis surveyor who combined rectilinear building plots with a curvilinear street plan. As a visitor to the community noted in 1869:

> The design here, everywhere manifest, is to aid Nature by Art and not to bully her, which a great many men in Western towns seem to have aimed to do and succeeded in accomplishing. . . . Winding graveled walks, smooth green lawns, rustic bridges—spanning ravines; gigantic vases—blooming with floral treasures; summer houses, rustic chairs and couches, and grand old trees present a scene the eye is never weary of.[10]

The initial proprietor of the 1,400-acre community was the Presbyterian Church, and a large parcel of land was appropriately reserved for a denominational university (now Lake Forest College). The suburb's most important period of growth began in the 1890s, when the Chicago business aristocracy—the McCormicks, Cudahys, Palmers, Piries, and Swifts—discovered the beauties of the north shore. The surnames of Lake Forest residents, who married with incestuous loyalty, formed a litany of the city's rich and powerful. When manufacturer Prentiss Loomis (1908), Reuben H. Donnelly of R. H. Donnelly and Sons Printers (1907); Edward Ryerson of Ryerson Steel (1906); and Ogden Armour of the meatpacking firm moved their households from the great city to the leafy suburb, they established a social pattern of genteel elegance. F. Scott Fitzgerald's first love was Lake Forest debutante Gineva King (later Mrs.

John T. Pirie), whom he immortalized as Gatsby's inamorata, Daisy Buchanan.[11]

Westchester County

No metropolis in the world, Chicago included, was as well served by railroad commuter lines at the turn of the century as New York. Although Pennsylvania Station would not open until 1910, hundreds of trains from northern and eastern New Jersey poured into Hoboken every morning, discharging passengers who immediately embarked on the ten-minute ferry ride to Manhattan. From southern and eastern Brooklyn and from Nassau County on Long Island steam locomotives brought passengers to Atlantic Avenue and easy access to the financial district. Before 1910, however, Grand Central Terminal at 42nd Street and Park Avenue had no rival as a commuter hub, and it was the rail lines to Westchester County that enabled the first large suburban area in the nation to develop.[12]

Westchester had been a summer and weekend retreat since early in the nineteenth century. "Lyndhurst," the most notable of baronial edifices, was constructed in 1838 for the merchant William Paulding on the banks on the Hudson River, about twenty-five miles north of the city. A quarter of a century later financier Jay Gould bought the structure and made it as elaborate as a castle. When Gould chose to spend the night there rather than at his Manhattan home, he either commuted down the Hudson aboard his yacht or flagged down the New York Central, which would make a special stop. The opposite side of Westchester County was also prime country house territory for nineteenth-century millionaires, especially after the New Haven Railroad pushed through the region at mid-century. In 1848 hotelier Simeon Leland built a castellated, turreted, and gabled castle of sixty rooms in New Rochelle; in 1852 another ultra-Gothic castle was built for millionaire William P. Chapman. These Westchester castles were ultimately overshadowed by the elaborate 4500-acre domain of John D. Rockefeller and his children at Pocantico Hills just east of Tarrytown.[13]

Westchester's importance in the history of American suburbanization, however, derives from the upper-middle-class development of Scarsdale, New Rochelle, Rye, Mount Vernon, and a dozen other villages scattered among its several hundred square miles of hills and lakes. With the opening of Grand Central Terminal in October 1871, the weekly *Westchester News* noted in its lead editorial:

> Great improvements are being made to utilize the splendid physical and commercial resources of Westchester County by making its unsurpassed

> harbors available for commercial purposes and its excellent sites for . . . suburban homes accessible to settlers. The reasons for these enterprises are the inevitable certainty that Manhattan Island, upon which is located New York City, will soon be entirely inadequate to provide homes for its increasing population.

The prognosis proved correct as tens of thousands of executives and businessmen from Manhattan chose well-kept lawns and an opulent overall ambience rather than the twenty-four-hour polyglot atmosphere of the great city. By 1898 the three major passenger railroad lines running up the Hudson River, the Harlem River Valley, and Long Island Sound, as well as several smaller routes, were disgorging 118,000 daily commuters into Grand Central Station, as the county's population more than doubled between 1850 and 1870, doubled again between 1870 and 1890, and doubled again between 1890 to 1910, when it had 283,000 residents. Most of the development was actively encouraged by the railroads, which developed communities, advertised suburban advantages, and offered frequent and reliable service. Indeed, there was more and better commuter service from most of the city's northern suburbs in 1880 than there was in 1980. The importance of the iron horse was recognized in an editorial in the *Railroad Gazette* on September 2, 1904:

> The extent to which suburban territory is dependent for its development upon a thoroughly harmonized transportation service is nowhere better shown than in the country which the New Haven road now proposes to open up by the improvement of its Harlem branch between the Harlem River at Port Morris and New Rochelle. The suburban territory along the New York and Harlem road . . . has been highly developed for a great many years, like the suburban territory across the Hudson River and that along the main line of the New York Central as far as Yonkers and beyond, but the towns reached by the inadequate service of the old Harlem branch of the New Haven road have scarcely developed at all, although situated within a radius of ten miles from New York. The service was inconvenient, involving a change of cars at the Harlem River and the use of the elevated line from that point into the city, and hence resulted in the curious anomaly of an undeveloped district in the midst of a thickly populated suburban territory.[14]

Of the nineteenth-century Westchester railroad suburbs, the prototype of the high-status, high-prestige community was Bronxville. Located on a craggy, breathtaking square mile, twenty-eight minutes (and fifteen miles) by train from Grand Central Station, this village of Italianate, Romanesque Revival, and Tudor estates was the brainchild of William Van Duzer Lawrence. A Gilded Age millionaire who could buy pretty much what he wanted, Lawrence reasoned that Westchester was directly

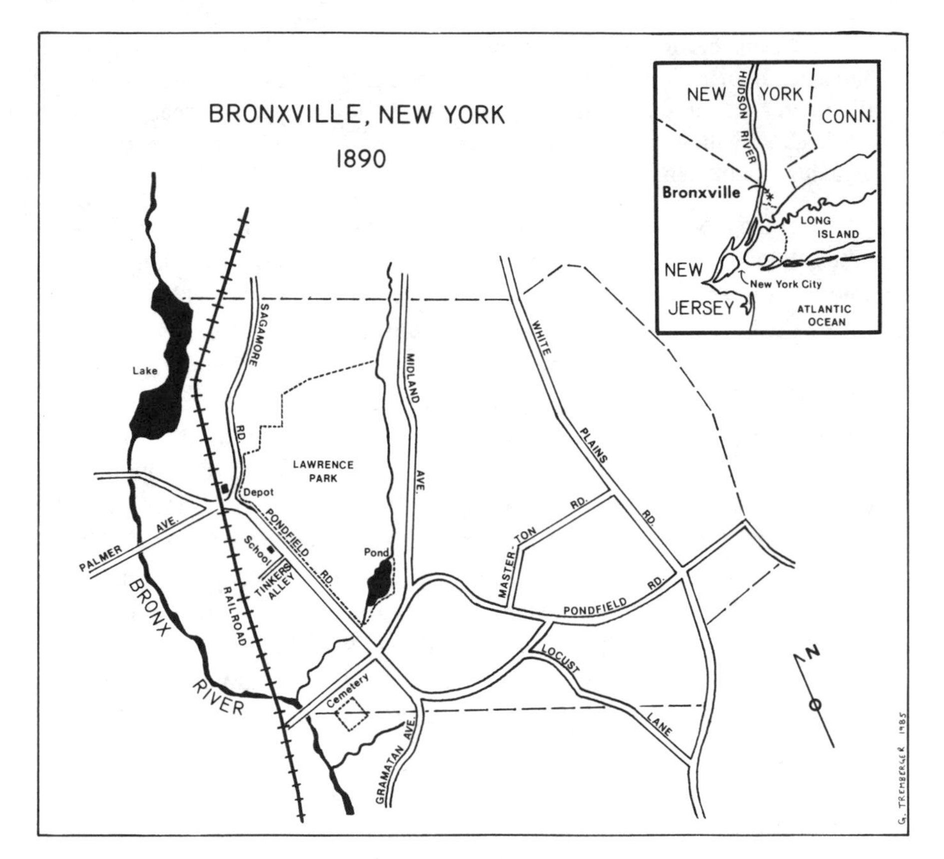

FIGURE 5-1.

in the path of New York's suburban growth. In 1889 he decided to develop an exclusive preserve for the very prosperous and the very artistic on a rugged, 86-acre estate near the house of his in-laws. He ordered his road builders to follow the cow paths and to construct curvilinear streets rather than the more familiar grid ones. He retained the raw slabs of rock, the fruit trees, and the wildflowers; forbade the use of fences; and insisted on architectural diversity, if not elegance. Architect William A. Bates designed many of the houses as well as the Hotel Gramatan and the Lawrence Arcade, a complex of stores adjacent to the railroad station that served as the town center. The success of Lawrence Park was immediate, and its expensive standards soon enveloped all of Bronxville, which was incorporated as an independent village in 1898. Many New Yorkers came out to the hotel for a leisurely weekend and eventually bought houses in Lawrence Park. Within a generation, Bronxville had earned a reputation as a "suburb endlessly copied and never matched."[15]

It remains today as similar to its origins as it is possible to be, and it looks much as it did in the 1930s, when four decades of development were all but completed. The 1980 census showed that the largest income category in this heavily Republican, largely Protestant enclave of the wealthy was "$75,000 or more."

Bronxville was copied in Westchester as often as anywhere else. The county was particularly suited for high-income residential development because of its attractive wooded and shoreline topography and because no ferry transfer was necessary for the commute into Manhattan. Its new communities, eighteen of which were incorporated between 1865 and 1898, were built at a time when there was little fashionable suburban development elsewhere in the country, and they set the basic pattern for many of the railroad suburbs in the entire nation (see TABLE A-6).

The Country Club and the Organization of Suburban Leisure

In his *Theory of the Leisure Class,* published in 1899, Thorstein Veblen described recreational activities in America as the private preserve of an elite group that engaged in unproductive games and sports purely to define themselves as superior to the working class. It had not always been so. As late as 1780, the prevailing view of intellectuals and men of culture was that hunting was the pastime solely of boorish backwoodsmen; not until after the Civil War did the sport take on profoundly aristocratic connotations as it evolved into a highly organized social activity of the gentry. In almost every case, the English preceded Americans by a decade or so in giving their sports organized form. Not until the 1850s did the British upper crust get caught up in clubs of cricket, lawn tennis, yachting, and eventually golf. Americans, however, quickly emulated their betters and established a new phenomenon—the country club— to give social sanction to outdoor recreation.[16]

The growing acceptance of physical activity and of sports was closely associated with the expansion of upper-class railroad suburbs in the late nineteenth century. Clubs for gentlemen dated back at least to 1792, when the Belvedere Club in New York offered its thirty-three members the convenience of rooms for public entertainment as well as ornamental gardens and a bowling green. Similar downtown clubs like the Knickerbocker (1871), the Union (1836), and the Century (1847) in New York; and the Somerset (1851) in Boston were the forerunners of the country-club movement. The members, however, were not initially well disposed toward the ostentation and pretense of social games. In the 1880s the older dues payers of the New York Athletic Club angrily protested the decision of the board to "go social," because the action departed from the intentions of the founders. No longer would the club's primary focus be upon athletics. "The social element in the clubs is like dry rot," explained Frederick W. Janssen. "It very soon causes them to fail in the purpose for which they are organized." According to the insurgents, the expense of palatial clubhouses could be better spent beautifying athletic grounds, improving tracks, and increasing public interest in sports more generally.[17]

Country clubs represented a new method of making spacious fields and restful shade available to the city resident while adding a few social amenities to suburban life. Initially, their members hailed more often from inner cities and used the clubs as a recreational destination on weekends. Increasingly, however, the country club came to be the focus of suburban social life and a spur to residential movement toward the periphery.

Although the Myopia Hunt Club of South Hamilton, Massachusetts, claims that it was founded in 1875 as a baseball club, and although the Larchmont Yacht Club was launched in 1880, the first of the modern genus was the Country Club of Brookline, which was established on one hundred picturesque acres about six miles from the State House in Boston. "The Country Club," as it is reverently known, was one of the primary agencies for preserving the cohesiveness of Boston society and for making suburban residence the social equal of the older neighborhoods. As one local historian said of its members: "They played with each other, not with others; they competed with each other, not with others; above all, they married each other only."[18]

By 1895 such venerable institutions as the Westchester Country Club, the Meadowbrook Country Club on Long Island, the Essex Country Club outside Newark, and the Philadelphia Country Club had been established. Most found their raison d'etre in the encouragement of outdoor sports in the company of social equals. With stables and kennels for the hunters and dogs, with graceful stone bridges to shimmering pools and green polo fields, they openly practiced class, ethnic, and racial exclusion. Only those with wealth could idle away time on games that demanded expensive equipment, immense acreage, and a staff of assistants. Indeed, in some communities, like Morristown, New Jersey, and Tuxedo Park, New York, membership in "the club" was a prerequisite for residence in the suburb. As George Birmingham, an Englishman, said of the Tuxedo institution, "The club exists primarily as a social center of Tuxedo. It is in one way the ideal, the perfect country club. It not only fosters, it regulates and governs the social life of the place."[19]

Although country clubs developed elaborate activities for women only or for whole families to go along with the polo, tennis, and yachting;

golf was the common denominator of the country club, the aristocratic pastime par excellence. First played in the United States at St. Andrews Golf Club in Westchester County (Yonkers), the game conformed to nearly all Veblen's canons of conspicuous waste and offered many opportunities to advertise pecuniary status. It was extremely time-consuming, and its dress and equipment costs could be inflated to very impressive proportions. With a few exceptions, a reporter wrote in 1898, golf "is a sport restricted to the richer classes of the country." Exclusive clubs such as the Onwentsia in Lake Forest and Shinnecock Hills Golf Club in Southhampton, Long Island, became famous for notable courses. In the first two decades of the twentieth century those men who stood a notch or two below the rich began to seek release, status, or relaxation on the links. By 1917 there were 472 golf courses in the United States, and by 1930 the total had jumped to 5,856, more than 90 percent of which belonged to private clubs.[20]

The country club exemplified the trend in which suburbs moved later in the twentieth century, when sport became almost an end in itself. The picturesque ideal fostered by Andrew Jackson Downing and Catharine Beecher emphasized a varied and significant life, but as leisure time increased compulsive play became an accepted alternative to compulsive work. Real nature was forgotten in the midst of manicured greens and all-weather tennis courts.

The Socioeconomic Composition of Railroad Suburbs

Although railroad suburbs obviously differed enormously, some general characteristics were so widely shared at the turn of the century that it is possible to construct an idealized model of the commuting village. Certainly, the country-club image was not appropriate for most of the inhabitants. Unlike post-World War II suburbs, which are relatively homogeneous socioeconomically, those of the tracked city were not restricted to a single economic class. There was diversity behind the posh Main Line stereotype. Even the richest communities were dotted with the small dwellings of those who furnished the support a grouping of large mansions required. In most railroad suburbs, about 30 percent to 50 percent of the heads of households in the late nineteenth century were affluent businessmen who traveled at least five miles to work and whose families were devoted to the pursuit of culture and recreation in the company of social equals. Similarly, most such towns had a larger and poorer group of citizens whose function was to provide gardening, domestic, and other services for the wealthier class.[21]

The Boston region offers excellent examples of the socioeconomic split that characterized most railroad suburbs. In the 1850s the Chestnut Hill section of Newton was already on its way toward attracting "the best class of residents" and becoming "the most select of all Boston's First Family suburbs."[22] Throughout the final decades of the century the town maintained an almost feudal image, and observers noted only the brilliant sectors of Chestnut Hill life. A careful analysis of Newton tax lists and city directories for the period between 1850 and 1890, however, reveals that about a third of the bread-winners (a maximum of 39 percent in 1875 and a minimum of 23 percent in 1891) in Chestnut Hill consisted of laborers, gardeners, and domestic servants. As it evolved into an upper-class community, it became a neighborhood of economic extremes. Although the average taxable income and property assessment were unusually high and increasing, almost half the population paid only a head tax or no tax at all. This situation occurred precisely because the community was so exclusive that the barons of Chestnut Hill regarded the close proximity of a poor servant class as an advantage.[23]

Another Boston railroad suburb with grandiose pretensions was Brookline, the self-styled "richest town in the world." As the most prominent of the region's independent suburbs, Brookline attracted a galaxy of the rich and famous in the latter part of the nineteenth century—architect Henry H. Richardson, landscape gardener Frederick Law Olmsted, financial publisher Henry V. Poor, and poet Amy Lowell, among them. Alongside these prosperous members of the upper crust, however, lived a poor, blue-collar, first-generation Irish Catholic community. In 1870, when 10 percent of Brookline taxpayers controlled more than 70 percent of the assessed property, less than 25 percent of the populace was in executive, professional, or managerial positions. The unskilled Irishmen found plenty of work hauling lumber, grading streets, tending lawns, and digging trenches. Moreover, Brookline was the home of several small industries after 1850, a circumstance that would be unlikely in a similar high-status suburb in the 1980s. In 1879 the *Brookline Chronicle* opined, with only slight exaggeration: "Brookline, unlike any other town in America, has, substantially, no middle class. We have the opulent class who occupy the country houses, on the one hand, and on the other, the laboring class and editors." In fact, the socioeconomic split in Brookline was duplicated in many other Main Line suburbs.[24]

The Pattern of Main Line Settlement

The gap between the rich and poor in railroad suburbs was reflected both in the journey-to-work of the inhabitants and in the distribution of houses within the villages. The cost of commuting to such outlying communities was always high and always beyond the reach of the average working man. Although the railroads often experimented with various types of yearly tickets and other discount fares, the annual cost of travel via such steam-powered trains ranged between $35 and $150, depending upon local circumstances, distance traveled, and promotional vagaries. Even the lower sum would have been a heavy burden for a laboring family, especially since about half the railroad commuters also had to use a local form of public transportation after arriving at the station.

Relative to other forms of travel in the late nineteenth century, railroad commuting was not only expensive but also time consuming. This was true because steam engines were difficult to both start and stop; unlike the horsecar or later the electric streetcar, the steam engine generated speed slowly. The practical result of this limitation was that railroad suburbs were usually discontinuous and separated by at least a mile or two of open space or greenbelt from each other. The typical pattern was for them to develop like beads on a string; the towns themselves were connected by the railroad line but were not initially contiguous either to each other or to the central city. Within the suburb itself, the natural limit to the spread of houses was walking distance to the railroad station. Only the very wealthy, who could afford a horse and driver, penetrated into open country. The towns therefore remained small in size and quite distant from the central city. Scarsdale, for example, was more than twenty miles from the business district and less than two miles square in 1920.[25] As Lewis Mumford noted in *The City in History,* "As long as the railroad stop and walking distances controlled suburban growth, the suburb had form."

Unlike their employers, the lower-paid workers who serviced the large homes in the railroad suburbs could not afford the cost of regular commutation. Typically, they lived either in servant's quarters within the estate grounds or in the more modest dwellings that surrounded the railroad stations. In Brookline, for example, the 40 percent of the population that was Irish almost inevitably lived in small, inexpensive houses on the low-lying land near the town center and station. Thus, the physical appearance of the railroad suburbs tended to duplicate the class-related spatial patterns of the core cities, with the poorest inhabitants living closest to the tiny business districts and the more affluent residents living in commodious homes on landscaped grounds of a quarter acre or more.[26]

The rapid growth of railroad suburbs after the Civil war and their substantial distances from large cities were reflected in the growing disjuncture between work and residence. As noted in Chapter 2, the tendency for a privileged few to move away from close proximity to their businesses began in the largest American cities about 1825, and by 1860 a commute of a mile or more had become a common experience of the urban elite. Among attorneys with offices in Manhattan, for example, the 40 percent who lived outside the island traveled an average of 8½ miles to their jobs in 1888, a harbinger of twentieth-century patterns among less affluent persons in smaller cities (see Appendix).

The nineteenth century was the century of steam, when the commuting railroads created suburbs of a new type in North America—distant from the city, distinguished by an elite dominant class, semirural in orientation, and mixed socioeconomically. As they concentrated industry at the rail junctions in the cities, as they blighted nearby neighborhoods with factories and switching yards, and as they shattered the quiet, they also made possible outlying communities of beauty, tranquility, and prosperity. Places like Lake Forest, Bronxville, and Brookline enabled the select few—and their supporting minions—to enjoy the benefits of American production and resources without bearing the burden of living near the noxious fumes, deafening noises, and poor peole who made the prosperity possible in the first place. The elite railroad suburb was not the solution for all. Not only was it expensive, but as it grew the rural ambience gradually eroded. But the well-known villages did not decline with urban development; they simply attracted white-collar newcomers who maintained community prestige. Most importantly, the railroad suburbs, although small in size and number, stood as a model for success. In the nineteenth century the image of suburbia as an affluent community of railroad commuters was set, and the image remained until the interstate suburbs developed in the 1960s. Such suburbs reached their apex about 1920, when Grand Central Terminal alone was handling more than 680 commuter trains per week, when Pennsylvania Station was even larger, and when every large city took special pride in the affluence and distinction of its railroad suburbs.

The phenomenon occured in other countries, but it assumed a different form. In Sweden, upper-class residential suburbs like Djursholm, Sundyberg, and Saltsjöbaden were developed on the new Stockholm-to-Västeras railroad after 1876. They featured large houses on spacious, well-landscaped grounds, as did those in America. But unlike the situation in the United States, suburban areas in Sweden were more commonly the dumping ground of the ill-housed. Beginning in 1904 the Stockholm city government began buying thousands of acres of land on the periphery for the housing of the working class. In Germany, Holland, and France, the pattern was similar.[27]

6

The Time of the Trolley

If the generation after the Civil War was notable for its great fortunes and its conspicuous consumption, it was equally noted for the wave of invention that swept over the United States and transformed the ways in which middle-class Americans worked and played. Among the more important developments were the air brake (Westinghouse, 1868), the telephone (Bell, 1876), the phonograph (Edison, 1877), the electric light (Edison, 1879), the fountain pen (Waterman, 1884), the adding machine (Burroughs, 1885), the linotype (Mergenthaler, 1885), the small camera (Eastman, 1888), the pneumatic tire (Dunlop, 1888), and the zipper (Judson, 1891). Some insight into the nature of the momentous changes involved is suggested by contrasting the ages of steam and electronics: heat, sweat, and grime gave way to the cool glitter of the metropolis at night, to the glistening, mathematically perfect shapes of mass-produced alloys, to the button-pushing leverage of new sources of energy. No invention, however, had a greater impact on the American city between the Civil War and World War I than the visible and noisy streetcar and the tracks that snaked down the broad avenues into undeveloped land.[1]

The Cable Car

The electric streetcar was not the first successful application of mechanical power to the problems of intraurban transport. In 1867 a maverick New York City inventor, Charles T. Harvey, developed an overhead vehicle connected by a releasable grip to a constantly moving cable and installed a primitive prototype over a three-block run in Greenwich Village. The effort ultimately failed, however, and it was left to Andrew

Smith Hallidie, a Scottish immigrant who had found wealth in San Francisco as a wire-rope manufacturer, to attempt an urban duplication of the English mining technique of hauling coal cars by large cables. Passenger vehicles ran along tracks similar to those of the horse railways, but the power came from the giant steam engines that moved the cable. Easily adaptable to the broad, straight avenues of American cities—as opposed to the narrow and sinuous streets of European urban centers—the cable car was particularly suited to Nob Hill and other perilous inclines of the city by the Bay. Chicago, however, quickly developed the world's most extensive cable system, particularly to its South Side, and by 1894, the middle western metropolis boasted of more than 1,500 grip and trailer cars operating over eighty-six miles of track. Philadelphia opened its first cable line in 1883, followed by New York and Oakland in 1887. By 1890 when cable transportation reached its peak, there were 283 miles of track in twenty-three cities carrying 373 million passengers per year.

Cable-car operation, at least in theory, was simple. When the conductor wanted to go forward, he pulled a lever that engaged a grip into the moving cable. When he sought to decelerate or stop, he released the lever and thus disengaged the car. The cable itself, and almost all of its supporting machinery, was underground, and it moved at a constant speed. The principle was very similar to that of escalators or people-movers in the late twentieth century.

Hallidie's invention had a number of advantages over the only available alternative—the horsecar. The cable car was cleaner (no dirt or cinders and no stable stench), quieter (most of the gear was underground), and more powerful (it could easily carry heavy loads up the steepest grades). As an added benefit, it involved no cruelty to man's faithful friend the horse. It was also swifter than animal-powered locomotion. As one satisfied customer noted:

> In the morning the business man finds it very convenient to linger at the breakfast table from a quarter to half an hour longer than before, and still by means of the cable car reach his office on time. Its unlimited power and tireless energy make possible the operation of cars of generous size affording ample seating capacity and better light and ventilation. On every warm summer evening the open cars are crowded with people riding solely to enjoy the cool and refreshing breeze.[2]

Because of these advantages, the cable car encouraged real-estate development. In Chicago, for example, the construction of the cable lines along Clark and Wells Streets in 1885 was coincident with the 1885–1894 building boom in Lincoln Park.[3]

Over the long term, however, the disadvantages of the cable car were more weighty than its advantages. The most serious was the initial capital cost; the minimum expenditure per mile of cable line in the 1880s was $100,000, and figures of twice that much were common. The fact that construction costs were several times those of the horsecar and later the trolley meant that cable operations had to be restricted to the very heavily traveled routes where passenger revenues would be sufficient to recover enormous investments.

A second major problem of the cable car was its inefficient and sometimes difficult operation. More than 95 percent of the power was expended to move the cable alone exclusive of the cars and the passengers. While this made the climbing of steep hills a simple proposition, it was an extraordinary waste of energy over relatively flat routes. In addition, the speed of the cable could not be adjusted during rush and slack hours.

Conductors also experienced with annoying frequency the inability to disengage their grips. On such occasions there was no way to stop the runaway car, so the conductor had to clang desperately on his bell, warning pedestrians, vehicles, and other obstructions to clear the track. Respite came only when someone was able to notify the power station to shut down the entire cable.

The expense and inefficiencies of the cable car meant that its popularity waned quickly. Most cities in fact remained with the horsecar, especially on the less-traveled streets and on the crosstown routes. As late as 1887, more than four hundred horsecar companies were operating in three hundred towns and cities and carrying about 175 million passengers per year. In smaller cities and towns, cable cars were never even installed.

Only San Francisco retained Hallidie's invention, primarily for nostalgia and tourism. In 1955 the San Francisco city charter was amended to provide that cable cars "shall remain in service in perpetuity," and in 1964 the remaining 10½ mile system was declared to be a national monument. Even so, the Municipal Railway declined sharply in size and profitability. In 1900 San Francisco had 600 cars operating over 110 miles of cable line. By 1980, when the system was losing $1.5 million per year, only forty cars operated over the short routes remaining.[4]

Disenchantment with the Horsecar

Horsecars were never as pleasant as legend would later make them out to be. As William Dean Howells noted from his perch in suburban Boston in 1870:

> I make my way to Bowdoin Square, and in the conscientious spirit of modern inquiry, I get aboard the first car that comes up. Like every other car, it is meant to seat twenty passengers. It does this, and besides it carries in the aisle and on the platform forty passengers standing. The air is what you may imagine, if you know that not only is the place so indecently crowded, but that in the center of the car are two adopted citizens, far gone in drink, who have the aspect and smell of having passed the day in an ash heap. The citizens being quite helpless themselves, are supported by the public, and repose in singular comfort upon all the passengers near them. . . . But they are comparatively an ornament to society till the conductor objects to the amount they offer him for fare, for after that they wish to fight him during the journey, and invite him at short intervals to step out and be shown what manner of men they are.[5]

Drunken and disorderly behavior could not easily be eliminated by even the most conscientious management, but the major problem of the horsecar lines was the inability of the companies to maintain the standards of clean and reliable service that were contracted for in their charters. As the *New York World* editorialized in 1886

> For filth, dilapidation, and a general appearance of squalor and slovenliness some of the streetcar lines of this city cannot be surpassed in the civilized world. Ladies and gentlemen are compelled to sit down on seats sticky with nastiness, breathe loathsome air, and look out of cracked windows that are splashed with dirt from one end to the other. Some of these cars are washed only by rain.[6]

Perhaps the most serious debility of the horsecar had to do with its mode of locomotion. In 1885 the nation's streetcar lines were dependent upon the muscle of more than 100,000 horses. Hills were a problem. Extra teams had to be stationed at steep grades, and the hitching and unhitching often consumed a full ten minutes. The sturdiest beast had to spend at least twelve hours per day in a stall, which meant that each car had to be assigned at least two and possibly as many as four horses. Even with such rotation, pedestrians and passengers were daily appalled by the spectacle of overworked animals straining under the weight of crowded cars and being beaten by drivers anxious to remain on schedule. Death for the animals was common. Horses often stumbled and fell, in which case they were destroyed where they lay. Among the unfortunate beasts in the 1880s, an estimated 15,000 horses died each year on New York's streets alone, and as late as 1912 Chicago experienced 10,000 such deaths annually. Carcasses were left in the gutters, there to collect flies and to give off a stench that but for an adjustment peculiar to the olfactory sense would have driven human life from the town.[7]

Horses shared the maladies of other living organisms. Occasional epidemics of zoogenic diseases illustrated the vulnerability of the horsecar systems to the vagaries of infection. In the 1870s and 1880s particularly serious outbreaks resulted in the virtual halting of service, despite hurried attempts to substitute gangs of men for sick and dying horses. More objectionable were the solid wastes. The average droppings of a drayhorse amounted to ten pounds per day, and much of it was deposited in the middle of the street. Professor Joel Tarr has estimated that in a city like Milwaukee in 1907, with a horse population of 12,500, this meant 133 tons of manure every day. In Rochester health officials calculated in 1900 that, if all the manure produced by the city's 15,000 horses were gathered in one place, it would rise 175 feet over an acre of ground and would breed sixteen billion flies.[8]

Slow, inefficient, polluting, and subject to disease, the horsecar companies awaited a better method. Little wonder that the electrified streetcar, and later the automobile, were looked upon as the salvation of the city.

The Invention of the Electric Streetcar

Although inventors in Europe were seeking similar electrified solutions to the urban transit crisis, the trolley was born in New Jersey. In 1880 Thomas Edison experimented in Menlo Park briefly and unsuccessfully with a half-mile-long electrified railway. Three years later, Leo Daft, an English immigrant living in New Jersey, produced the forerunner of the streetcar, a self-propelled vehicle called the *Ampere*. On August 5, 1885, Daft tested his electric locomotive over a three-mile stretch of a Baltimore horsecar line. Although this was the first electric railway in regular operation in the world, the third-rail system had an unfortunate tendency to electrocute small animals unaware of the hazard. The Baltimore line subsequently returned to four-legged power.

Meanwhile, in 1884, Edward Bentley and Walter Knight were experimenting in Cleveland. Using an underground wooden conduit carrying two copper-wire conductors, the car drew its current by means of a small plow connected to the trench. However, a myriad of mechanical breakdowns and technical difficulties caused the East Cleveland Railway to abandon the system in 1885.

The world's next trolley line opened in Montgomery, Alabama, in April, 1886. Designed by Charles J. Van Depoele, a Belgian-born Detroit furniture manufacturer who had first demonstrated an electrical traction system at the Toronto Exhibition of 1884, this method used overhead wires instead of underground conduits. Each car contained a heavy electric motor which transmitted power from the wire to the wheels. Old residents initially feared the attempt to move large vehicles by "lightning," and the *Montgomery Advertiser* had to assure its readers, "There is more real danger . . . in a Texas mule's heels than in all the electric motor system." Although their suspicions were allayed and the city experienced a real-estate boom as a result of the experiment, Van Depoele could not secure enough financial backing to improve his electric motors before another inventor made the breakthrough that launched the streetcar era.

The most successful of the early trolley ventures and the one that would spawn a revolution in urban transport was that of Frank Julian Sprague. An Annapolis graduate who had tried, unsuccessfully, to devise a way for warships to be illuminated by electricity rather than by oil, Sprague resigned from the Navy in 1883, worked briefly as an assistant to Thomas Edison, and founded the Sprague Electric Railway and Motor Company in 1884. After experimenting on a 200-foot length of track in a New York City alley, he entered into an agreement in 1887 with the Richmond Union Passenger Railway to devise a system that would serve Virginia's largest city in its entirety. Justly regarded as the "father of electric traction," Sprague overcame the manifold technical requirements for a feasible operation. He used twelve miles of track and forty cars, which got their power—and their name—from a little four-wheeled carriage connected to the cars by a flexible overhead cable. This carriage was called a "troller" because it was pulled or trolled along the wires. A corruption of the word produced "trolley," the universal designation of the electric streetcar.[9]

Although the Richmond undertaking cost Sprague more money than he made on it, the well-publicized attempt to move many cars simultaneously by means of overhead wires demonstrated the feasibility of transit electrification on a broad scale. Thereafter, Richmond became a temporary mecca for railway investors and operators. By the turn of the century, half the streetcar systems in the United States were equipped by Sprague, and 90 percent were using his patents.[10]

The typical trolley resembled a nineteenth-century railroad car. It had metal wheels underneath, open platforms front and rear, and large windows all around. About half the size of a modern bus, it swayed and clanged down the small railroad tracks that were specially designed for its use. With its constantly humming motor controlled by a driver in a glassed-in cubicle, the vehicle ordinarily had no front or back because it could not be turned around at the end of the line.

Pollution-free electric traction possessed many advantages. Faster than either the cable car or the horse-drawn tram, it raised the potential speed of city travel to 20 miles per hour (the average was 10–15 miles per hour), and was capable of additional acceleration in low-density or undeveloped areas. Similarly, it achieved substantial economies over alternative forms of transit. It required neither the extensive underground paraphernalia of the cable car, nor the heavy investment in animals, feed, and stables of the horsecar. Because the trolleys themselves tended to be larger than the horse trams, the cost per passenger mile was reduced by at least 50 percent, and the average price of a fare dropped from a dime to a nickel.[11]

The Robber Barons of the Street Railways

Even as the fares dropped, the streetcar lines provided the basis for substantial fortunes. In New York City, for example, Peter Widener, William Whitney, and several others formed the Metropolitan Street Railway and profited to the tune of $100 million between 1893 and 1902.[12] As with the omnibus and the horsecar, the exclusive right to operate an electric railway line along particular city streets was typically granted by municipal governments to private companies in return for certain guarantees of service. Bribery and political favoritism were the most common requisites of successful applications. Unlike the situation in Europe, however, where government control was pervasive, American cities granted lengthy franchises, potentially worth tens of millions of dollars, on the basis of only vague statements about rates and fares, and with virtually no limitation on the value (shareholders' equity) that the company could assign to its properties. Profitability was the main criterion for the establishment of any route.

Initially numerous smaller companies provided service over only two or three streets. Consolidation into a few large firms, however, took place in most American cities before the turn of the century. In Philadelphia, sixty-six different street railway companies were incorporated between 1854 and 1895. By the latter year, most of them had combined to form the giant Union Traction Company. In Massachusetts there were 142 operating consolidations between 1890 and 1915, the most important of them in Boston, where steamship mogul Henry M. Whitney formed a syndicate that merged six streetcar companies. Although this strategy led to heavy debt from buying out competing routes and from enormous outlays to install and to repair the new technology over the short run, electrification and consolidation generated enormous operating profits later, especially in the 1890s, when the cost of both wages and equipment declined. And such mergers were usually popular with the public because they generated more free transfers from one route to another, enabling passengers to travel farther on a single fare.[13]

The greatest street railway tycoon was Charles Tyson Yerkes, who organized transit enterprises all over the Middle West. Born in Philadelphia in 1837, he became a wealthy commission broker and dealer in municipal securities before losing everything during an 1871 panic on the Philadelphia exchange caused by news of the great Chicago fire. In 1872 he went to prison for embezzlement, but served only seven months of his term. Within two years he had recovered most of his fortune.

Yerkes moved to Chicago in 1881 and established a grain and brokerage office. With borrowed money, he got an option on the North Chicago Street Railway and slowly gained control of different lines. By the end of the decade, most of the companies operating on the north and west sides, including those in the suburbs, had been consolidated into the Yerkes group.

In a purely technological sense, Yerkes gave Chicago a superb transportation system. He replaced horsecars with cable or electric traction, increased surface lines by five hundred miles, and built the renowned downtown Loop, consisting of elevated lines around the central business district. By 1895 Chicago had more miles of track, more routes, and a longer average ride than any other city on earth.

However, Yerkes also had a legendary contempt for the public and for politicians who honestly represented them. "Whatever I do," he boasted, "I do not from any sense of duty, but to satisfy myself." Newspaper criticism of overcrowded rush-hour coaches seemed not to bother him. As he reminded his stockholders, "It is the people who hang to the strap who pay you big dividends."[14] Unpopular in the extreme, Yerkes was the target of Theodore Dreiser's fictionalized trilogy, *The Financier, The Titan,* and *The Stoic.*

Yerkes's "Chicago Traction Tangle," as it was derisively called, was finally beaten in 1897, when he sought to extend his franchises for fifty years. A compliant state legislature, aided by under-the-counter payments to influential party leaders, passed the Allen Law giving Chicago the authority to make such an agreement. Unfortunately for Yerkes, however, public disgust for him was coupled with a strong sentiment for municipal ownership. When the usually venal city aldermen met to give Yerkes what he wanted, a mob carrying guns and ropes surrounded City Hall and so menacingly threatened their representatives with bodily injury that the extension failed. Disappointed but undaunted, Yerkes promptly sold his Chicago holdings and moved to England, where he was in control of the London Underground at the time of his death in 1905.

The Spread of Electrified Traction

Despite the greed, and occasional callousness, of men like Whitney, Widener, and Yerkes, the American people embraced the trolley with extraordinary rapidity and enthusiasm. In 1890, when the federal government first canvased the nation's rail systems, it enumerated 5,700 miles of horsecar track, 500 miles for cable cars, and 1,260 miles for the trolley. By 1893, only six years after Frank Sprague's successful Richmond experiment, more than 250 electric railways had been incorporated in the United States, and more than 60 percent of the nation's 12,000 miles of track had been electrified. By the end of 1903 America's 30,000 miles of street railway was 98 percent electrified. Meanwhile the proportion of cable mileage had fallen from 6 percent to 1 percent and that of horsecar mileage from 69 percent to 1 percent. It was, as a distinguished transportation historian has written, "one of the most rapidly accepted innovations in the history of technology." By comparison the automobile, which was invented at about the same time, was a late bloomer.[15]

Heavy transit use extended down to smaller cities as well. In Massachusetts alone, 211 "towns" were served by street railways in 1910. The trolley represented progress and technological achievement; no community that thought well of its future could afford to be thought backward and unpromising. The electric streetcar was a source of pride; the very symbol of a city. In Georgia, where a device called the county-unit system extended rule control over the state legislature to any gubernatorial election, Governor Eugene Talmadge's most widely known political boast was that he never bothered to campaign in a county that was large enough to have a streetcar.

The rapidity of the American adoption of the trolley was especially striking in comparison with Great Britain and Europe. In 1890 for example the number of passengers carried on American street railways (including cable and elevated systems) was more than two billion per year, or more than twice that of the rest of the world combined. In cities of more than 100,000 inhabitants, the average number of rides per person each year was 172, a figure that included children and other persons who traveled scarcely at all. Berlin, which had the best system in Europe, would have ranked no higher than twenty-second in the United States. At the turn of the century, when the horsecar had virtually disappeared from American streets, it remained the dominant form of urban transport in Britain. In Tokyo the electric streetcar did not make its first appearance until 1903, and in 1911 the 190 kilometer system was less than one-tenth as large as that of New York City.[16]

Tying the City Together

Aside from acting as a major spur to suburbanization, a topic to be considered in more detail in the next chapter, the trolley was the major instrument by which ordinary citizens broke out of the casements of their blocks and began to explore other parts of the city. Almost from the introduction of streetcar service in the late 1880s, patronage on weekends and holidays exceeded that of regular work days by a wide margin. Because the trolley was relatively inexpensive, the public took advantage of the mobility it offered to explore other neighborhoods that had previously been as foreign to them as another land. During the warmer months, the companies put on open-air cars, and whole families took advantage of the opportunity to have an outing or just to marvel at the changing urban tapestry.[17] Particularly adventurous souls purchased little pocket guidebooks which explained how to undertake more elaborate journeys by transferring from trolley line to trolley line.

The streetcar companies encouraged pleasure-riding by establishing race tracks, beer gardens, parks, beaches, and resort hotels at the end of the line. Their greatest stimulus to recreational travel was the amusement park. Usually located on the edge of the city and at the end of a trolley route, such parks were physical expressions of the new importance of leisure in the life of urban families; at the same time, they provided an escape to a fantasy world that was far removed from the humdrum existence of everyday life. Although Coney Island, which developed by 1900 into a massive fairyland of lights and side shows at the extreme southern edge of Brooklyn, was the largest of the new amusement parks, every city provided examples of essentially the same phenomenon. In 1902 the trolley systems of Massachusetts owned thirty-one pleasure parks, virtually all of them on the distant edges of urbanized areas. Chicagoans had a choice of routes to the thrill rides of Riverside; Memphians took the Crosstown trolley to the Fairgrounds and East End Park. In Oakland "Borax" Smith and his Realty Syndicate built Idora Park as a lure to riders, and of course as an inducement to potential customers who would pass by land for sale on the way to the outing. The description of Smith's park by a California Press Association brochure in 1903 would have applied almost equally to many other such places: "One of the finest theaters in Oakland is one of the principal attractions at Idora. . . . The Tea

Garden is always crowded because of its beauty. Then there is the Merry-Go-Round, the scenic railway, the laughing gallery, and a dozen other attractions.''[18]

The Electric Streetcar and the Spatial Distribution of Economic Activity

In the nineteenth-century metropolis, industrial location was largely determined by steamships and railroads. Where they met, and only there, could factories assemble the masses of coal—their driving power—and the basic raw materials necessary to fuel industrial enterprise. And only from the railroad-navigable waterway juncture could factories ship products to distant markets. The trolley did not affect this equation, except to make it easier for workers to reach the plant from more distant residences. This simply created more favorable conditions for more factories, a point noted by Edward Ewing Pratt in his 1911 study of *Industrial Causes of Concentration of Population in New York City*.[19]

As more steam-railroad lines were built late in the nineteenth century, however, more intersections of freight-hauling tracks were created. At these nodal points, new factory locations became possible, and more regional industrial activity moved away from the center city. Olivier Zunz has carefully detailed this process in Detroit, showing how the manufacturing enterprises of the Motor City tended toward the crossing of railroads in the years between 1880 and 1920. A few older concerns continued to reflect the industrial design of the past and remained downtown, but those industries that required a great deal of space and those dealing with warehousing activities moved away from the core, where there was room to expand.

The central business districts of large cities thrived during the time of the trolley. There were many centralizing forces unrelated to the electrification of public transport. The steel-frame skyscraper, developed in the last two decades of the nineteenth century, was the perfect physical embodiment of the heavily centralized city. The telephone and the elevator made life in tall buildings bearable and possible, as did the electric light bulb. But the extraordinary prosperity and vitality of most urban cores between 1890 and 1950 cannot be understood without reference to the streetcar systems. Unlike the railroads, the streetcars penetrated to the very heart of the city. The tracks radiated out from the center like spokes on a wheel, tying residential areas far distant to the heart of the metropolis. Because the routes almost invariably led downtown, with only an occasional crosstown or lateral line, the practical effect was to force almost anyone using public transit to rely on the central business district.

The relationship between the electric streetcar and the spatial distribution of commercial activity is perhaps best illustrated by the development of the department store. Such giant emporiums of material abundance originated before the Civl War, most notably in the enormous dry goods establishment of Alexander T. Stewart in New York. By 1870 he had amassed a fortune of several million dollars, had built one of the most elaborate mansions ever seen on the continent at Fifth Avenue and 34th Street (across from the present site of the Empire State Building), and had launched the suburban development scheme of Garden City.[20]

The man most intimately connected to the rise of the great department store, however, was John Wanamaker of Philadelphia. In 1874 Wanamaker took a calculated risk by purchasing an old railroad freight depot at 13th and Market Streets. Although the structure was much larger than required by Wanamaker's commercial needs and was far removed from what was then the retailing hub of the city, he gambled. In order to fill his enormous space with customers, Wanamaker refined four merchandising concepts that had been developed by himself and others in recent years: low competitive prices, departmental organization, money-back guarantees, and large-scale advertising. By 1900 his store was the commercial center of downtown Philadelphia.

John Wanamaker, and the other merchandising kings who followed—Gimbel, Rich, Marcus, Hudson, Goldsmith, Woodward, Lothrop—could not have succeeded without the aid of an efficient transit system. It was the streetcars that delivered the hordes of shoppers to the huge selling spaces, and the streetcars that took them home again. The department-store owners learned to locate at the intersection of the busiest transit lines. At those points of highest pedestrian accessibility were found the highest land values. In 1910, for example, the one-half square mile of the Loop in downtown Chicago represented almost 40 percent of the total assessed land value of the entire 211-square-mile city. The accessibility of the sites, made possible by mass transit, accounted for the rise in value.[21]

Suburbanization at the Turn of the Century

In the period between 1888 and 1918, when the automobile was still a novelty and a toy, the electric streetcar represented a revolutionary advance in transportation technology. Radiating outward from the central business districts, the tracks opened up a vast suburban ring and enabled electric trains to travel as fast as fourteen miles per hour, or four times faster than the horse-drawn systems they replaced. By the turn of the century, a "new city," segregated by class and economic function and encompassing an area triple the territory of the older walking city, had clearly emerged as the center of the American urban society. The electric streetcar was the key to the shift. So important and pervasive had the trolley become by 1904 that its inventor Frank Sprague could reasonably claim: "The electric railway has become the most potent factor in our modern life."

7

Affordable Homes for the Common Man

> Your Home Is A Retreat—from the elements, the frustrations of daily life . . . where you can shut out the world and its traumas. It's a haven your children can run to, not run away from . . . where worries are calmed and weary hearts and minds renew their strength and courage.
>
> —ETHAN ALLEN advertisement, 1982

The quick and enthusiastic American response to the trolley was more than just a respect for speed and novelty and more than just traditional Yankee fondness for the latest, most complicated gadgetry. Rather, as Joel A. Tarr has indicated, an important part of the reception was the expected "moral influence" of the electric streetcar. In 1890, despite a process of population deconcentration that had been underway for half a century in some neighborhoods, congestion in many sections remained frightfully high, and on some streets, such as the immigrant Lower East Side of Manhattan, it was moving upward toward the highest levels ever recorded on earth. As land values rose, the temptation to exploit every inch of ground proved irresistable. Yet because the poor by the 1890s were concentrated in central areas, where the demand for space was greatest, the people who could ill afford even the barest necessities of life were in fact living on tiny fragments of extraordinarily valuable real estate. The infamous "dumbbell" tenement in New York, which occupied 90 percent of its plot, which crowded twenty-four families (not including boarders) into its meager rooms, and which offered residents only the vista of a narrow airshaft, was simply one result of the inflation of land values in urban areas that were accessible to places of employment.[1]

The obvious solution was to spread out the population toward open land and the relatively inexpensive housing on the periphery. The streetcar was thus seen as a safety valve against further overcrowding. The working classes were not expected to jump automatically at the chance, however, because some people seemed to enjoy the teeming circumstances of their tenement neighborhoods. As *Harper's New Monthly Magazine* put it in 1882:

> Myriads of inmates of the squalid, distressing tenement-houses, in which morality is as impossible as happiness, would not give them up, despite their horrors, for clean, orderly, wholesome habitations in the suburbs, could they be transplanted there and back free of charge. They are in some unaccountable way terribly in love with their own wretchedness.[2]

Most observers saw the suburbs as a panacea, however, and looked forward to the day when cheap fares would enable even the poorest classes to surround their habitations with open spaces. Adna Ferrin Weber, the leading student of nineteenth-century urbanization, thought that electrification made possible "a solution of the problem of concentration of population," because quick transit offered "the solid basis of a hope that the evils of city life, so far as they result from overcrowding, may be in large part removed." In Cincinnati Mayor Henry T. Hunt suggested in 1912 that better streetcar lines to the suburbs would reduce disease and death rates and combat all social ills by decentralizing the inner city population. As a leading Los Angeles reformer noted in 1907:

> The laying out of subdivisions far out beyond the city limits makes cheap and desirable home sites obtainable for a multitude of working men, where they are able to build their bungalows or California houses. . . . The family unit, the desire of the sociologist, can be recovered, when, by rapid transit, giving a fare of from five to seven cents for a 30 minute ride, the working man can be induced to locate his family far from the noisy city. No work for civic betterment is worth more than this.[3]

It had of course been the dream of Andrew Jackson Downing in the 1840s to resettle "honest workingmen" in the distant open spaces by relying on the speed of modern transportation. There they could build the cottages that would promote family stability, peace of mind, patriotism, and moral character. His idea, and that of many other visionaries, was not to duplicate the European experience of the few landlords and the many tenants.[4]

Relative to the Old World, homeownership was easily attainable in the American Republic. An 1838 inquiry in Bristol, England, found that only one-third of one percent of manual workers owned their own homes. In the United States, by contrast, property ownership was within the reach of the lowliest laborer. In his pioneering study of social mobility, Stephan Thernstrom reported that "real estate was strikingly available to working class men who remained in Newburyport for any length of time." After twenty years of residence, the percentage reporting property holdings ranged from 63 to 78 percent. It was a slow and expensive task to discharge the financial burden of these cheap houses, and such purchases often took place at the expense of their childrens' education, but the families owned rather than rented. Similarly, in Detroit, Olivier Zunz has shown that in 1900 fully 55 percent of the Germans, 46 percent of the Irish, and 44 percent of the Poles owned their own homes. A striking feature of these housing statistics is that immigrants owned their homes proportionately even more than middle-class, native white Americans. This conclusion has been confirmed by the extensive research of Carolyn and Gordon Kirk, whose analysis of the general pattern of homeownership in all American cities of 100,000 or more inhabitants at the turn of the century reveals that the proportion of immigrants owning their own homes ranged from a low of 11 percent in New York City to a high of 58 percent in Toledo. Native American rates of ownership varied somewhat less: from 15 percent in New York to a high of 40 percent in Los Angeles. Obviously, variation by city and by ethnic group was enormous, but from an international perspective what is most important about these statistics is that it was not a native-American, or middle-class, or urban phenomenon, but an American phenomenon.[5] Moreover, except for orthodox Jews, whose strict religious observances encouraged group cohesion, and blacks, who encountered discrimination on a scale unknown by other minorities, all ethnic groups participated in the suburbanization process by moving toward the peripheral areas of the city.[6]

The Trolley and Suburbanization

The electric streetcar was vital in opening up the suburbs for the common man. In a pioneering study of Boston, Sam Bass Warner, Jr., has shown that, beginning in the 1870s, the introduction of improved street railway lines made possible a continuing outward expansion of the city by ½ to 1½ miles per decade. In practical terms, this meant that the outer limits of convenient commuting (by public transit as opposed to steam railroad) stood at about six miles from City Hall in 1900 as compared to two miles in 1850. That was the distance that one could reasonably be expected to traverse in one hour or less.[7]

Medford, Massachusetts, which had been a popular place for wealthy Bostonians to build their summer residences in the 1850s, illustrates the trend. With the opening of a double-track streetcar line to Scollay Square, the quiet town became attractive to the "middling sorts," who sought permanent suburban houses. Medford's great estates were divided up into house lots and streets, and its population swelled from 11,000 in 1890 to 23,000 in 1905.[8]

Two policies of the streetcar entrepreneurs were especially important in facilitating the outward movement of population. The first was the practice of extending the lines beyond the built-up portion of the city and into open country. As will be discussed later, this had the practical effect of enabling heads of households to see that a convenient transportation mode would be available from their homesite.

The second essential policy of the trolley companies was the five-cent fare. Unlike the European streetcar systems, which depended upon high prices instead of high passenger volume and adhered to the practice of the zone fare—or payment according to distance—American firms usually adopted a flat fee with free transfers, thus encouraging families to move toward the cheaper land on the periphery. The cheap fare thus served the social purpose of preventing congestion and of reducing the necessity for tenement dwellings.

The pattern was as follows. First, streetcar lines were built out to existing villages, like Tenally Town in the District of Columbia, Hyde Park in Chicago, Idlewild in Memphis, and Weequahic in Newark. These areas subsequently developed into large communities. Second, the tracks actually created residential neighborhoods where none had existed before. In 1886, when President Grover Cleveland bought a house northwest of Georgetown between Wisconsin and Connecticut Avenues in Washington, the area was undeveloped countryside. Ten years later, the trolley had helped turn the section into thriving Cleveland Park.

The best evidence on the relationship between mass transit and urban growth comes from the work of Sam Bass Warner. Examining Boston, Warner found that Henry Whitney and his associates regarded the endless expansion of ridership as the key to profit. To this end, they laid 238 miles of track in Boston, Lynn, and Cambridge. Theoretically, as more areas had access to the city, more people would have a valid reason to ride the cars.[9]

This is not to say that a trolley line could itself determine the pace of change; as a matter of fact transportation is and was only one of a number of variables influencing development. As Max Foran has demonstrated with respect to Calgary, however, the presence of a street railway was the dominant factor in predicting growth. In every case he examined, the areas most popular with prospective home builders were those close to the streetcar routes. This relationship has long been recognized by scholars. In 1903 economist Richard Hurd became the first person to compute formulas for this equation. He assumed that the speedier service of the electric streetcar was saving the average rider fifteen minutes in each direction, thus making possible a journey of three more miles in the same time span. In 1903 dollars, his estimate of the increased values within the three-mile zone of a semi-circular city like Chicago was $456 million. Hurd's precise calculations are unimportant, but he raised to a science what transit entrepreneurs already knew and what the public was beginning to learn. As the Massachusetts Street Railway Commission noted in 1918: "It is a well known fact that real estate served by adequate street railway facilities is more readily saleable and commands a higher price than real estate not so served."[10]

Trolley Tracks and Suburban Developers

The close American relationship between land speculation and the construction and location of streetcar tracks can be demonstrated through an examination of three cities in which transit tycoons were less interested in the nickels in the fare box than they were in their personal land development schemes. They learned quite early that transit access would make undeveloped farmland attractive to potential commuters and thus raise its value. Their goal was simply to reap the land-speculation profit by dictating in their corporate chambers the direction and extent of transportation lines.[11]

Oakland

F. M. "Borax" Smith was such an investor. After making a multimillion-dollar fortune by marketing borax as a cleanser, Smith bought a controlling interest in several electric railway companies in the East Bay area in 1893. Over the next ten years he added other companies to his portfolio, and by 1903 Smith's Oakland Transit Company was operating seventy-five miles of track and carrying thirty million riders per year.[12]

Smith cut costs with a vengeance. Previously, the various companies had rendered service on tracks of different guages which could not be adapted to each other. On one street alone there were six sets of tracks, not to mention duplicating power plants. Smith eliminated unprofitable lines and franchises, especially those that replicated existing services, and he achieved economies of scale by consolidating eight previously separate managerial and accounting operations.

The absence of yards even for the homes of the wealthy in the era of the walking city can be seen in this group of row houses on East 11th Street in New York City. Built by the same developer in 1845, they were built flush with the street and with zero lot lines, a common practice in all big American and European cities of the time.

Andrew Jackson Downing published plans for many rural cottages during his short lifetime. This particular photograph is of a somewhat more elaborate dwelling, but it illustrates Downing's firm belief that the best environment for healthy and happy families was a comfortable home surrounded by gardens and trees and sunlight.

(Opposite) Nineteenth-century prints, lithographs, and paintings were important elements in shaping a broad national consensus in favor of suburban living. This lithograph by Nathaniel Currier was produced in 1855, two years before he began his long association with James M. Ives. Labeled "American Country Life," it was enormously popular and illustrated the sentimental and idealized portrait of rural America that has long been common in the United States. *Courtesy of the Library of Congress.*

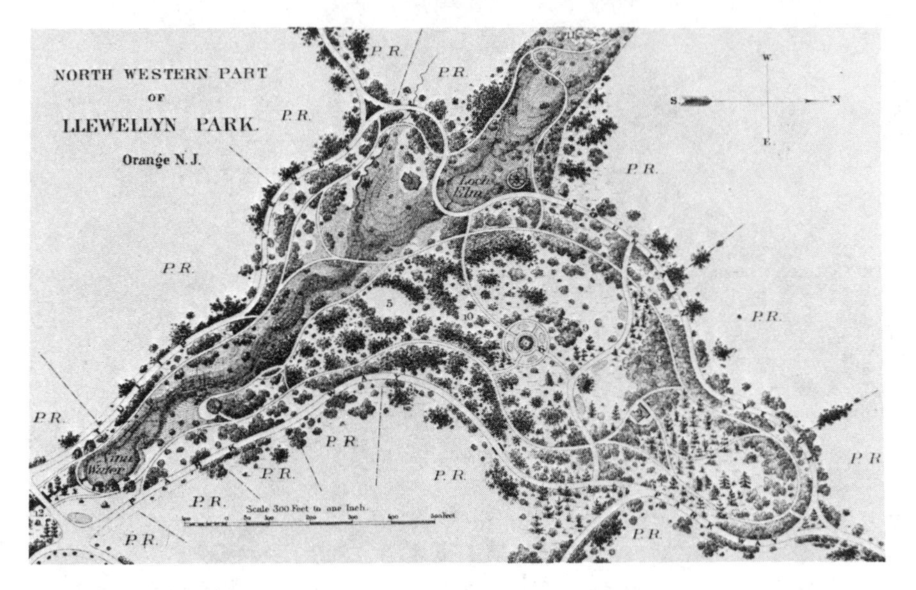

The original plan of the northwestern part of Llewellyn Park, New Jersey. Note the attempt of the developers to use curvilinear roadways and to retain as much of the natural beauty of the site as possible. *Courtesy of Carol Willis.*

The privacy and exclusivity of Lewellyn Park, New Jersey has been preserved by this gatehouse since the 1850s, and in 1985 a private police force continues to man this structure on a 24-hour basis. The idea of a controlled-access community was adopted by only a few communities—Tuxedo Park, New York, and Sea Gate in Brooklyn among them before 1970. In recent years, however, many retirement and resort communities, as well as a few extremely rich towns in the South and West, have resorted to the Llewellyn Park precedent of the gate house and have restricted access to the complex to invited guests. *Courtesy of Carol Willis.*

Organized in 1896 in Lake Forest, Illinois, the Onwentsia Club was typical of the ''country institutions'' that made suburban life fashionable and socially acceptable by the turn of the century. Indeed, the Onwentsia Club claimed many of Chicago's business leaders among its members. *Courtesy of Gottscho Collection, Avery Architectural Library, Columbia University.*

The Lawrence Park Golf Club in Bronxville, New York, shown here in the 1920s, offered it members the pleasure of outdoor sports in the company of social equals. Golf required expensive equipment and clothing, spacious and well-kept grounds, and the ability to leave the workplace for several hours during the day. It thus satsified all of Thorstein Veblen's requirements of ''conspicuous consumption.'' *Courtesy of Lawrence Investment Company.*

The leading symbols of suburban identity are the separate post office and the village hall. In Bronxville, New York, these two institutions were housed together in the building at the center of this 1935 photograph. Although the village itself was only a few miles from the Bronx and New York City, it very carefully emphasized its separateness and worked hard and successfully to retain an exclusive image. *Courtesy of Lawrence Investment Company.*

(Opposite) These two photographs illustrate the dramatic difference between American building techniques and those of Europe. In the first photograph, which was taken in Palos Verdes, California in 1976, the balloon-frame construction method can be clearly seen against the backdrop of recently completed homes, all of which have a core of two-by-four-inch wooden studs. In the second photograph, taken in a suburb of Cologne in West Germany in 1981, a similar picture reveals the almost complete absence of wood. Instead, the Germans make heavy use of masonry, as do most other European peoples.

Size 22' 6"x36' 0" **Design 12681-B** 6 Rooms and Bath

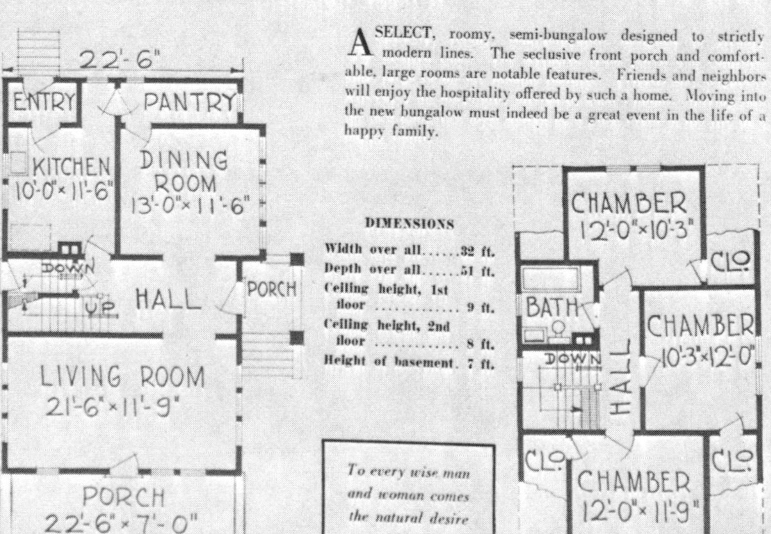

A SELECT, roomy, semi-bungalow designed to strictly modern lines. The seclusive front porch and comfortable, large rooms are notable features. Friends and neighbors will enjoy the hospitality offered by such a home. Moving into the new bungalow must indeed be a great event in the life of a happy family.

DIMENSIONS

Width over all	32 ft.
Depth over all	51 ft.
Ceiling height, 1st floor	9 ft.
Ceiling height, 2nd floor	8 ft.
Height of basement	7 ft.

To every wise man and woman comes the natural desire to own a home.

44

This American pattern book of the early twentieth century offered its purchasers a selection of houses of varying price. With each pattern went a picture of the finished structure and a room-by-room diagram complete with exterior and interior dimensions. Interested families could then purchase finished architectural plans for a home from the publisher or the architect for as little as five dollars. *Photograph by Carol Willis.*

Pattern books offered American families an inexpensive way to build an architect-designed home without the expense and uncertainty that accompanied one-of-a-kind-construction. Typical of such books was this selection from a Rhode Island lumber company of "practical homes of moderate price." *Courtesy of Carol Willis.*

Without water connections, electric utilities, roads, and sewers, the subdivisions of the United States would not have developed as they have. This 1924 view of sewer construction in Brooklyn would not have been duplicated in the outlying areas of most European cities until the 1950s. *Courtesy of Municipal Archives, Department of Records and Information Services, City of New York.*

Smith's land-development plans were closely tied to his trolley operations. In 1895 he associated himself with several other local businessmen to form the Realty Syndicate. This consortium promptly went about the task of purchasing thirteen thousand acres of undeveloped land extending from Mills College to Berkeley, as well as farm properties inaccessible to existing towns. These tracts were then held until the growth of the towns necessitated new subdivisions. The Realty Syndicate would then divide the larger parcels into small building lots and arrange for a trolley line to extend only to that section while bypassing the land of other real estate concerns. The Realty Syndicate not only installed the streets, sidewalks, and sewers, but also provided financing for home builders, an unusual practice available only to heavily capitalized firms. Land that was accessible to an electric railway, namely that owned by the Realty Syndicate, was usually the first choice of middle-class families who depended on public transportation to Oakland to get the breadwinner to work.[13]

So important was real estate to Smith's overall money-making plans that in 1903 he created a commuter line called the Key System that was never intended to make money. Directly competitive with the Southern Pacific Railroad in the Bay Area, the Key System was instead designed to sell real estate in the manner of Smith's other ventures. Smith expected that the profits that the Realty Syndicate would earn on land would more than offset the operating losses of the transit system. In the words of one of Smith's senior executives: "The matter of operation or relations between the two companies (the Key System and the Realty Syndicate) has been very similar to the relation between two pockets in the same man's trousers."

The Southern Pacific Railroad was too large and powerful for Smith, who had hoped that his long-distance competitor would simply concede the commuter traffic to his company. For ten years, the Key System absorbed enormous operating losses; the red ink finally overwhelmed the Realty Syndicate in 1913, and Smith lost control of his empire. But the stamp of his operations would remain on Oakland for the rest of the century.[14]

Los Angeles

Los Angeles provides the premier example of the confluence of street railway entrepreneurs and real-estate development. Land speculation was part of the tradition of the City of Angels from the northern conquest in 1847. Properties inflated 200 percent between 1865 and 1866 and another 500 percent in 1868. During the railroad boom of the mid-1880s the process repeated, fueled in part by Robert Maclay Widney, Francis P. F. Temple, and their associates, who organized the first streetcar line in 1874. Other powerful Los Angeles real-estate moguls soon organized other lines. Although land prices collapsed in 1888–1889, and although over $14 million in property values disappeared in Los Angeles County, transit moguls, especially Moses H. Sherman and Eli P. Clark, bought more properties in the 1890s.[15]

The greatest of Los Angeles transit barons was Henry E. Huntington. At the turn of the century, even as the Southern Pacific was battling "Borax" Smith in Oakland, one of its former owners was himself engaged in almost exactly the same land-transit enterprise six hundred miles to the South. Between 1890 and 1910 Huntington amalgamated several shoestring transit lines into the potent Pacific Electric Railway Company, and he put down tracks throughout the Los Angeles Basin, from Santa Monica to San Bernardino and from Pasadena to Balboa. Technically called "interurbans" because the cars were larger and faster than those which operated solely within city limits, Huntington's "big red cars" soon became as familiar as the orange groves through which they swayed and clattered.[16]

Like "Borax" Smith, Huntington concentrated on short-haul passenger traffic, and like Smith, he was more interested in selling land than streetcar tickets. In any case, Pacific Electric had too many branch lines to be a big money-maker. Deciding that the best financial opportunities were in the manipulation of property values, Huntington formed a land company in 1901 (which was composed of Moses Sherman and Otis and Henry Chandler, about whom more later, among others) to select and promote residential sites. He studied weekend transit patronage to determine which areas appealed to riders, and he focused his real-estate promotions there. Laying out rectangular blocks upon which row upon row of tidy houses could be built, Huntington even operated water companies, usually at a loss, in order to encourage development in the desert-like region. Most importantly, Huntington connected the widely separated towns of the Los Angeles area with one of the nation's best transit systems, and as much as any single person, he initiated the southern California sprawl that still baffles visitors[17]

Washington, D. C.

On the opposite side of the continent, a more exclusive type of suburb was growing up near the nation's capital as the brainchild of Senator Francis G. Newlands of Nevada. A longtime Washington resident who married the daughter of wealthy Senator William Sharon (Nevada, 1875–1881), Newlands was among the first investors to perceive the speculative implications of Frank Sprague's invention.[18] Newlands dreamed of extending Connecticut Avenue into Maryland as a corridor for high-class suburban expansion. In 1888 his Chevy Chase Land Company received a charter to run a trolley line along the Avenue that would connect with the regular District of Columbia transit system. With the Congressional assistance of Senator William Morris Stewart, who himself bought $300,000 worth of the first issue of the Chevy Chase Land Company stock, Newlands not only received a favorable charter for the street railway, but he also arranged for the creation of Rock Creek Park. As Roderick S. French has noted: "Not only did the presence of the park raise the value of nearby properties owned so largely by the Land Company, but at the same time, as Stewart so candidly expressed it, the action took '2,000 acres out of the market.' "[19]

Meanwhile, Senator Newlands and his associates, most notably Colonel George Augustus Armes and realtor Edward J. Stellwag, had been systematically, and at first secretly, purchasing as much farmland as they could along the proposed route. By 1890 they had put together 1,712 acres; landowners who held out for what Newlands regarded as excessive profit were bypassed by a shift in the direction of the road and the trolley. This is why Connecticut Avenue today changes direction somewhat at Chevy Chase Circle.

After completing his purchase, Senator Newlands determined the route of Connecticut Avenue; built and graded a broad, 150-foot-wide right-of-way; and then deeded the thoroughfare to Maryland and the District of Columbia. The action was hardly philanthropy, for Newlands well knew the impact the glamorous road would have on the attractiveness and accessibility of Chevy Chase.

Newlands's aim was to build a totally planned residential area that would serve as a "home suburb" for the national capital. Chevy Chase would be a model of elegance and planning for the entire nation. Alleys were proscribed, and the wide new streets were given appropriate English and Scottish names. Even the trees and shrubs were carefully selected to represent the best in contemporary style and taste. The first subdivision opened in 1893—a depression year—on 250 acres just beyond the District of Columbia line. Newlands said he wanted "a community where every residence would bear a touch of the individuality of the owner," but that individuality had to be expressed within very specific limits. No commercial ventures and no apartments were to mar the serene landscape. Lots were sold with the understanding that no home on Connecticut Avenue could cost less than $5,000, and no dwelling on a side street for less than $3,000. Individual properties had to be at least sixty feet wide, and houses had to be set back at least twenty-five feet from the street. And to give prospective purchasers some idea of the pretentiousness that was envisaged, the land company built the first four houses itself.

Although most streetcar suburbs had a proletarian or middle-class image, Senator Newlands was successful in creating a different tone for Chevy Chase. More interested in quality than in rapid growth, he did not create an instant suburb; in 1900, when the streetcar line made the trip to the White House in thirty-five minutes, only fifty families were living in the town. And Chevy Chase was to remain a tightly controlled enclave of upper-income Americans long after its great growth in the first half to the twentieth century.

"Borax" Smith, Henry Huntington, and Francis Newlands were linked by their common perception of the impact of mass transit on land values. All used the electric streetcar to generate customers for their building lots. All pretended to be operating as independent entrepreneurs in the best traditions of a democratic society. All in fact manipulated government agencies and employed political favoritism in order to use public streets and to gain public franchises for their private ends.

Although few eyebrows were raised over the way politics and business were mixed in the development of American suburbs, such tactics were unique to the United States. In Great Britain and on the European continent, transit owners were not allowed to speculate in real estate served by their lines, and landowners were not given streetcar franchises. In this country, by contrast, no such prohibitions existed, and many entrepreneurs recognized that the opportunities for fast profit in real estate multiplied with the development of mass transit. Even in Boston, where small operators dominated the residential development of the suburbs, the process was helped along by the self-interest of the transit leadership. The great consolidator of the Boston streetcar system, Henry Whitney, was a big speculator in Brookline real estate. With advance information on the planning and layout of Beacon Street and of the trolley line, his West End Land Company bought farms along the right of way at a fraction of their true worth. After title had passed and the lots were ready for sale, the West End Railway brought willing customers to the property.[20]

The Balloon-Frame House

The development schemes of Smith, Huntington, and Newlands were made possible by a new method of residential construction that revolutionized building in the United States. Indeed, the balloon frame was as important as mass transportation in making the private home available to middle-income families and even to those of more marginal economic status. Initially, the earliest European settlers in the colonies possessed few sophisticated tools or machines, and their rude structures were sub standard even in comparison with those they had left behind. By 1650, however, pit saws and other precision instruments had been imported, and the seaport towns hugging the Atlantic Coast had achieved a built form dominated by houses of heavy-timber framing. This type of domestic construction, which would remain common until 1840, was almost identical to that used in medieval England. In contrast, log cabins, with their solid bearing walls of hand-hewn logs interlocked at the corners, were unknown to seventeenth-century Englishmen. The log cabin, which probably developed in the forest culture of northern Europe during the Bronze Age, was part of the native traditions of Scandinavian, Finnish, and German settlers, and it was initially popular only in Delaware and New Jersey. With its distinctive horizontal look, the lob cabin ultimately became the classic woodsman's home in the primitive American wilderness, and it spread over much of the continent.[21]

The more traditional New England frame house, the most common in the colonies, was usually made of oak, a heavy hardwood of great strength, following a post-and-beam construction method. This meant that the weight of the building rested on thick horizontal beams held up by bulky vertical posts. Crosspieces, known as "knee braces," were set in each joint to strengthen the house. The timbers were either hand hewn or power sawed to a rectangular shape and interlocked with a straight butt joint, or later, a mortise-and-tenon joint with oak dowels or hand-wrought nails. In other words, heavy 8-by-8-inch beams were shaped at the ends to fit into slots in adjoining beams. Typically, the frame members were assembled on the ground, then raised into position and joined to create the standing frame, which then supported the remainder of construction. The result was a home of rugged durability.

While it was a reasonably simple and sturdy method of supporting a house, and served the colonists fairly well, the heavy timber frame had many drawbacks. Its name implies the leading fault: the use of thick posts and beams was unwieldy and required much labor in construction. Then again, the specific joining and carpentry techniques, while uncomplicated in principle, were difficult in practice and required specialized knowledge, so that house-building was limited by the availability of experienced craftsmen.[22]

In 1833 in brand-new Chicago—later to become the birthplacc of the other great American architectural innovations, the skyscraper and the prairie house—a type of building appeared that would radically alter the face of built America. Mockingly derided as the ''balloon frame,'' it made possible the new suburban neighborhoods that would absorb most of the population growth of the United States over the next one hundred and fifty years. While even today Europe retains the exterior masonry wall as the primary support for the walls and roof, American houses of every type—brick, stucco, wood, or stone—use an interior wood framing as their primary support.[23]

Solon Robinson, the tireless advocate of the Western immigrant, gives the honors for the invention of the balloon frame to George W. Snow (1797–1870), a jack-of-all-trades in early Chicago who owned a lumber yard. More recently, Christopher Tunnard and Henry Hope Reed have assigned credit to carpenter-architect Augustine Deodat Taylor. In either case, the first building in the new construction was St. Mary's Catholic Church, the initial sanctuary of that denomination in the Windy City.[24]

The balloon frame spread across Chicago, then the Ohio Valley, and finally the populous East, where it was called Chicago construction. It was characterized by the substitution of thin 2-by-4-inch studs, nailed together in such a way that every strain went in the direction of the wood (i.e. against the grain), for the heavy beams and posts held together by mortise and tenon. Abandoning entirely the bulky members of the New England braced frame, this easy method of permanent construction was like a box. Unlike the timber frame, it required no heavy corner post for stability, and it had lateral as well as vertical integrity, which meant that it could withstand heavy wind loads. The weight rested on the 2-by-4-inch posts 16 inches apart and on the floors, which acted as platforms. By spreading the stress over a large number of light boards of a few sizes, the balloon frame had a strength far beyond the seeming capacity of the wood studs. The wall units could be framed on the ground without much carpentry ability and simply raised into place.

Because the balloon frame reduced construction to a few basic hand and tool techniques, the new structure could be erected more quickly by two men than the heavy timber frame by twenty. It was in fact quite common for urban workers in America to build their own homes, a practice that would have been virtually impossible in European cities. This partially explains the findings of the Kirks and of Olivier Zunz that many poorly paid immigrant groups had homeownership rates as high, and in Detroit higher, than more affluent native-white Americans. The immigrants were nailing their homes together themselves, while the local elite employed professional builders and carpenters to put up their more elaborate residences.[25]

Part of the balloon frame's success lay in its generous use of the machine nail, forever banishing complicated mortise-and-tenon joints. Before 1817 nails were hand-wrought and costly (twenty-five cents per pound in 1825). Even with the introduction of improved nail-making machinery, their use was held back by tradition: it was thought that they would not be able to hold the beams in place. But the balloon frame house survived wind and rain, demonstrating the strength and durability of these inexpensive joining devices, which cost only 15 percent as much as the old wrought-iron nails.[26]

Within a generation, home building was transformed from a specialized craft into an industry. As new transit developments made commuting easier, speculators bought large tracts of farmland adjacent to the city and carved them into lots. Individual families and small contractors then built houses modeled upon a common design. Entrepreneurs developed kits that could be delivered to any railroad depot in the nation; no longer was it necessary for the home builder to be able even to cut his studs to the right size; that was done for him. His only task was to follow the instructions and assemble the house. By 1872 prefabricated stores had been introduced, complete with windows and doors.[27]

At about the same time, beginning with Andrew Jackson Downing's *The Architecture of Country Houses* (1850), Gervase Wheeler's *Houses for the People* (1855), Calvert Vaux's *Villas and Cottages* (1857), and Henry Hudson Holly's *Country Seats* (1866), magazines and pattern books began to include house plans for people to emulate. *Godey's Lady's Book,* for example, published about 450 model-house designs between 1846 and 1898. Other successful pattern-book authors and compilers prior to 1876 were William Ranlet (1806–1865), Samuel Sloan (1815–1884), George E. Woodward (1829–1905), and Amos J. Bicknell. Most saw their writings as a way to gain credibility as professional architects. At a time when there were few hard-and-fast distinctions between the architect and builder, at least in the public mind, literary accomplishments could set the architect apart as a cultivated and educated practitioner. A pattern book that included an account of historical styles as well as original designs conferred a certain cachet on its author. Unlike builders' guides, which consisted primarily of ornamental and structural details drawn to scale, the pattern books were composed almost exclusively of designs for suburban and country houses. Because the totality of the house and its integration into the natural surroundings were the paramount concerns, the illustrations were perspective views. Such pattern books appealed to non-professional readers who found it difficult to picture a house from the details, plans, and elevations commonly found in the builders' guides.[28]

Pattern books, whatever their author's intentions, helped to reinforce the public's suspicion that an architect's fee was an unnecessary and luxurious item in a family budget. Detailed specifications for labor and materials and precise cost estimates were also included. Potential home builders could thus copy standardized plans, adding whatever variation they desired. With the initiation of publications such as *House and Garden,* new styles swept across the country with more rapidity. The result was the development of a specifically suburban type of architecture that combined the requirements for servantless domesticity with the ideal of independence and privacy. Unlike the farm house or the manor house, the suburban dwelling was not economically tied to the land. And whether it was in the shingle or colonial style of the East or the mission style of the West, the suburban house represented a distinct visual example of the attempt to take the city to the country. Thus began the American tradition by which most residential structures were put up by builders who took the plans out of a portfolio.

In Europe methods of construction did not change, rates of homeownership remained miniscule, and Engels continued to advise that this was in the best interests of the worker. Writing in *The Housing Question* in 1872, he cited the cottage-owning peasants in rural Germany who worked for starvation wages: "For our workers in the big cities freedom of movement is the first condition of their existence, and land ownership could only be a hindrance to them. Give them their own houses, chain them once again to the soil and you break their power of resistance to the wage cutting of the factory owners."[29]

In the United States, however, as towns all across the nation turned to the balloon-frame house and as detached dwellings came within the reach of many people previously unable to afford such a luxury, the look of the country was changed. As Solon Robinson told a group of New Yorkers in 1855: "If it had not been for the knowledge of the balloon-frames, Chicago and San Francisco could never have arisen, as they did, from little villages to great cities in a single year." And the new method of building was as unique to the United States as the idea of low density suburbs itself.[30]

Cheap Land/High Wages

The rapid suburbanization of the United States cannot be viewed in isolation from the material prosperity of its people and the sheer abundance of its land. By the middle of the nineteenth century, Americans were already a "people of plenty." The wages of the working man, no matter how meager, were almost invariably higher than those of his counterparts elsewhere in the world. Geographically, the amount of space potentially available to each citizen was also staggering. Other countries—Canada, Australia, and Russia—were as large as the American republic, but much of their space consisted of treeless desert or frozen tundra. The United States, by contrast, was composed substantially of heavily forested or grass-covered ground, most of it habitable by human beings. In comparison with other countries of the world, the real estate of North America was almost literally endless.

In urban areas, and especially in developing suburban sections, this translated into land that was cheap by international standards. The average price of a lot suitable for building ranged widely from city to city, but a figure of more than five hundred dollars would have been high before 1900, and prices of $150 were common. The sales target was often the "little man," the working-class clerks, mechanics, and struggling businessmen who aspired to more security and space for their families and who were sensitive to slight variations in price. As an advertisement for the Morris and Southwick Company in the *Louisville Courier-Journal* promised on April 18, 1871:

> $50 CASH WILL BUY A LOT
>
> $50 cash on each lot, remainder in 1, 2, 3, 4, and 5 years. The community, and in particular clerks, mechanics and laboring men who are paying nearly one-half of their earnings for rent are invited to this sale.
>
> BE ON HAND, BUY A LOT
>
> and in the course of a few years you will be rid of house rents and exacting landlords.

Affordable property was in part a function of the continental size of the nation and in part a function of the speed of mass transit. This occurred because the streetcar took advantage of a law of geometry to the effect that the area of land increases with the square of the radius from the center of the city. Thus, by simply doubling the radius, the amount of land available for development would be quadrupled. In economic terms, the marginal cost of transportation did not rise as fast as the price of land fell with distance from the city center. In other words, the least expensive housing option for middle-class families that could afford a commute was to move outward.

The relative affordability of land was greatly aided, especially in the three decades after Appommatox, by a sustained agricultural depression. Between 1865 and 1896, the prices of most commodities fell in real terms, as the American farmer paid more and more and received less and less. This meant that the value of land was falling for agricultural purposes, even as acreage within commuting range of a city was becoming desirable for residential use. On the edges of large communities, therefore, farming could be economically justified only by ignoring the potential value of the property for building lots. Most farmers responded rationally and eventually sold their land; some to speculators and others to individuals. In either case, the amount of suburban-style real estate increased to meet the demand.

The process of land conversion was also speeded by the pre-Civil War introduction of building and loan associations. Even before 1860, realtors claimed "Long Credit and Low Terms." The long credit meant the note was payable in six years, and the low terms were interest due semiannually at 6 percent. Building and loan associations smoothed the lending process. First organized in Philadelphia in 1831, they enabled an individual of modest means to invest his savings in shares of an association and ultimately to borrow against the value of those shares at low interest. An essential part of the loan contract was that the associations provided for a method of repayment.

Most building and loan associations—by 1874, there were four hundred in Philadelphia alone—adamantly preferred small, balloon-frame homes in the suburbs, where typical prices ranged between $1,000 and $4,500. Although such associations were less pervasive in the Middle West, there were thirteen within Louisville by 1875 and eighteen more by 1892. Everywhere, they fostered the view that proper households could and should purchase their own dwelling and that people of moderate means could benefit from the institution of private property.[31]

The role of the savings and loan associations was especially crucial in the nineteenth century because before 1916 most commercial banks were prevented by national legislation from providing long-term credit for real-estate loans. Beginning in the 1890s, such institutions began cautiously to issue a few short-term mortgages for non-income producing residential property, and by 1915, most important banks had set up real-estate departments. Although they acted conservatively and rarely extended large loans to builders, their change in policy during World War I stimulated suburbanization and eased the transition of many families to home-ownership.[32]

The Provision of Urban Services

New balloon-frame communities usually developed in conjunction with the establishment of essential services. Except for immigrant neighborhoods, where improvements were often resisted in order to hold down taxes, massive public investments in roads, storm sewers, street lighting, curbs and gutters, playgrounds, and schools were necessary for growth, and real-estate advertisements usually emphasized the availability of upgraded city facilities. Sewers, for example, were absolutely essential to most native white American neighborhoods; they were usually paid for by public works departments rather than by the developers, especially after the 1890s. And not only were such improvements financed at general taxpayer expense, but their low-density sprawl involved numerous inefficiencies.[33]

Streets are instructive in this regard. Before the Civil War, streets were paved or widened when owners of a certain percentage (usually three-fourths) of the property facing the right-of-way petitioned the city to do so. To finance such improvements, property-owners "abutting and directly affected" paid special assessments. The municipal government played a limited role: the basic decisions as to when and how to pave were made by private individuals. Because the owners would presumably benefit from the increased value of their land after the street was opened, the system had a certain logic and justification. Where the cost of new services would be a financial problem for working-class homeowners, the residents simply delayed the paving.

In the final decades of the nineteenth century, however, a second method of financing became more common, one that passed the cost of peripheral street improvements on to the municipality as a whole. As horsecar lines enabled upper- and middle-income families to move away from downtown, outlying residents placed increasing pressure on the city to build smooth pavements at public expense. By the 1890s engineering publications were hypothesizing that well-paved roads, paid for by the city, would reduce the cost of freight-handling, thus encouraging new businesses and reducing the tax rate. Reformers added their voices to the chorus in the belief that suburbanization and the improved housing it promised would alleviate the evils of tenement districts. By 1900 the changeover was effected. The centralization of street administration meant that all city dwellers subsidized those who moved to the edges.[34]

Tenement house laws are another example. The 1867, 1879, and 1901 New York City ordinances, which influenced similar measures across the nation, did not so much alleviate conditions in immigrant neighborhoods as insure that the worst abuses would not be reproduced in newly developing sections of Brooklyn and the Bronx. The important 1901 requirements dealing with light, ventilation, fire protection, and water closets could not easily be retroactively applied to the crowded buildings of the Lower East Side and Harlem, but they could be enforced with regard to new construction.

Methods of constructing schools and sewers exhibited the same pattern of creating the best environment on the edges, and if necessary, paying for it by taxing the entire city. Howard L. Preston, for example, has documented the way in which Atlantans within the city limits, one-third of whom were black, subsidized the necessary municipal services of unincorporated Fulton County residents, most of whom were white and on the whole better able to pay for schools, police, and street repairs. Preston discovered that more than 50 percent of the local tax revenue collected in 1937 and earmarked for public schools in Atlanta suburbs came from Atlanta taxpayers. This was true for highway maintenance, health costs, and police expenditures in unincorporated sections of Fulton County.[35]

In the twentieth century, the availability of light, heat, and power became a prerequisite for suburban living after electrical appliances and central heating became necessities of the American middle class. The expansionist policies of the private utility companies, which extended power and sewer lines to new subdivisions in advance of actual construction, meant that suburbanization was not stalled by the absence of essential services. Instead, the provision of light and power helped to create a city of islands—one industrial; a second of working-class residents dependent on coal, wood, oil, and ice; and a third of affluent sections on the rim enjoying gas and electric kitchens. This pattern contrasted sharply with Germany, France, and Great Britain, where utility lines were not available in suburban areas until the 1950s, and then only on a piecemeal basis.[36]

Individual Effort and Homeownership

Cheap land, inexpensive construction methods, favorable peripheral taxing policies, and the rapid expansion of public utilities only partially explain high homeownership rates in American urban areas. In the last quarter of the nineteenth century, working-class people were offered free horsecar rides, lunch, and entertainment at sales of land through auction. Under the banners of "Get a Slice of the Earth" and "Own Your Own Home," ethnic arrivals from the Piedmont, Lombardy, Holland, Germany, and Ireland built modest homes along the streetcar lines. They aspired to and achieved the detached house and garden that had been so far beyond their reach in the Old World.

As Olivier Zunz, Roger Simon, and Stephan Thernstrom have noted, immigrant groups managed to purchase homes at rates equal to or above those of native Americans because of extraordinary personal and family sacrifice. As Simon has demonstrated for the 14th Ward of Milwaukee, Polish newcomers purposefully delayed the expense of paved streets and city sewers, eagerly rented first floors or basements to tenants, and did without leisure and material comforts in order to buy modest bungalows. In Simon's judgement, they adapted the physical environment to their needs, for as he says, "the new housing stock and public policy permitted an ordering of priorities in which financial security won out over convenience and public health.[37]

As in so many other aspects of national life, black Americans did not share in the homeownership boom. Their migration from the plantation South to the urban North led to gains in civil rights, but the pattern of the ghetto—residential segregation, underemployment, substandard housing, disrupted family life, inferior education, and disease—separated the black experience from that of white ethnics. Because of racial discrimination, blacks were unable to enter the housing market on the same terms as other groups before them. Thus, the most striking feature of black life was not slum conditions, but the barriers that middle-class blacks encountered in trying to escape the ghetto.

The Process of Suburban Land Conversion

At the heart of all suburban growth is land development—the conversion of rural or vacant land to some sort of residential use. The process involves property owners, speculators, banks, private lenders, builders, and buyers. As land values at the center of the metropolis rise, individual parcels either produce the higher yields to hold their place, or, in the course of a few years, more profitable businesses move in on the site. By the same token, if much the same yield can be earned at a peripheral site of lower value, there is little incentive to remain in and around the central business district. The pattern of urban land investment affects the value of outlying farmlands, which either increase yields by more profitable crops, or, as is usually the case, they give way to more lucrative subdivision and real-estate developments.

A single model cannot possibly describe the entire suburban experience. Before World War I, however, rarely did a single individual or firm buy land, lay out streets, build houses, and finance sales to the ultimate occupants. Instead, a landowner typically hired a civil engineer to determine streets and lots, and then, depending upon local circumstances, either pressured the municipal government to extend pavement at public expense or brought in private crews to construct roads. The land was subsequently sold, often at auction in the nineteenth century, to numerous buyers who would either build houses for their own occupancy for sale, or would retain vacant lots for speculation. The subdivider often retained some of the land for his own use and built his own home there. Some real-estate syndicates operated in the larger markets, but the predominant force was the small developer.[38]

In contrast to the heavy governmental involvement in the use of land in Europe, residential development in the United States has largely been the work of private interests. Indeed, urban real estate was the single most important source of leisured wealth in the nineteenth century. The dynastic Goulet, Rhinelander, and Schermerhorn families in New York and the Palmers in Chicago were all founded by men who invested heavily in city property. In James Fenimore Cooper's 1838 novel, *Home As Found,* the main character returns to a Manhattan world in which quiet respectability has given way to incessant real-estate speculation. Speculators were active in most suburban developments, and advertisements usually stressed the advantages for the investor. "A SPLENDID OPPORTUNITY for speculative investment," noted a *Louisville Courier-Journal* ad on May 20, 1871: "No other locality promises to be so profitable." One month later another company was even more explicit: "If you wish to make a fortune come to this great sale for so certain and great an opportunity to do it by the investment of a small amount of capital was never offered before. . . . The man who invests $10,000 at this sale can take $50,000 for it in five years."

Appropriately, the richest man in the United States before the Civil War was John Jacob Astor (1763–1848), a "self-invented money-making machine" who came to this country at age twenty with five dollars in his pocket. Although Astor's initial successes were in the China and the fur trades, he aggessively bought cheap Manhattan farmland north of Canal Street after 1810. Sensing a great city in the making, he purchased additional tracts at reduced prices during the Panic of 1837, and he ruthlessly foreclosed on hundreds of property owners who were temporarily unable to keep up with their mortgage payments. By 1840 the greatest source of his $20 million estate was in the increased value of his lands in the city, and shortly before he died he remarked: "Could I begin life again, knowing what I now know, and had money to invest, I would buy every foot of land on the island of Manhattan." The wisdom of his judgement remained apparent in 1980, when a corporation offered to purchase St. Bartholomew's Protestant Episcopal Church in midtown Manhattan for $125 million, the highest value yet put on an acre of land in the United States.[39]

The Subdivision and the Real-estate Specialist

The basic unit of development in the nineteenth-century suburb was the subdivision. In 1843, for example, lots were auctioned in the twenty-acre Linden Place subdivision in Brookline, where a few years later the 300-acre Longwood development was opened, this time without an auction so that the residents could more easily select desirable neighbors. In the Bronx (then part of Westchester County), subdivisions date from 1850, when Morrisania first experienced a suburban boom. Within a decade, nearby Fordham and Tremont were also laid out, and by the end of the Civil War their combined population approached 20,000. They were typical of early subdivisions in that they were adjacent to railroads which established commuter stations at their centers.[40]

Just as the auction system was not universal, so also was there no single pattern for subdivision development. In cases where a large residential section emerged from a single rural property and a single investor, the developer usually had full responsibility for the street system. Where contractors lacked the capital to construct more than a few houses at a time, they were in no position to question the city engineer's imposition of the grid street plan. Thus, in many peripheral areas within cities between 1875 and 1945, uniform, narrow rectangular lots defined the houses before they were built.

Whether their subdivisions were large or small, real-estate specialists were more active in the city building process than anyone else. The theory that early suburbs just grew, with owners "turning cowpaths and natural avenues of traffic into streets," is erroneous. Subdividers lobbied with municipal governments to extend city services, they pressured streetcar companies to send tracks into developing sections, and they set the property lines for the individual homes. Each city and most suburbs were created from many small real-estate developments that reflected changing market conditions and local peculiarities. And even when the area was developed by large operators, ultimately the land filtered down to the private buyer.[41]

One of the nation's biggest nineteenth-century developers, for example, was Samuel Eberly Gross of Chicago. Between 1880 and 1882, he laid out 40,000 lots, developed 16 towns and 150 subdivisions, and built and sold more than 7,000 houses, some of them priced below $1,000 and all of them below $5,000. Gross, however, did not try to control the building process nearly so completely as the Levitt organization would do so effectively two generations later. And Gross was an exception on the gargantuan side. More typical of the residential experience in the United States was the Boston area, where Sam Bass Warner has reported the prominence of small contractors and individual homeowners. In the three residential areas that he examined—Roxbury, West Roxbury, and Dorchester—approximately 22,500 dwellings were constructed between 1870 and 1900, and no one company or individual was responsible for more than 3 percent of them.[42]

For the first time in the history of the world, middle-class families in the late nineteenth century could reasonably expect to buy a detached home on an accessible lot in a safe and sanitary environment. Because streetcars were quick and inexpensive, because land was cheaper in suburbs than in cities, and because houses were typically put up using the balloon-frame method, the real price of shelter in the United States was lower than in the Old World. Increased homeownership was reflected in new magazines. In the 1890s *Good Housekeeping, Careful Builders, The American Home,* and *Sanitary News* offered illustrated articles about style and domestic fashion. As Clarence Cook, a well-known New York art critic wrote of the period: "There never was a time when so many books and magazines written for the purpose of bringing the subject of architecture—its history, its theory, its practice—down to the level of popular understanding were produced as in this time of ours."[43]

The general availability of "affordable homes for the working man" did not bring universal joy. In the decades after the Civil War, Charlotte Perkins Gilman and a few kindred spirits denounced widely spaced private dwellings as "bloated buildings, filled with a thousand superfluities." They did not see the evolving suburbs as instruments of female or family liberation but as lace-curtain prisons.[44]

A more common criticism of the burgeoning peripheral subdivisions was that their very popularity undermined the results incoming families hoped to achieve. Privacy and solitude disappeared as the populace streamed outward, blighting the rural charm that had lured them in the first place. The environment they found was never quite as open or as isolated as theorists had wished. The typical streetcar suburb featured one- or two-family homes on lots of about three thousand to six thousand square feet (about one-tenth of an acre). Such dimensions were generous by the standards of the walking city, but much more cramped than Andrew Jackson Downing had envisaged. This compactness occured because the streetcar, the common man's mode of transit par excellence, required a certain volume of traffic for profitable operation, and because the very availability of a transit line tended to raise the price of land. Thus the trolley, which was supposed to provide space for the middle class, was not feasible unless the residential neighborhoods were closely packed.

Nevertheless, the electric streetcar, and the land developers who were so quick to take advantage of its possibilities, had created a new kind of metropolis by 1900, one that was very different from the walking city of a century earlier. By 1900 the center of the city had become an area of office and commerical uses that was almost devoid of residences. Nearby were the grimy factories, and just beyond them the first tenement districts of the poor, the recent immigrants, and the unskilled, persons unable to afford even the streetcar fare and forced to compete for housing space where real estate was the most expensive and housing the least desirable. Along these same streets, the well-to-do had lived only two generations earlier.

Beyond the compact confines of the walking city lay the new streetcar suburbs, the essence of the American achievement at the turn of the century. The residential structures that filled them were not elegant, but they were spacious and affordable by European standards and they represented an attainable goal.

Farther out, the railroad commuters lived in houses that sprawled in ample yards, thick with trees and shrubbery behind iron or wooden fences. These residences represented a new American ideal. Unlike the intown dwellings of the Old World wealthy and unlike the country houses of the English gentry, these structures were uniquely American and, with their pseudo-Gothic towers and cupolas and mansard roofs, they set a suburban rather than an urban standard for achievement-oriented Americans.[45]

A number of models have been developed to explain residential patterns in relationship to the two most important factors: the quality and cost of housing and the convenience, speed, and cost of transportation. The well-known model of urban growth by Burgess assumed that concentric zones of residential areas develop around the central business district in an ascending hierarchy. William Alonso has added to Burgess's conception the idea that in general the price of land decreases with increasing distance from the center. In order to earn a return comparable to that from other uses, residential development near the center has to be dense and compact. The lower-income groups live near the center in the United States because the factors of centrality and cost are more important than the quality of housing. Among the middle and upper classes, the importance of centrality decreases because of the sinking importance of transportation costs. Thus, affordable housing means that the middle class could set as a priority the quality of the dwelling unit, preferring to live in suburbs with a low population density far outside the city.[46]

8

Suburbs into Neighborhoods: The Rise and Fall of Municipal Annexation

Get your children into the country. The cities murder children. The hot pavements, the dust, the noise, are fatal in many cases, and harmful always. The history of successful men is nearly always the history of country boys.

—Wilmington, Delaware, real-estate advertisment, 1905

As new developments in home construction and in urban transportation encouraged American families to move away from their old neighborhoods to new residences on the periphery, the most basic of questions involved the provision of schools, sewers, utilities, and police and fire departments. Four different approaches were possible: (1) cities could simply expand their boundaries by annexing newer sections into the municipal corporation, (2) new municipalities could be created within the suburban ring, (3) special taxing districts could be established to provide for one or more important functions, and (4) county governments could expand their powers by becoming more like cities themselves.

Throughout the nineteenth century, the first alternative was predominant as American cities annexed adjacent land and grew steadily larger in area and in population. Of course, there were always a few small communities here and there that lost population, but most—and certainly all the larger ones—added residents between each decennial census. Historically, city fathers tended to be concerned with the *rate* of growth and with the relative standing of their community and rival cities.[1]

In the second half of the twentieth century, Americans have learned that urban population growth is not inevitable and that city boundaries do not indefinitely expand. Boston, for instance, reached its maximum

TABLE 8-1
Territorial Growth in Square Miles of the Twelve Largest American Cities That Gained Population Between 1950 and 1980

City	1870	1890	1910	1930	1950	1970	1980
Los Angeles	29	29	85	440	451	455	465
Houston	25	9	16	72	160	453	556
Dallas	NA	9	16	42	112	280	378
San Diego	74	74	74	94	99	307	323
San Antonio	36	36	36	36	70	183	267
Phoenix	NA	NA	NA	10	17	247	325
Indianapolis	11	11	33	54	55	379	379
Memphis	4	4	19	46	104	217	290
San Jose	NA	NA	NA	8	17	117	157
Columbus	12	14	23	39	39	114	184
Jacksonville	1	10	10	26	30	827	841
Seattle	11	13	56	69	71	92	92
Totals	203	209	368	936	2161	3671	4257

SOURCES: *Municipal Year Book,* 1981; Various City and County Data Books, various United States Census Reports, and Roderick D. McKenzie, *The Metropolitan Community* (New York, 1933).

population in 1950, and in the next three decades had a net loss of 238,000 inhabitants. And the Hub is not unusual. New York City lost 800,000 in the same period; Chicago, 600,000; Detroit, 465,000; Philadelphia, 400,000; and St. Louis, 400,000. The process of annexation, as we shall see, ended even earlier.

Declining cities are typically old and congested, and except for Birmingham, San Francisco, New Orleans, and Atlanta, located in the East and Middle West. It is often alleged, in fact, that the characteristics of these cities are themselves the reasons for their decline. In a youthful, mobile, and affluent society that seems to be drifting toward the South and the West, one would expect aging and crowded surroundings to be unpopular. But cities that are losing population share another more significant characteristic: their boundaries have not expanded in the last half-century. The municipal area of the dozen largest American cities that experienced a net decline in permanent residents between 1950 and 1980 has increased by less than one percent since 1930. Individual cities have expanded even less (See TABLE 8-2). Philadelphia, San Francisco, and Buffalo have not absorbed additional land since before the Civil War.

Those who have grown up in the suburbs or central cities of the East or Middle West find nothing very strange in all this. There is a Brookline as well as a Boston, an Evanston as well as a Chicago, a New Rochelle as well as a New York, and no one is likely to argue successfully

TABLE 8-2
Territorial Size in Square Miles of the Twelve Largest American Cities That Lost Population Between 1950 and 1980

City	1850	1870	1890	1910	1930	1950	1980
New York	22	22	44	299	299	299	304
Chicago	10	36	169	185	207	223	228
Philadelphia	2	130	130	130	130	128	128
Detroit	6	13	22	41	138	138	138
Baltimore	13	13	30	30	79	79	79
San Francisco	5	42	42	42	42	45	46
Washington	60	60	60	60	62	61	61
Milwaukee	NA	13	17	23	41	54	97
Cleveland	5	12	28	46	71	81	81
Boston	5	13	39	39	44	46	50
St. Louis	14	61	61	61	61	61	61
Pittsburgh	2	23	27	40	51	52	55
Totals	144	438	669	996	1225	1267	1328

SOURCES: *Municipal Year Book,* 1981; Various City and County Data Books, various United States Census Reports; Roderick D. McKenzie, *The Metropolitan Community* (New York, 1933); and *The Book of American City Rankings.*

that they should share a single municipal government. Yet the very fact that these dual communities continue to exist represents a break with the earlier urban tradition. When cities first experienced explosive growth in the nineteenth century, they expanded outward as well as in density. If the earlier pattern had continued, Boston would probably encompass the entire area circumscribed by Route 128, New York City would reach to White Plains in Westchester County and at least to the Suffolk County line on Long Island, and Chicago would stretch half the distance to Milwaukee. Those who find such assertions fantastic are reminded that dozens of American cities, including all those that boast high population growth rates since World War II, have expanded their boundaries in just such a fashion.

The Nineteenth Century

Without exception, the adjustment of local boundaries has been the dominant method of population growth in every American city of consequence. If annexation (the addition of unincorpoated land to the city) or consolidation (the absorption of one municipal government by an-

TABLE 8-3
Territorial Size in Square Miles of the Twenty Largest American Cities in Area, 1980

City	*Square Miles*
1. Anchorage	1955
2. Jacksonville	841
3. Oklahoma City	650
4. Honolulu	604
5. Houston	556
6. Nashville	533
7. Los Angeles	464
8. Dallas	378
9. Indianapolis	375
10. Phoenix	325
11. San Diego	323
12. Kansas City	316
13. New York City	304
14. Memphis	290
15. Lexington	283
16. San Antonio	267
17. Virginia Beach	259
18. Fort Worth	250
19. El Paso	240
20. Chicago	228

SOURCE: *Municipal Year Book,* 1981.

other, usually adjacent) had not taken place, there would now be no great cities in the United States in the political sense of the term.[2] Only New York City would have grown as large as one million people, and it would have remained confined to the island of Manhattan.[3] Viewed another way, if annexation[4] had not been successful in the nineteenth century, many large cities would have been surrounded by suburbs even before the Civil War.[5] For example, the cities of New York and Philadelphia contained in 1980 less than one-half the opulation of their metropolitan areas; in St. Louis, Pittsburgh, and Cleveland the proportion was much less than one third (Appendix A-12). Their boundaries have not been altered for the past three-quarters of a century, and these cities are now extreme examples of core areas being strangled by incorporated suburbs. A St. Louis school administrator complained a decade ago that suburbanites had "erected a wall of separation which towers above the city limits and constitutes a barrier as effective as did those of ancient Jericho or that of the Potsdamer Platz in Berlin."[6] Yet if these cities had been unable to add territory before the Civil War, their central areas would have contained about the same percentage of the metropolitan population in 1850 as in 1980.

If we consider the dozen largest American cities that lost population between 1950 and 1980, we find that they had a very different experience in the nineteenth century.[7] Taken as a group (see TABLES 8-2 and 8-4), they expanded their boundaries by more than 500 percent through the addition of more than 800 square miles of land between 1850 and 1910. In percentage terms, the decade of greatest gain was the 1850s; in absolute terms, the premier decades were the 1890s and the 1880s. At no time between 1850 and 1930, however, did these dozen cities in the aggregate annex less than an average of 100 square miles in a decade.

Appropriately, the most significant annexations in the nineteenth century involved the nation's three largest cities: New York, Chicago, and Philadelphia. Philadelphia's mammoth consolidation of the city with Philadelphia County in 1854 is still, in percentage terms, the largest single annexation in American history. In one move, the City of Brotherly Love quadrupled its population, expanded its area from 2 to 130 square miles, and, until Paris annexed its outer *arrondissements* in 1859, became the largest city in the world in terms of area.[8] Included in the new city of Philadelphia were the formerly independent suburbs of Spring Garden, Northern Liberties, Kensington, Southwark, and Moyamensing, which in 1850 ranked as the ninth, eleventh, twelfth, twentieth, and twenty-eighth largest cities in the United States.[9] In terms of relative growth of population and area, the impact would be less today if Philadelphia were to annex the equivalent land area and population of Los Angeles, Detroit, and Boston.

Chicago's largest annexation took place in 1889, when 133 square miles and most of what is now the far South Side were added. The addition included pleasant residential villages between 35th and 71st Streets like Hyde Park, Kenwood, and Woodlawn, as well as peripheral industrial communities in the Calumet Region like Grand Crossing and South Chicago and the famed model town of George Pullman. At the time of the annexation, only about 225,000 people lived in the area, and large stretches of the tract were rural or sparsely settled. Within thirty years, however, there were more than one million people on the land added in 1889.[10]

The most important municipal boundary adjustment in American history occurred in 1898, when Andrew Haswell Green's lifelong dream of a Greater New York City was realized. Brooklyn, which at the time was the fourth largest city in the United States, joined Manhattan, as did Queens (with a portion withheld as the soon-to-be-created Nassau County), Staten Island, and additional parts of Westchester County, which came to be known as the Bronx. The size of the city increased from about forty-four to about three hundred square miles, and the population grew by almost two million, most of it as a result of the consolidation. Strangely enough, the impetus for the move came from a Republican governor and a Republican state legislature, who presumably meant to dilute the influence of Tammany Hall in the governance of the metropolis by adding middle-class voters in outlying boroughs to the city electorate. There was relatively little public debate on the issue, considering the rhubarb that currently attends even minor zoning changes in most cities, and only in Brooklyn was an 1894 advisory vote close—65,744 to 65,467, a plurality of only 277 in favor of consolidation in a total vote of 131,000.[11] The state legislature went ahead with the vast merger, which became effective on January 1, 1898. A new city was created, and until the population spilled out for great distances into New Jersey, Westchester, and Connecticut, the government of the nation's largest city was a unique form of metropolitan organization.

Although smaller cities did not match the square-mile additions of Philadelphia, Chicago, and New York, every large city shared in the expansion boom. St. Louis increased its area from 4.5 to 14 square miles in 1856 and to 17 square miles in 1870. The biggest change came in 1876, when city voters overwhelmed the opposition of rural St. Louis County, raised the municipal area to 61 square miles, and removed the city from the county, making St. Louis one of the nation's very few examples of a completely independent city.[12] Boston added about 15 square miles by joining with Roxbury in 1868 and Dorchester in 1870, while New Orleans absorbed Carrolton in 1876 to give the Crescent City most of the area it occupies today. Baltimore more than doubled its size in 1888.[13]

Many of America's great cities, such as Minneapolis, Cleveland, Cincinnati, and Pittsburgh, expanded their boundaries through a series of small additions rather than through the single massive change that characterized New York, Chicago, and Philadelphia. Detroit provides a good example of such a pattern. Outlying areas were constantly integrated into the city as they became settled, so that vigorous independent communities did not develop to compete for a share of the expanding population. The city annexed almost constantly between 1880 and 1918, absorbing large portions of the townships of Greenfield, Springwells, Hamtramck, Gratiot, and Grosse Pointe, and of whole villages like Fairview, Delray, and Woodmere.[14]

Motives for Annexation

American cities have been especially susceptible to the notion that "bigger is better." During the taking of the 1890 federal census, partisans of Minneapolis and St. Paul each accused the other community of falsifying the returns in order to appear larger. Investigators found that the whole enumeration was a frightful tangle, and a recount was made. The new census revealed that Minneapolis had enrolled the dead, while St. Paul's standing had been enhanced by the listing of hundreds of inhabitants who evidently lived in depots, barber shops, and dime museums.[15]

If counting the dead was frowned upon, annexing populous suburbs was a perfectly respectable method of fueling the municipal booster spirit. Brooklyn gloried in its rise to third place among the nation's cities when it absorbed Bushwick and Williamsburg in 1855, and Chicago took pride in its second-place status after its massive annexation of 1889. In fact, it was partly the fear that Chicago would become the nation's largest city in 1900 or 1910 that prompted various factions in New York to agree to the consolidation of 1898.[16] Not only would a city gain additional residents by expanding its borders, but the fact of growth often inspired citizens with renewed confidence in a community's future and spurred them to greater efforts in civic development. As a leading Philadelphia newspaper commented with regard to that city's successful consolidation of 1854: "All of us may feel today that we are citizens of a new city. The Philadelphia we have known heretofore . . . has undergone a transformation which at once not only magnifies it immensely in physical proportions, but invests it with a social spirit hitherto unknown in its experience."[17]

The desire to annex was inspired not only by the booster spirit, but also by the business idea that a large organization was more efficient than a small one and that substantial economies would accrue from a consolidation of municipal governments.[18] According to this view, even when suburbs were honestly governed, their management was inefficient; large cities, on the other hand, could be run by highly paid experts.[19] In what businessmen regarded as a typical and laudable development, the Philadelphia police department was reduced from 850 to 650 men after consolidation, presumably with no loss of effectiveness. On other occasions, annexationists pointed out that competing communities could sometimes offer mutal advantages after joining together. Thus, Los Angeles provided the resources and San Pedro the location for a new port; Cleveland provided the financing and Ohio City the site for a waterworks.[20]

In many cases, the cry for efficiency was a mask for the desire to exploit and to control; it might be termed the local or downtown brand of urban imperialism.[21] Often the large merchants and businessmen of the central business districts sought to eliminate neighborhood governments that in their view inhibited progress. In Philadephia, where wharfage taxes, railway rights, and water prices were among the issues of contention between the city and its suburbs, the supporters of consolidation were overwhelmingly middle- and upper-income residents of the core. Suburban supporters of the proposal tended to be well-heeled commuters to the central business district. Neither group was representative of the laboring and farming constituency of the outlying areas to be added.[22]

The business community also sought to regularize the relationship between Cleveland and Ohio City, two communities on opposite sides of the Cuyahoga River. Prior to their consolidation in 1854, partisans of each city disrupted trade by tearing down bridges thought to be advantageous to the other.[23] Brooklyn and New York fought for control of the lucrative ferry trade on the East River, Memphis and South Memphis for commercial trade on the Mississippi. Through annexation, the strongest political unit could organize the government of a large area for its benefit.[24]

The desire to regularize the economy was further buttressed by the felt need for greater social control. In New Orleans the so-called "Spanish riot" of 1851 made clear to the city fathers the difficulty of police operations when law enforcement authority was divided among three separate communities.[25] In Philadelphia business leaders thought that a unified police force would eliminate undesirable conditions in districts then beyond the city limits. Riots in 1838, 1844, and 1849 had reduced city and suburbs to a garrison, and a group of prominent Philadelphians proposed consolidation so that "the peace of our community will be preserved and the prosperity of its citizens protected without the unpleasant necessity of a resort to armed force."[26] Because the local constabulary of Southwark, Moyamensing, and the Northern Liberties proved unable to control the volunteer fire companies and the roving bands that moved back and forth across municipal boundaries, the *North American* said in 1850: "Philadelphia never before needed a stronger government nor ever possessed a weaker one."[27] According to a nineteenth-century historian of the Pennsylvania metropolis, "The miserable system of a city with adjacent districts each independent of each other was a protection to the disorderly and encouragement of them to unite together for the purpose of showing their disregard for the law."[28]

Land speculators also supported annexation, but they usually worked behind the scenes, and their precise role is difficult to measure. In a pattern familiar enough over the last century, real-estate promoters purchased large tracts of rural land in the expectation that the advancing horsecards, steam railroads, and trolleys would make the area attractive to urban families. In the absence of decent sewerage, water, and educational systems, land speculators looked to annexation as a sort of guarantee to potential buyers that the suburb would eventually possess the comforts of the city. The desire to turn a fast buck was undoubtedly an important reason why nineteenth-century urban boundaries were usually set far in advance of actual settlement. In Memphis, in Baltimore, in Cleveland, in Chicago, and in other cities, the municipality sometimes included land that had not even been surveyed, let alone laid out into streets.[29] The Philadelphia consolidation of 1854 brought into the city agricultural areas that remained working farmland for two generations.[30]

The Nineteenth-Century Success

What is most important about annexation in the nineteenth century is not motivation; rather, the important aspect is the single, overwhelming fact of area growth. With the exception of Boston, the thrust of municipal government was imperialistic, and the trend was clearly toward metropolitan government.

To the extent that historians have bothered with the history of annexation, their tendency has been to credit its early success to the sense of community or mutality that supposedly existed between residents of the core and residents of the periphery.[31] Certainly, one could not deny that many suburbanites did regard the city as their achievement and were willing and even eager to be joined with it. But probably more important than such lofty notions were pragmatic, mundane considerations of sewers, schools, water, and police.[32]

As we have seen, the early nineteenth-century residents of outlying areas were more likely to be poor than affluent, and their residential areas normally could not offer a level of public services comparable to that of the core.[33] Such considerations were apt to be important, particularly to newcomers who might have bought a house without water intakes or sewer outlets, on land that was overrun by snakes and rabbits, on a street that was neither paved nor served by storm drains nor watched over by the police.[34] In Detroit the Fairview Village annexation of 1907 was precipitated because Fairview badly needed a sewer system which the villagers could not afford.[35] Similarly, Roxbury joined Boston partly to gain relief from an intolerable sewerage situation;[36] Hyde Parkers looked to Chicago for better fire protection and cheaper gas rates; and residents of Kensington, Spring Garden, and Germantown were able to share more fully, and at a better price, in a Philadelphia water system that was rated among the best in the world.[37]

As the centrifugal movement of the middle class gathered force after 1865, suburbia gradually shook off its reputation for vice and squalor. Slowly, many peripheral residents came to believe that local autonomy could mean better public services. Education was among the first suburban institutions to be upgraded. As early as the middle of the nineteenth century, Ralph Waldo Emerson had boasted of Concord: ''We will make our shools such that no family which has a new home to choose can fail to be attracted hither as to the one town in which the best education can be secured.'' In subsequent generations, such suburban school systems as Newton, New Trier, Scarsdale, Great Neck, Bethesda, Chappaqua, and Ladue became synonymous with educational excellence.

As suburban services and self-consciousness became stronger, the desire for absorption into the metropolis waned, and fewer annexations were unopposed; some took place over the objections of 90 percent of those concerned.[38] But in the nineteenth century, success depended much less on public than on legislative approval. Legally, of course, a city is a corporation that receives from the state government special powers of regulation over the residents of a precisely defined geographic area. Thus it normally remains within the power of the state to change the boundaries of governmental units under its jurisdiction.[39] In the nineteenth century, states tended to exercise this power without the advice of those who would be affected; that is, rarely were public referendums held on the issue. And when a vote was taken, it was often ignored if it was negative.

The predominant view in the nineteenth century was the doctrine of forcible annexation. No small territory could be allowed to retard the development of the metropolitan community; the most important consideration was simply the greatest good for the greatest number. The most recent articulation of this view came in 1917, when Judge Harlan of Maryland overruled the objections of Baltimore suburbanites and approved an annexation that tripled the area of the city. He declared:

> Those who locate near the city limits are bound to know that the time may come when the legislature will extend the limits and take them in. No principle of right or justice or fairness places in their hands the power to stop the progress and development of the city, especially in view of the fact that the large majority of them have located near the city for the pur-

pose of getting the benefit of transacting business or securing employment or following their profession in the city.[40]

Examples of forced annexations in the nineteenth century are numerous. In 1854 the consolidation of Philadelphia was approved not by suburbanites, who in fact sent delegation after delegation to oppose it, but rather by lawmakers in Harrisburg. Local referendums were not held on the San Francisco Consolidation Act of 1856 or on any one of the frequent annexations to Chicago and Baltimore prior to the 1880s.[41] Lawmakers added to both St. Louis and Boston three times before 1860, but the local electorate in each city rejected annexation measures submitted in 1853. A vote to reconsolidate the three municipalities of New Orleans met a popular defeat in 1850, only to be forced by special legislation in 1852. Louisiana also gave Carrolton to the Crescent City in 1876 without seeking the approval of the aroused residents.[42] The Ohio General Assembly added to Cleveland in 1829 and 1834, but when merger with Ohio City was first submitted to the voters in 1851, the total of the votes of the larger community was 1,098 to 850 against consolidation.[43] Some annexations did meet with popular approval, but it was in the legislative halls that the annexationists won their most important nineteenth-century victories.[44]

The Twentieth Century Failure

For various reasons, then, the addition of peripheral land to cities was a normal process of urban growth in the nineteenth century. To most people, it seemed entirely logical and even inevitable that cities would add to their boundaries to accommodate a spreading and increasing population, even if the affected families were themselves resistant to the idea. In 1899 a suburban Chicago newspaper admitted, falsely as it turned out, that "the time may and doubtless will come when Oak Park will be swallowed up by the great city."[45] Several rash people even predicted that Minneapolis and St. Paul would come together into "one great city."[46]

But something has happened—or more precisely has failed to happen—in the twentieth century. For many cities, and particularly for the older ones in the East and Middle West now losing population, metropolitan government is a phenomenon of the past. Quite simply, these cities are no longer able to annex or to consolidate in order to keep pace with the overflow of population beyond established boundaries (See TABLES 8-2 and 8-4).

TABLE 8-4
Comparison of the Number of Square Miles Annexed of Twelve Growing and Twelve Declining Cities, 1850–1980

Period	*Declining Cities*	*Growing Cities*
1850–1870	294	NA
1870–1890	231	6
1890–1910	327	159
1910–1930	229	568
1930–1950	42	1510
1950–1980	61	2096
Total Area in 1980	1328	4257

SOURCE: TABLES 8-1 and 8-2.

The growing rejection of municipal expansion through annexation and consolidation can be seen in several ways. For the dozen largest cities in the United States now losing population, there was a marked percentage drop in the amount of new territory added after 1870 and a sharp absolute drop after 1930. Although these dozen cities added to their area by 400 percent in the last half of the nineteenth century and by almost 200 percent between 1870 and 1930, they have added less than 10 percent since 1930.[47]

The first really significant defeat for the consolidation movement came when Brookline spurned Boston in 1874. Starting in 1868, the Hub doubled its area by annexing, in turn, the cities and towns of Roxbury, Dorchester, Charlestown, West Roxbury, and Brighton. But Brookline, the self-styled "richest town in the world," voted against union by a vote of 706 to 299. They were not rejecting growth or development, but were expressing a determination to control the physical and social environment in which they lived.[48]

After Brookline spurned Boston, virtually every other Eastern and Middle Western city was rebuffed by wealthy and independent suburbs—Chicago by Oak Park and Evanston, Rochester by Brighton and Irondequoit, and Oakland by the rest of Alameda County.[49] Some consolidation proposals, such as those of St. Paul in 1924, Cleveland in 1925, and Boston in 1931, never even gained constitutional or legislative approval. Others, such as the proposals for Birmingham and Louisville, were defeated by statewide public referendum.[50] And, as the suburban trend gained momentum, state legislators became increasingly reluctant to override the wishes of the voters concerned.

Because large cities sometimes felt the need for financial retrenchment, the core areas themselves occasionally rejected consolidation. In

1902 Mayor Carter H. Harrison of Chicago contended in his annual message that the city was too large to be administered efficiently. "An attempt to increase this territory," he said, "should meet with instant and emphatic discouragement. The ideal city is compact. With its area fully occupied, the care of all branches of administration can be applied to all sections expeditiously and well."[51] Particularly during the depression years of the 1930s, cities were not prepared to make the enormous capital expenditures annexation usually entailed.[52]

But the inability of some of the older cities to grow in the twentieth century has not generally resulted from a lack of will on the part of the core area. In fact, the well-publicized consolidation attempts of St. Louis and Pittsburgh in the 1920s failed despite enormous campaigns, led by the local business elite, on their behalf. The Greater St. Louis Conference attempted to enlarge borders that had been frozen since 1876, and it prophesized, correctly as it turned out, that failing such action, the Missouri city could only shrink in comparison with other cities of the nation and the world. The proposal won by a big majority in the city, but was rejected in the outlying areas by more than two to one.[53] The Pittsburgh vote came in 1928, following passage of an enabling amendment to the Pennsylvania constitution authorizing a federated city of Pittsburgh. It was approved in the statewide vote, but it failed to win the required two-thirds vote in a majority of communities. Like St. Louis, Pittsburgh did not expand in the 1920s and has not expanded since; it has fallen from the nation's twelfth largest city in 1950 (with 677,000 residents) to the thirty-first largest in 1980 (with 424,000 residents).[54]

There are basically three reasons why America's older cities are now ringed by incorporated suburbs that emphasize their distinctiveness from rather than their relationship with the metropolis: (1) sharper racial, ethnic, and class distinctions, (2) new laws that made incorporation easy and annexation unworkable, and (3) improved suburban services. Most important was the changing reality and image of the periphery and the center, particularly with regard to demographic characteristics. With the vast increase in immigration in the late nineteenth century, the core city increasingly became the home of penniless immigrants from Southern and Eastern Europe. And of course, in the early years of the twentieth century, increasing numbers of Southern blacks forsook their miserable tenant farms for a place where, they hoped, "a man was a man." In the view of most middle-class, white suburbanites, these newcomers were associated with and were often regarded as the cause of intemperance, vice, urban bossism, crime, and radicalism of all kinds. And as the central city increasingly became the home of the disadvantaged, the number of white commuters rose markedly. These recent escapees from the central city were anxious to insulate their neighborhoods from the "liquor power" and other pernicious urban influences. An independent community offered the exciting promise of moral control. As the Morgan Park *Post,* a suburban Chicago weekly, remarked in an antiannexationist editorial on March 9, 1907: "The real issue is not taxes, nor water, nor street cars—it is a much greater question than either. It is the moral control of our village. . . . Under local government we can absolutely control every objectionable thing that may try to enter our limits—but once annexed we are at the mercy of the city hall."[55]

Some people felt that the suburbs could best serve the city as a moral force if the central city annexed the suburbs and then used the additional middle-class votes to crush the liquor and vice interests. As Zane L. Miller has noted of late-nineteenth-century Cincinnati, the successive enlargements of city boundaries kept the wealthy, highly educated, and sophisticated residents of the Hilltop suburban fringe firmly engaged in the city's affairs. And annexationists in Boston predicted that their city would share the fate of ancient Rome if the middle class, which had earlier provided reform leadership, was to separate itself from active involvement in municipal affairs.[56]

But even when annexations occurred, the political machines they were designed to unseat proved to have remarkable staying power. After all, argued the antiannexationists, how could the small suburbs possibly overcome the great city? "What influence would Oak Park have as the tail end of the 35th Ward?" asked one man at a suburban Chicago meeting. "About as much as the hair on a dog's tail," shouted a citizen in the audience.[57]

Middle- and upper-income suburbs like Oak Park had a much better chance of preserving their independence than did the low-status peripheral communities of the nineteenth century because of changes in incorporation and annexation laws. Although each state conformed to a slightly different pattern, pre-Civil War villages were typically chartered in the same way that cities were chartered, and by about 1820 the village had become a recognized step in the development of a city. A certain area of a township that wished to become a city in the future would incorporate as a village to secure the advantages of individual local services and improvements that might not be of interest in the township as a whole but that might improve the possibility of a community's development into a city. Chartering by state legislatures slowly became a more simple process.

After the Civil War, however, the traditional role of the village as a prelude to the formation of a city began to shift. Small areas that had little hope and less interest in growing to become large cities began to incorporate under the easily met provisions of state incorporation laws. Rather than a first step towards becoming a city, the new wave of village incorpoations was essentially a protective action, an attempt to save certain communities from annexation by larger urban units. Weschester County, New York, offers a good example of the pattern. Directly north of the nation's greatest metropolis, it was a prime target for growth during the last quarter of the nineteenth century. In the 1870s and again in the 1890s, New York City expanded into the county, gobbling up once-quiet villages and transforming them into wards of a massive city. Not surprisingly, village incorporations in Westchester County cluster around the years of the New York City threat. A number of villages in the county were founded immediately before and after the 1874 annexation of Morrisania, West Farms, and Kingsbridge. A second clustering coincided with the joining of Wakefield, Williamsbridge, the town of Westchester, and parts of the towns of Eastchester and Pelham to New York in 1895. Bronxville, for example, incorporated in 1898, when it had fewer than five hundred residents, and Scarsdale incorporated in 1915, when White Plains was expanding in its direction (Appendix A-6). In Westchester, as in other suburban areas around the country, the village had become an alternative to urban growth rather than its precursor, a method by which suburbs could protect their reputation, status, and independence.

In addition to easier incorporation, or "villagification," affluent suburbs like Brookline, Newton, Evanston, Beverly Hills, and Shaker Heights, in cooperation with rural areas, were able to move state legislatures away from the doctrine of forcible annexation. With some notable exceptions in the South and West, where cities can sometimes annex without a popular referendum, it is now commonly held that annexation should be a voluntary affair that must gain the approval of the residents of an affected area. Even so, rigorous procedural and substantive requirements block the way, and special acts calling for annexation have been defeated by antiurban state legislatures. Where annexation is provided for in a state constitution, as in the case of San Francisco, the relevant provision seems intended to thwart rather than to promote the process.[58] Conversely, some states even require that central city services must be provided to newly incorporated communities at central city rates.[59]

A third factor causing the breakdown of annexation as a process has been the improvement of suburban services. Since World War II especially, county governments have grown out of their old rural orientation and have begun to offer municipal-type services. This method has been publicly associated with Nashville, Indianapolis, and Miami, but the trend has been almost everywhere apparent. Even more important in upgrading suburban lifestyles has been the increased use of special service districts. This type of governmental structure was first used as an alternative to annexation in Philadelphia, where after 1790 special districts were established to administer prisons, schools, public health, and port administration.[60] Nineteenth-century examples were the New York Metropolitan Police Board (1857), the New York Metropolitan Board of Health (1866), the Massachusetts District Commission (sewerage, 1889; parks, 1893; water, 1895), and the Chicago Sanitary District (1889). To a man such as Andrew Haswell Green, these regional institutions emphasized the logic of complete consolidation.[61] But to most suburbanites, special-service districts were an alternative rather than an avenue to metropolitan government. By bringing together suburbs that individually lacked the resources to provide high-quality sewerage, water, educational or law-enforcement services, the special-service district enabled suburbanites to have their urban amenities without the urban problems. In 1915, for example, only 45 percent of Chicago's Cook County environs had a public water supply; by 1934 85 percent of the municipalities within a fifty-mile radius of the Loop had the service.[61]

Professor Jon C. Teaford's recent analysis of city-suburban relationships has concluded that the group that most vigorously supports metropolitan government are elite businessmen who live in the suburbs but whose companies and livelihoods are associated wtih the larger city. By contrast, those people most opposed to metropolitan government are blue-collar homeowners in the suburbs who fear racial change and higher taxes.[62]

Annexation is no longer a viable process for most of the old Eastern and Middle Western cities of the United States. Where once they moved their boundaries outward in a consistent pattern, they now lie surrounded by unfriendly suburbs. Their metropolitan populations often continue to grow, but their central cities decline in numbers of residents and in wealth because they are prevented from enlarging their boundaries.

Centralization and size are no longer seen as desirable objectives. "We haven't had a consolidation or annexation in decades," said Panke Bradley, Atlanta's city planning director in 1984. The Georgia capital lost 14 percent of its population in the 1970s, and the status quo seems likely to continue. According to Bethel Minter, an Atlanta economist: "Blacks who predominate in the city don't want to give up politcal control. Suburbanites don't care to share the problems of the city. Everybody, for different reasons, is satisfied with the way things are. Annexation is pretty much a dead issue."[63]

In fact, during the 1960s the issue of the dispersal of power at all levels became popular in the United States. In large metropolitan areas especially, "Power to the people" became a common goal for blacks who sought greater control of their inner-city neighborhoods, for middle-class whites who wanted to protect a suburban way of life, and for professionals and intellectuals who feared that respect for institutions and authority was being eroded by mindless bureaucracies. In 1967 the borough president of Staten Island established a commission to study the possibility of secession from New York City, and in 1985 the separation proposal was given added impetus when Staten Island lost its co-equal status on the city's powerful Board of Estimate. In 1968 the United Black Front of Roxbury, a predominantly black community, demanded independence from Boston. In 1969 small, middle-class property-owners in Glen Park looked to secession from Gary as necessary for the protection of their housing investments. And in 1970 the Association of the Bar of the City of New York suggested the formation of up to forty-five new units of local government in the metropolis.

While the movement for neighborhood control gained adherents in some communities, the tendency in a few places was to pursue the opposite course—that is, to consolidate a larger geographical area and a greater number of people under a single autonomous government. In Texas, where a Home Rule Law allows large cities to annex unincorporated territory without a popular referendum, spatial expansion has been the rule rather than the exception. In 1940 Dallas, for example, was a compact city of about 45 square miles, or about the size of San Francisco. Under the guidance of a determined business community, Dallas followed an aggressive postwar annexation policy, and it took in a huge reservoir of unincorporated county land in anticipation of rapid suburbanization. Between 1940 and 1960 its population increased 131 percent to almost 680,000 while its area increased by 507 percent. Over the next twenty years, the Dallas population rose to more than 900,000 and its area increased to 350 square miles. And Dallas was hardly unusual. Each major Texas city stood at least ten times larger in 1960 than it had in 1900, and Houston was twenty times as large.[64]

Texas was not the scene of the most dramatic changes. Indianapolis and Oklahoma City added huge expanses of land to their corporate limits. No city could match the achievement of Jacksonville, however. In 1968 Jacksonville became the largest municipality in the continental United States in area, and it joined the twenty-five largest cities in population when its government and territory were consolidated with almost all of Duval County.

These two generally opposite tendencies are simply recent manifestations of a dilemma that has confronted American cities for two hundred years. On the one hand, democracy seems to call for government to remain small and close to the people; on the other hand, efficiency and the regional character of many contemporary problems point to the necessity of government that is metropolitan in authority and planning.

Professor Kingsley Davis of the University of Southern California has suggested that annexation is not a terribly important factor in the study of urban demography and ecology because the expansion of boundaries by politcal annexation tends to approximate the physical spread of the city. Thus, a community annexes because it becomes more populous. While this is to some degree a self-evident proposition, the data presented here suggest that cities also become more populous because they annex, and if they do not annex they will not grow. One could hardly argue that Hominy Pot, New Hampshire, can become a metropolis by annexing five hundred square miles of New England woodlands. But the whole of the population growth of Memphis, Houston, Indianapolis, Phoenix, and many other cities has come from annexation. Within their 1940 boundaries they lost population between 1940 and 1980.

If nothing more than civic pride were at stake, it would make little difference whether a given city were eight or forty-eighth on the list of large cities. But it makes a great deal of difference where the city limits are placed. In New Jersey, for instance, most suburbs flourish and try to ignore the fact that Newark and Camden, both seriously depressed and geographically small, must struggle with the whole range of contemporary urban problems. The rich have long since departed; the middle class is almost gone. Professor Scott Greer is correct when he says that the decline of American cities is really an optical illusion; only a small part of the city is suffering while most of it is relatively prosperous, particularly those parts on the edges. But in Newark the area of decline is practically the entire city because annexation has taken place only on a tiny scale, and the city does not have a substantial middle-class zone. Assimilation is often more difficult for blacks than for other minorities because movement from the ghetto involves movement into another governmental jurisdiction rather than simply movement into another neighborhood.

Resistance to annexation is symptomatic of the view that metropolitan problems are unsolvable and that the only sensible solution is isolation. Elite suburbs are communities encapsulated from the crises of urban capitalism, yet able to benefit and enjoy the system's largesse. The result has led Lewis Mumford to comment:

> In the suburb one might live and die without marring the image of an innocent world, except when some shadow of its evil fell over a column

in the newspaper. Thus the suburb served as an asylum for the preservation of illusion. Here domesticity could flourish, forgetful of the exploitation on which so much of it was based. Here individuality could propser, oblivious of the pervasive regimentation beyond. This was not merely a child-centered environment; it was based on a childish view of the world, in which reality was sacrificed to the pleasure principle.[65]

9

The New Age of Automobility

> The ordinary 'horseless carriage' is at present a luxury for the wealthy; and altho its price will probably fall in the future, it will never, of course, come into as common use as the bicycle.
>
> —*Literary Digest*, October 14, 1899

The *Literary Digest* was no more perceptive about the automobile than it later was about the 1936 Presidential election, when it predicted that Landon would defeat Roosevelt. At the turn of the century, however, its assessment probably mirrored that of the population as a whole. Initially considered a curiosity and a toy, the car was more an offspring of the bicycle than a successor to the horse-drawn carriage. In 1898 there was only one automobile in operation for every eighteen thousand Americans, each of them a hybrid creation secured by crossing a bike with a buggy and installing in the product a noisy, sputtering little engine. Two years later there were only eight thousand motor vehicles in the United States, and half of those were of European manufacture. The primary means of intraurban movement, therefore, remained the electric trolley, which was a ubiquitous and commonplace feature of everyday American life at the turn of the century.[1]

Although the horseless carriage remained a rarity in 1900, it was in fact older than the streetcar. As early as 1860, Etienne Lenoir produced a crude prototype of an automobile, and in 1876 Nicholas Otto unveiled his four-stroke engine. The pace of mechanical invention had quickened by 1884, when Karl Benz built a self-propelled, three-wheeled vehicle in Germany. Working independently, another German engineer, Gottlieb Daimler, introduced an improved internal combustion engine the following year. Ten years later, in 1894, the French, supported by Daimler's company, brought out the Panhard, the first car in the world to be marketed successfully. Soon thereafter, French words like *garage, chassis, automobile,* and *chauffeur* became internationally accepted motoring terms. The *New York Times* expressed skepticism on January 3, 1899:

> There is something uncanny about these newfangled vehicles. They are unutterably ugly and never a one of them has been provided with a good or even an endurable name. The French, who are usually orthodox in their etymology, if in nothing else, have evolved 'automobile,' which being half Greek and half Latin is so near indecent that we print it with hesitation.[2]

In the United States, most mechanics were unaware of these inventions, or even of the fact that the word "automobile" had been coined by the French Academy to describe the steam omnibuses that were lumbering around Paris by 1875. North America thus lagged behind Europe in motoring development, and horseless carriages were manufactured in the 1890s by only a few Americans, among them Charles and Frank Duryea, Ransom Olds, Elwood Haynes, and Alexander Winton. The Duryeas, who won most of the road races in 1894, 1895, and 1896, dominated the field, and their production did not reach ten vehicles per year until 1896.[3]

In contrast to the trolley, therefore, which was in widespread use in American cities within ten years of its invetion, automobile adoption moved much more slowly. One problem was legislative. On the theory that lumbering automobiles frightened horses and raised dust, many states followed British precedent and passed laws limiting self-propelled vehicles to four miles per hour and requiring that each be preceded by a man on foot carrying a red flag.

Another problem was the virtual absence of highways. Prior to 1920, when the railroad was smoother, faster, and more economical than any other form of overland travel (even in 1920 most Americans relied on rail transportation for travel away from home), there were almost no paved surfaces between cities in the United States. Many roads were nothing more than dirt paths cleared of obstructions. Recalling his boyhood in Michigan, Edmund G. Love told of being stuck in the mud eight times in one ten-mile stretch between Lapeer and Imlay City and of being hopelessly lost on a detour to Owesso, only twenty miles from home.[4]

Getting lost was not difficult because neither roads nor towns were easily identifiable. There was no highway-numbering system; thus Rand-McNally and other firms published small, illustrated, auto-touring books which guided the motorist by detailing the twists, turns, and landmarks necessary to negotiate every bridge and fork in the road. Towns and villages along the route did not welcome the traveler with signs announcing name and population; after all, the distinction between city and country was still sharp, and the local folk knew where they lived. Adding to the challenge was the scarcity of gasoline stations and service facilities.[5]

Because the early motorcars were little more reliable than the highway system, driving was an adventure that required ingenuity and daring. In 1910 the *Brooklyn Eagle* called automobiling "the last call of the wild" and the "world's most exciting sport," and as late as 1918 the War Industries Board could label automobiles "among the least essential manufactures" and regard the shutdown of the entire industry as a mere inconvenience.[6]

More perceptive observers realized that a transportation revolution was taking place, and that the United States was becoming the world leader in automotive technology. Unlike European manufacturers, who concentrated on expensive motorcars for the rich, American entrepreneurs early turned to economical vehicles that could be mass-produced. Charles Duryea enunciated this idea on January 31, 1896, in a letter to *Horseless Age:* "We require in the horseless carriage a mechanism so simple as not to get out of order easily or give trouble to the unskilled operator, and a carriage arranged as to be comfortable to use, viz. it should be clean, free from objectionable odor, vibration or possible danger. If it is simple in construction, it will, in all probability, become cheap in cost."

In 1900 insisting that an uncomplicated, inexpensive, and utilitarian car was needed, Ransom E. Olds began assembling a single model from parts bought in quantity from other manufacturers. He priced his horseless buggies below five hundred dollars, and by 1905 more than sixty-five hundred of the curved-dash models were on the road. In that year, Gus Edwards wrote, "In My Merry Oldsmobile," the best-known song ever written about the automobile:

Come away with me, Lucille,
In my merry Oldsmobile,
Over the road of life we'll fly,
Autobubbling you and I,
To the church we'll swiftly steal,
And our wedding bells will peal,
You can go as far as you like with me,
In our merry Oldsmobile.

By 1908 twenty-four American companies were producing simply constructed automobiles at low prices. They made it possible for the common man to aspire to ownership and thus disproved Woodrow Wilson's 1906 prediction that automobility would bring on socialism by promoting envy of the rich. By 1913 there was one motor vehicle to every eight people, and by the standards of a single decade earlier, every one of them was a marvel of silence and service.[7]

The Model T

Henry Ford did not invent the gasoline-powered engine, and he made no important technological contribution to early automotive technology. He did not even originate the idea of an economical car for the average man. But Henry Ford was alone in sticking to it with a grim persistence, and he became the most important and successful of the industry's pioneers. A Michigan farm boy who migrated to Detroit in 1879 at the age of sixteen, Ford worked variously as a machinist, a watch repairman, and an engineer, all the while tinkering with internal combustion engines and auto buggies in his back-yard shop and in a tiny brick building on Bagley Avenue. In 1896 he produced his first "flexmobile," and for the next six years he tried, mostly unsuccessfully, to market at a profit improved models of the vehicle. His big break came in 1902, when his racing car won several nationally publicized races and gave Fords a reputation for toughness and reliability. With $28,000 invested, he formed the Ford Motor Company in 1903, and over the next five years he sold his automobiles for an average price of $1,600, considerably more than a dozen major competitors, including Cadillac.[8]

Within the next generation, Henry Ford became a legend—the very symbol of modern industrial technique. In 1908 he introduced a boxlike vehicle which was easy to operate, simple to repair, and dependable even under trying conditions. Dubbed the Model T— and popularly known as the "Tin Lizzie"—it remained unchanged in outward appearance for the next two decades. In describing his early plans, Ford revealed both his common touch and his hard-headed business acumen:

> I will build a motor car for the great multitude.
>
> It will be large enough for the family, but small enough for the individual to run and care for. It will be constructed of the best materials, by the best men to be hired, after the simplest designs that modern engineering can devise. But it will be so low in price that no man making a good salary will be unable to own one—and enjoy with his family the blessings of hours of pleasure in God's great open spaces.[9]

Ford's genius lay in his ability to reduce the cost of his popular Model T even while increasing the wages of his employees. In 1914, at a new factory in suburban Highland Park, Michigan, the "Flivver King" initiated the moving assembly line—perhaps the most important contribution to manufacturing technology since the introduction of the principle of interchangeable parts in the eighteenth century. In 1919, with the opening of his enormous River Rouge complex, the making of automobiles entered the stage of giant enterprise. With greater industrial efficiency and constant attempts to reduce the work process to the simplest possible steps, Ford dropped the price of his Model T from $950 in 1910 to $290 in 1924. This occurred during a period of rising wages and prices. Whereas it took the average worker twenty-two months to buy a Model T in 1909, by 1925 the same purchase would have required the labor of less than three months.[10]

Ford managed his feat even while revolutionizing American industry on January 5, 1914, by announcing, unilaterally, unexpectedly, and in the midst of a business recession, a raise in the minimum daily wage of his employees from $2.30 to $5.00 per day. Realizing that boredom and monotony were major causes of employee concern, Ford simply paid his workers more for their time. In the process, he stabilized his work force, increased the pace of his assembly lines, and created more potential customers for his product.

Ford also undercut the American labor movement. On the one hand, he instituted tyrannical discipline in his factories and used spies and armed thugs to resist attempts at unionization. On the other hand, his cheap cars weakened the Marxist claim that laborers "had nothing to lose but their chains." As an elderly man explained to Robert S. Lynd in Muncie, Indiana, "The Ford car has done an awful lot of harm to the unions here and everywhere else. . . . As long as men have enough money to buy a secondhard Ford and tires and gasoline, they'll be out on the road and paying no attention to union meetings."[11]

By 1925 Ford was turning out nine thousand cars per day, or one every ten seconds. That such unprecedented production earned the "Flivver King" $25,000 per day and made him a billionaire did not detract from his image. In *Vanity Fair,* he was described as "the Colossus of Business, an almost divine Master-Mind." A study in contradictions, Ford was a salesman whose product destroyed vast areas of traditional small-town life, and who, at the same time, devoted a considerable amount of his fortune and his spiritual energies in rebuilding models of old-fashioned villages and promoting old-fashioned square dancing. At the end of the 1920s college students were asked to rank the greatest people of all time. Henry Ford came in third—behind Jesus Christ and Napoleon Bonaparte.[12]

When Model T production finally ceased in 1927, the ownership of an automobile had reached the point of being an essential part of normal middle-class living (TABLE 9-1). Almost sixteen million "Tin Lizzies"

TABLE 9-1
Automobile and Truck Registrations in the United States at Five Year Intervals, 1905–1975

Year	*Automobiles*	*Trucks*	*Ratio/Trucks to Cars*
1905	8,000	1,400	1 to 55
1915	2,332,426	158,506	1 to 15
1925	17,481,001	2,569,734	1 to 7
1935	22,567,827	3,919,305	1 to 6
1945	25,793,493	5,079,802	1 to 5
1955	52,135,583	10,302,987	1 to 5
1975	106,713,000	25,755,700	1 to 4

SOURCE: *The Statistical Abstract of the United States: From Colonial Times to the Present* (Stamford, Conn., 1966), 462 and 470E.

had rolled off the assembly line, and every second car on earth had a Ford nameplate. Although much cheaper than most European vehicles, the American car was more able to withstand abuse, was easier to repair, and had more horsepower. It was an all-purpose family machine that was designed for the average driver rather than for the professional chauffeur.

Led by Ford, American automobile registrations climbed from one million in 1913 to ten million in 1923, when Kansas alone had more cars than France or Germany, and Michigan counted more than Great Britain and Ireland combined. By 1927 when the American total had risen to twenty-six million, the United States was building about 85 percent of the world's automobiles, and there was one motor vehicle for every five people in the country. For the next three decades, only Canada could join the United States in that claim. Most American blue-collar workers were unable to afford private transportation until after World War II, but on a comparative basis the automobile was ubiquitous only in the United States in the 1930s (TABLE 9-2). In that decade Mickey Rooney starred in a series of *Andy Hardy* movies about a high-school boy with a car. Sixteen-year-old students elsewhere in the world could not even have dreamed of such a possession.

The rapid rise in motor-vehicle registrations created a booming optimism, a national faith in technological progress. Ransom Olds said, "The automobile has brought more progress than any other article ever manufactured," and Charles Mott asserted in 1923, "It would be hard to name a branch of human activity . . . not made to function more smoothly and more effectively in the service of humanity, because of the . . . automobile." Industry spokesmen harped on this theme of improve-

TABLE 9-2
Number of Inhabitants Per Registered Passenger Vehicle in Major Western Nations, 1905–1970

Year	*United States*	*United Kingdom*	*Germany*	*France*	*Sweden*	*Italy*
1905	1,078	2,312	983	1,850	NA	NA
1920	13	228	1,017	247	277	1,206
1930	5	42	135	37	59	225
1940	5	32	54	22	NA	163
1950	4	22	116	24	28	139
1960	3	9	15	8	6	25
1970	2	5	4	4	4	5

SOURCE: Calculations made from data in B. R. Mitchell, *European Historical Statistics, 1750–1970* (London, 1975), 350–4.

ment. "When I sold a car," Roy Chapin remarked in 1926, "I sold it with the honest conviction that I was doing the buyer a favor in helping him to take his place in a big forward movement." As Sinclair Lewis's popular 1922 novel *Babbitt* indicated, the private car had become no longer a luxury, but a necessity of the American middle class.[13]

The Road-Building Revolution

Although the motorcar was the quintessential private instrument, its owners had to operate it over public spaces. What would be the reaction of government? One solution would have been to levy heavy user fees to reimburse local treasuries in full for the cost of streets, traffic maintenance and police services. Another possibility was to rely on general taxation to support private transportation. That the latter course was adopted is testimony both to the public perception of the benefits of automobility and to the intervention of special interest groups.

A massive investment in roadways was necessary because city streets at the turn of the century were notoriously dilapidated and rutted. Typically, municipal governments had decided upon new routes and removed tree stumps and rocks from the rights-of-way before there were adequate funds to pave or maintain them. Alternately muddy and dusty and always accented with potholes, the streets were more nearly obstacle courses than thoroughfares. And while a horse might step nimbly over the yawning gaps, cars were unable to avoid all the holes.

The removal of horses from cities was widely considered a proper object for the expenditure of public funds. Indeed, the private car was initially regarded as the very salvation of the city, a clean and efficient alternative to the old-fashioned, manure-befouled, odoriferous, space-intensive horse. On the basis of the common good, many local governments applied general revenues to easing the way for the motorcar.[14]

The basic thrust for street improvement came less from idealism than from special interest, however. By the 1920s, a coalition of private-pressure groups, including tire manufacturers and dealers, parts suppliers, oil companies, service-station owners, road builders, and land developers were lobbying for new streets. Merchants argued that traffic congestion caused real-estate values to decline by increasing the cost of doing business downtown. They viewed highway building as a form of social and economic therapy, and they justified public financing for such projects on the theory that roadway improvements would pay for themselves by increasing property-tax revenues along the route. City planners dropped dutifully into line and added to the chorus of commercial and civic organizations.[15]

Elected officials bowed to private pressure and the public purse was opened to improve the quality of streets. Initial efforts focused on existing road surfaces. In the late nineteenth century, the most common pavement used in heavily traveled areas was cobblestone, which consisted of small, water-rounded stones set on end in a bed of sand. Readily available and exceptionally durable, cobblestones had been used by cities since the Middle Ages. The cracked surface generated low friction between wheel and road and prevented horses from slipping. But cobblestones were often uneven. By the late nineteenth century, American cities had begun to adopt three new types of surfaces. The first was a type of crushed stone—called macadam—that was popular in less-traveled residential and suburban routes. It was lightweight, however, and unable to resist the suction of automobile tires, which tore macadam to pieces if usage was heavy. A second road type was asphalt, a method developed in Europe. Contractors had to relearn the ancient Roman system of crushed-rock foundations because asphalt was dependent upon solid, well-drained underpinnings. Finally, concrete of predictable quality became available in the late nineteenth century.[16]

Changes in the construction and financing of public rights-of-way reflected not only new technology, but also new attitudes toward the function of streets. In the midnineteenth century, when row houses predominated, the street was the primary open space, and it performed an important recreational function.[17] By 1920, however, most urban residents and virtually all highway engineers saw streets primarily as arteries for motor vehicles. Speaking of Paris in 1924 the French planner LeCorbusier bemoaned the change:

> In the early evening twilight on the Champs Elysees it was as though the world had suddenly gone mad. After the emptiness of summer, the traffic was more furious than ever. Day by day the fury of the traffic grew. To leave your house meant that once you had crossed the threshold you were a possible sacrifice to death in the shape of innumerable motors. I think back twenty years, when I was a student: the road belonged to us then.[18]

The significance of the smooth asphalt and concrete roadways to the growth of automobile suburbs can be illustrated by Detroit. Six major arteries radiate from the Motor City toward Pontiac, Toledo, Lansing, Windsor, Ann Arbor, and Port Huron. The road to Pontiac, Woodward Avenue, is Detroit's main street; it begins at the Detroit River and follows the original Saginaw Trail of the Indians. The twenty-five-mile stretch between the city and Pontiac offers several examples of commuter suburbs that sprang up in response to a good transportation facility—the highway itself.[19]

As Woodward Avenue angles away from Detroit, it passes through communities such as Highland Park, the home of Ford's mammoth plant; Birmingham, which had a long, independent history before it became a commuter suburb; and Ferndale, a four-square-mile area that boomed in the 1920s as new houses surged across Detroit's northern tier, as well as villages such as Royal Oak and Huntington Woods, which developed in the twentieth century. The Woodward Avenue "family" includes Berkley, a poor neighborhood with homes built of whatever materials were handy, as well as Bloomfield Hills, one of the wealthiest communities in the United States. All share a dependence on the highway and a conviction that its function is that of an artery rather than an open space. As early as 1892, realtors were advertising the "repaving of Woodward Avenue" as an incentive to purchasers, and in 1923, when a Wider Woodward Project was started to provide an eight-lane concrete road all the way from Detroit to Pontiac, every town along the way joined in enthusiastic support. When the improvement was finished in the late 1920s, local boosters proclaimed the road the nation's best, and developers began to advertise land in terms of its distance from Woodward Avenue. The new facility enabled executives to travel fifteen miles to their offices in downtown Detroit in less than forty minutes.[20]

Because traffic clogged even wide thoroughfares like Woodward Avenue, most urban regions soon proposed "express" streets without any stop lights or intersections at grade. The first of the new type were called parkways because the land on either side of the travel ways was typically part of a park and because parkways followed the natural topography of the land, generally running along the banks of stream valleys.

Their distance was measured in miles, not blocks, and they were usually sinuous, not straight.[21]

The French engineer Eugene Henard is credited with inventing the grade-separated interchange *(le carrefour a voies superposées)* in a 1906 plan for Paris. In the years before World War I, the Italian futurist architect Antonio Sant'Elia produced a revolutionary scheme in his designs for "La Citta Nuova" (New City). His projected Central City Station for Milan (1913–1914) brought together seven levels of separated railway and motor traffic.

But the idea of an expressway was first systematically applied in the United States. William K. Vanderbilt's Long Island Motor Parkway (1906–1911) was the world's first thoroughfare restricted solely to the automobile, and especially designed for its needs. Made of innovative concrete, it featured open speeds, bridges and tunnels to separate it from local cross traffic, and limited access through its own toll gates. Even more significant was Westchester County's bucolic and meandering Bronx River Parkway, begun in 1906 and completed in 1923. There, the complete separation of crossing traffic from the parkway was accomplished cheaply. Because the park roadway ran through a valley, it could be bridged by crossing roads without massive earthwork. The result was an enormous aesthetic success. Running sixteen miles from Bruckner Boulevard in the Bronx alongside the New York Central tracks to White Plains, the beautifully landscaped road stimulated automobile commuting from Scarsdale, Mount Vernon, Bronxville, and New Rochelle.[22]

Within ten years, the New York area also witnessed the construction of the Hutchinson River Parkway (1928), the Saw Mill River Parkway (1929), and the Cross County Parkway (1931). The Henry Hudson Parkway, the first inner-city freeway (there was a toll across the bridge to the Bronx, however) to have limited access, no grade crossings, and service stations of its own, was begun along the West Side of Manhattan in 1934. Meanwhile, by 1929, New Jersey had started a thirteen-mile expressway between Jersey City and Elizabeth; Chicago had completed an "elevated drive" north of the Loop that could handle 60,000 vehicles per day; Boston had finished two main traffic arteries north and south of the city; and Philadelphia's Planning Federation had recommended the construction of seven highway routes to permit motorists to bypass city congestion. Bridges such as that over the Delaware River between Philadelphia and Camden (1926) and the George Washington Bridge, which connected New York City with northern New Jersey in 1933, also spurred suburban growth and increased the demand for feeder highways. Although most of the new thoroughfares were radial connections linking the core with what was then the fringe of the built-up area, many were carefully landscaped to accommodate the popular practice of pleasure driving. Indeed, most were designed more for recreation than for rush-hour journeys-to-work. All contrasted dramatically both in form and function with the German *Autobahn,* a massive roadway built primarily for military transport in the 1930s.[23]

Across the North American continent, California planned a statewide system of express-traffic arteries. The Arroyo Seco Parkway was first envisioned in 1911, but not until downtown Los Angeles department stores backed the project was the $1.3 million "speedway" launched. Opening on December 9, 1940, with a chain reaction rear-end accident involving three carloads of dignitaries, the 8.2 mile Arroyo Seco Parkway (later renamed the Pasadena Freeway) was the first section of what would later become the most extensive freeway system in the world. Intended to lure shoppers to the center of Los Angeles, it instead enabled city residents to move to the suburbs, and by 1960 it was carrying 70,000 cars per day, or 25,000 more than its planned capacity. Most importantly, it raised the value of Pasadena real estate to such an extent that developers and builders anxiously supported freeways elsewhere in the region.

At the national level, rural interests such as the National Grange Association lobbied vigorously for better roads between farm and market. As early as 1902, Congress considered, but did not pass, a highway bill that prohibited cities from receiving any assistance. Gradually, urban spokesmen became more insistent, joining with farm groups to promote a national highway program. The Federal Road Act of 1916 offered funds to states that organized highway departments; the Federal Road Act of 1921 designated 200,000 miles of road as "primary" and thus eligible for federal funds on a fifty-fifty matching basis. More importantly, the 1921 legislation also created a Bureau of Public Roads to plan a highway network to connect all cities of 50,000 or more inhabitants.[24]

Meanwhile, the adoption of gasoline taxes, beginning with a one cent per gallon tax in Oregon in 1919 (by 1929 every state had enacted a similar tax on motor fuel at the pump), provided the necessary state revenues for massive road-building programs. By 1925 the value of highway construction projects exceeded $1 billion for the first time; thereafter, it fell below that figure only during a few years of the Great Depression and World War II.[25] Even during the troubled thirties, however, state and federal funds were made available for roads because they employed many workers and could be planned quickly. During this period Robert Moses of New York became the nation's greatest builder as he combined a genius for getting things done with a zeal for large-scale projects. His parkways crisscrossed the huge metropolis and ushered in a new era of federal road subsidies. Moses was hardly typical, but his extraordinary success derived partly from the fact that his attitude toward the street was in harmony with that of a growing number of citizens. No longer a market place or the scene of informal social interaction, the street was becoming a place where movement was paramount and where the motorcar was King. Clay McShane has succinctly summed up the change:

> Thus, in their headlong search for modernity through mobility, American urbanites made a decision to destroy the living environments of nineteenth-century neighborhoods by converting their gathering places into traffic jams, their playgrounds into motorways, and their shopping places into elongated parking lots. These paving decisions effectively made obsolete many of urban America's older neighborhoods.[26]

Public Policy and Public Transit

While private transportation was flourishing with general taxpayer subsidies and the road was defined as a public good, mass transportation was floundering because of government decisions that the streetcar represented private investment and should "pay for itself." As early as 1910, American urban transit, once the envy of the world, had been overtaken by Europe. In Germany, Holland, and France, alternatives were developed to the ugly overhead wires that were a common feature of United States systems. And whereas streetcar ridership had been about four times as great as Europe on a per capita basis in 1890, the two were about equal in 1910.[27]

The major problem was fiscal. From the beginning, trolley promoters watered stock in a reckless and selfish manner. Moreover, the typical franchise agreements, dating from the 1890s, guaranteed the five-cent fare. The Boston Elevated Company, for example, agreed in 1897 to maintain the five-cent fare with free tranfers for at least twenty-five years. Prior to 1900, such an amount was usually sufficient to generate substantial profits. With the discovery of gold in Alaska in 1898, however, and with continuing inflation after the turn of the century, operating costs rose to such an extent that the nickel price was no longer adequate to maintain the existing system, much less provide a surplus for new equipment. During World War I, severe shortages almost doubled the cost of living again, but city officials, sensitive to the voters, refused the companies' requests for fare increases. Essentially, the street railways were being asked to provide rides for half price. The Boston Elevated Railroad, for example, never made a profit and declared bankruptcy in 1918. As company after company fell from profitability, the stock prices of the entire industry dropped, so that necessary money for modernization could not be raised. The result was a vicious cycle in which aging equipment and reduced services were accompanied by falling ridership. By 1925 buses were beginning to replace trolleys on the outskirts, where the long distances made the streetcars less economical, at least during a time of cheap gasoline.[28]

Three possibilities existed for the regulated private companies. The initial response of most firms was to increase the volume of traffic even further, especially in cooperation with land developers. Indeed, even in the era of the Model T, some of the most spectacular residential ventures were expressly tied to the fareboxes of streetcar lines. In Brooklyn, which led the nation in housing construction in both 1922 and 1923, large-scale development in Sheepshead Bay began only after the Brighton Beach line of the BMT was opened to the district in 1920. Throughout the giant borough, speculators were active in petitioning for transit routes near their housing ventures. The trolley extension along Nostrand Avenue from Avenue Q to Avenue U, for example, was completed only as a result of vigorous campaigning by realtors, who paid $10,000 of the $30,000 needed to construct the line.[29]

In Shaker Heights, southeast of Cleveland, the Van Sweringen brothers followed the late-nineteenth-century pattern of owning both the traction system and the land brought within commuting range. At a tract of several thousand acres—the site of an old Shaker religious colony—they developed an upper-middle-class community renowned for its preservation of parklands, its imaginative street plans, and its rigid architectural and construction standards. One English visitor visitor described it as "the finest residential district in the world." In 1920 the Van Sweringen electric rapid transit lines reached Shaker Heights. Because four miles of that track ran along the old right-of-way of a railroad, the brothers bought the entire 513-mile railroad to realize their dream. Thereafter, the town boomed. In the next decade population grew by 1,000 percent, and real-estate valuation increased more than five-fold.[30]

Successful streetcar suburbs, however, even those like Shaker Heights that were attractively packaged, were the exception after World War I. More typically, decisions to expand service were badly timed. After track mileage had been increased, heavier cars introduced, and service made more frequent, the additional patronage did not materialize, and the companies were simply saddled with larger debts. And the often-absentee owners paid little attention to cultivating ridership.

The second possible solution to the public transportation crisis was public subsidy. Precedent for municipal investment in mass transit came from Boston and New York, where the subway systems had been very largely financed by the city treasury. But unlike the road, which was defined as a public good and thus worthy of public support, mass transit was defined as a private business unworthy of aid. And when the companies took their case to the public, as in Detroit, Seattle, and Los Angeles, voters rejected their appeals for taxpayer financing of rail improvements. Thus, Americans taxed and harassed public transportation, even while subsidizing the automobile like a pampered child.[31]

A third solution was to raise the fare. But such a change required the elimination of regulation, which required political action, and the highly visible and unpopular streetcar companies were not objects of sympathy. The cities typically refused even to relax the regulations and restrictions which prevented innovation. The nickel fare remained until it was too late, and then the raises rarely kept up with need.

The electric streetcar systems might have survived the impersonal forces of the marketplace, if only because the trolley was as efficient as any alternative form of intracity movement. But the automobile industry did not leave the existence of competitive forms of travel to chance. Beginning in 1926 and continuing for the next thirty years, General Motors operated a subsidiary corporation to buy nearly bankrupt streetcar systems and to substitute rubber-tire vehicles for the rail cars. In New York City, for example, GM combined with the Omnibus Corporation to convert an enormous streetcar operation to buses. In 1919, their peak year, the five boroughs accounted for 1,344 miles of trolley track, and these surface lines carried more passengers than the famed subway and elevated trains. Yet by 1939, there were only 337 miles of streetcar track remaining in the city, and even those were scheduled for extinction. Manhattan led the way, with most of its routes ripped out during a single eighteen-month period during the 1930s, despite massive rider complaints and petitions.[32]

The details were remarkably similar in other cities, and by 1950 General Motors had been involved in the replacement of more than one hundred streetcar operations—including those of Los Angeles, St. Louis, Philadelphia, Baltimore, and Salt Lake City—with GM-manufactured buses. A federal grand jury ultimately found the giant corporation guilty of criminal conspiracy for this effort, but the total fine—$5,000—was less than the profit returned from the conversion of a single streetcar.[33]

The demise of the nation's rail-transit systems occasioned few protests because most people agreed with New York Mayor Fiorello LaGuardia that the automobile represented the best of modern civilization while the trolley was simply an old-fashioned obstacle to progress. In 1940 the Denver Planning Commission suggested that streetcars be removed from major thoroughfares "because (they) delay the faster vehicular traffic." In Detroit the chairman of the rapid transit commission himself spoke of the automobile as "the magic carpet of transportation for all mankind." In this regard, they were simply echoing what futurists had been saying for over a generation. "We declare," Marinetti wrote in 1909, "that the splendor of the world has been enriched by a new beauty—the beauty of speed. A racing car . . . racing along like a machine gun is more beautiful than the winged victory at Samathrace."[34]

The misguided and unfortunate result of such thinking was that Americans would no longer have transit options and that the car would become a prerequisite to survival, with disastrous consequences for energy consumption and traffic deaths. Far from supplementing electric-rail systems, the automobile became the single form available, and the suburbs became abjectly dependent on a vehicle that demanded ever-larger resources in terms of street space, parking facilities, and traffic patrols. As early as 1940, about 13 million people lived in communities beyond the reach of public transportation.

Without additional riders, public subsidies, or higher fares, and in the face of a concerted effort by General Motors to eliminate street railways, the nation's trolley systems could not last. The number of electric streetcars peaked in 1917 at 72,911, while total ridership crested in 1923 at 15.7 billion nationwide. Patronage slowly declined in the 1920s (to 14.4 billion in 1929) and precipitously in the 1930s (to 8.3 billion in 1940). By 1948 the number of streetcars in service had fallen to 17,911, and by 1985 the clang of the trolley could be heard on only a few lines in Boston, New Orleans, Pittsburgh, Philadelphia, and Newark.

Commuting by steam railroad survived automotive competition better than did the trolleys. Indeed, the "golden age" of railroad commutation was during the 1920s, when every major Eastern and Middle Western city benefited from inexpensive and frequent passenger rail service. The Depression years saw a sharp drop in patronage, but the demise of the once-magnificent passenger railroad system of the United States did not come until the 1950s, when bankruptcy and deteriorating service were the inevitable result of a national transportation policy that subsidized air and automobile travel and that taxed the railroads. By the mid-1980s, only a handful of cities—including New York, Boston, Chicago, and Philadelphia—could boast of impressive railroad-commuter traffic.[35]

10

Suburban Development Between the Wars

> To possess one's own home is the hope and ambition of almost every individual in our country, whether he lives in a hotel, apartment, or tenement. . . . Those immortal ballads, *Home Sweet Home, My Old Kentucky Home,* and *The Little Gray Home in the West,* were not written about tenements or apartments . . . they never sing songs about a pile of rent receipts.
>
> —PRESIDENT HERBERT HOOVER
> Address at Constitution Hall, December 2, 1931

If the automobile represented the future, it was not clear early in the century what shape that future might take. Although gasoline prices ranged from twenty cents a gallon in metropolitan areas to fifty cents a gallon in the desert, and although most cars were priced at well over one thousand dollars, the costs associated with driving fell between 1900 and World War II. In conjunction with better roads and the abundant fuel that became available following the 1906 Spindletop discoveries of oil in Texas, the motor vehicle lowered the marginal cost of transportation.[1] This meant that, once a household had purchased an automobile, short trips were cheap, and children and spouses could be transported without the additional fare that would have been necessary on a streetcar. Automotive families had neither to wait nor to walk. New possibilities in shopping, living, and working were opened. In 1925, Columbia University sociologists Robert and Helen Lynd completed the field work for their now classic study of "Middletown." In all six areas of social life in Muncie, Indiana—getting a living, making a home, raising the young, using leisure, engaging in religious practices, and participating in community activities—the private car played either a contributing or a dominant role. When President Herbert Hoover convened a commission to examine contemporary patterns of American life, the report said this about the car:

> In a considerable degree the rapid popular acceptance of the new vehicle centered in the fact that it gave to the owner a control over his movements that the older agencies denied. Close at hand and ready for instant use, it carried its owner from door to destination by routes he himself selected, and on schedules of his own making; baggage inconveniences were minimized and perhaps most important of all, the automobile made possible the movement of an entire family at costs that were relatively small. Convenience augmented utility and accelerated adoption of the vehicle.[2]

With all its convenience, however, it remained to be seen whether this dream machine would most benefit rural areas, cities, or suburbs. Initial speculation favored the countryside. As the United States became a nation on wheels, a disproportionate number of motorcars went to farm families, and rural spokesmen predicted that the life of the yeoman would regain lost luster. In 1907 a publicist forecast that the automobile would "remove the last obstacle to the farmer's success. It will market his product, restore the value of his lands, and greatly extend the scope and pleasure of all phases of country life."[3] By providing easy access to a wide area, thereby reducing the traditional isolation of rural existence, reliable cars and good roads were expected to stimulate a back-to-the-land movement. At the very least, they would give young people less reason to forsake the farm for the bright lights of the city. When a lonely woman was asked by a government agent in the 1920s why her family had purchased an automobile before equipping their home with indoor plumbing, she answered matter-of-factly: "Why, you can't go to town in a bathtub."[4]

The automobile undoubtedly improved the quality of rural life, but its permutations, such as the popular Fordson tractor, drastically reduced the amount of human labor necessary to produce a given crop. The back-to-the-land movement never materialized, as mechanization cut the farm proportion of the national population from 32 per cent in 1900, to 23 percent in 1940, to 3 percent in 1980.

Urban spokesmen were also deluded into thinking that the internal combustion engine would be a boon for their constituents. Municipal officials were outspoken advocates of road construction and improvement, and no less an authority than Allan Nevins argued that the car was the most potent stimulus to urbanization ever devised. It would reduce the problem of dead horses and mercifully eliminate manure from the streets. Moreover, an increase in motor travel initially meant that it was easier for customers to get downtown for serious shopping, and some city officials blissfully suggested that the automobile would actually decrease traffic congestion. In 1925, for example, Newark merchants boosted the attractions of their businesses: "Broad Street today is the Mecca of visitors as it has been through all its long history. They come in hundreds of thousands now when once they came in hundreds."[5]

In the short run, merchant predictions were accurate. Employment in central cities mushroomed, and downtown office space in the ten largest cities tripled between 1920 and 1930. Steel skeleton skyscrapers grew like weeds on the urban landscape; few people understood the prescience of Lewis Mumford's lament that the automobile was antithetical to the very meaning of the city.[6]

Soon, however, downtown sections became hopelessly congested, and the ease of moving about by car was cancelled out by the impossibility of finding a place to park. As early as 1930, more than half the daily commuters into medium-sized cities like Milwaukee, Washington, and Kansas City came by car. Parking lots slowly appeared, but they were expensive, and congestion worsened. To hasten traffic flow, New York began installing a block-signal, stop-light system in 1923, and the following year General Electric began production of timed electric lights. But clogged streets had become part of the rhythm of urban life by the late 1920s. In Manhattan average late afternoon travel speed on Fifth Avenue fell below three miles per hour in 1926. In Los Angeles the number of people entering the downtown area declined by 24 percent between 1923 and 1931 despite a population boom in the metropolitan area.[7]

The situation was almost as bad in smaller cities. One Atlanta drugstore owner, forced to close in 1926, prophesized: "The place where trade is, is where automobiles go. . . . A central location is no longer a good one for my sort of business." Indeed, farther out from the core the automobile quickly gained an advantage in speed over public transportation. A Kansas City study in 1930 revealed that even at rush hour, at seven and one-half miles from the downtown area the automobile had a 15-minute advantage over the primary trolley lines and a 35-minute advantage over the secondary ones.[8]

The Suburban Boom of the 1920s

Those who gained the most from the automobile were those living in the newer suburban areas. Undeveloped land on the metropolitan fringes became prime residential real estate. Housing reformers like Lawrence Veiller, Jane Addams, and Jacob Riis welcomed such a result because of their belief that poverty and crime were largely the handmaidens of overcrowding in the central cities. As Mark Foster has noted, thoughtful urban decision makers supported any project that promised to give slum dwellers access to open spaces. Frank Lloyd Wright and LeCorbusier embraced the automobile as a revolutionary liberating force,[9] while Henry Ford himself predicted, "The city is doomed," and "We shall solve the city problem by leaving the city."[10] The Flivver King took his own advice and moved to a 2,000-acre suburban estate, "Fair Lane," replete with an artificial lake, ten miles from Detroit. If other citizens could not share Ford's lifestyle, they could nonetheless experience the cheaper cars and better roads that made low-density living feasible. Motor vehicle registration statistics in the United States have consistently indicated that car owners are concentrated not in the city but in the area immediately surrounding it.[11]

The decade after the end of World War I was the first in which the road and the car had full impact. The 1920 census revealed that only 46 percent of American families were homeowners. In central cities, the proportion was even lower: 27 percent in New Orleans, 18 percent in Boston, 12 percent in New York. In the seven years between 1922 and 1929, however, new homes were begun at the rate of 883,000 per year, a pace more than double that of any previous seven-year period. Block after block of the American dream turned into one-story frame houses with chain-link fences, white ruffled curtains, and wrought-iron posts holding up small front porches. The fact that wages were rising and housing prices were falling helped the construction industry.[12] So also did real-estate tax exemptions, such as that passed by New York City in 1920. But automobile access was critical, and, as would so often be the case in the United States, the new roads were typically paid for not by the beneficiaries, but by general taxpayers. As Los Angeles planner Gordon Whitnall explained in 1924:

> When we faced the matter of subdivisions in the County of Los Angeles . . . we reached the conclusion that it would be absolutely necessary to go out and try to beat the subdividers to it by laying out adequate systems of primary and secondary highways at least, thus obtaining the necessary area for highways and boulevards.[13]

New suburbs sprouted on the edges of every major city. Of the seventy-one new municipal incorporations in Illinois and Michigan in the 1920s, two-thirds were Chicago, St. Louis, or Detroit suburbs. Between 1920 and 1930, when automobile registrations rose by more than 150 percent, the suburbs of the nation's 96 largest cities grew twice as fast as the core communities.[14] Statistics for individuals communities were particularly startling. Grosse Pointe, near Detroit, grew by 725 percent in ten years, and Elmwood Park, near Chicago, by 717 percent. Long Island's Nassau County almost tripled in population, while Connecticut's nineteen fastest growing towns of the decade were all suburbs. By 1922 about 135,000 suburban homes in sixty cities were already completely dependent upon cars for transportation, and in 1923 *National Geographic* did a special feature on the automobile industry. "Cities," it noted, "are spreading out:"

> Long Island is built up for half its length to accommodate those who make New York the metropolis of America; so is New Jersey from Morristown to Long Branch and Jersey City to the Empire State boundary at Suffern. Even Connecticut, as far as Stamford, Greenwich, and New Canaan, is peopled with those who work in Gotham by day and sleep in the country by night.[15]

Appropriately, the great novel of 1925, *The Great Gatsby,* stressed motion of all kinds and featured a yellow Rolls-Royce with which the hero hoped to woo Daisy Buchanan.

The impact of the automobile was even greater than suburban statistics and quotations indicated, however, because a large part of the growth *within* cities was in sections newly opened by the car. Los Angeles's immense annexation of the San Fernando Valley in 1915 gave the southern California metroplis an arena for residential growth that would not be filled in until the 1950s. In Chicago the great consolidation of 1889 brought into the municipal corporation areas like Beverly and Morgan Park that were suburban in everything but legal status. In New York City, Brooklyn gained 540,000 residents in the decade, but its growth was concentrated in Flatlands, Canarsie, Sheepshead Bay, and Bensonhurst. All except Bensonhurst were on the outer edges of the borough, where public transportation was less important than elsewhere in the city.[16]

The benefits of this suburban boom were not evenly distributed; the biggest gainers were usually those who owned close-in farms. Sometimes, long-time rural residents earned a bonanza, but more often, real-estate speculators and lawyers who had advance information about street openings bought such property at bargain prices. Every multi-lane ribbon of concrete was like the touch of Midas, transforming old pastures into precious property. Even the promise of a road was sometimes enough to stimulate development. In southeastern Brooklyn, promoters stressed that paved streets would soon be extended to their "conveniently situated lots," making commuting to Manhattan possible even in the absence of mass transit. They announced that the city expected to upgrade Kings Highway, which would run for seventeen miles across the borough, that Flatbush Avenue would become a direct artery between

Rockaway Beach and downtown Brooklyn, and that Nostrand and Bedford Avenues would be connected with the proposed Marine Park on Jamaica Bay. These promises had not been fulfilled by 1930, but by that time the developers had sold their houses and land and moved on to other areas.[17]

Not all speculators got rich, and some even lost money. Just before the Great Depression, overanxious subdividers platted more land than could be occupied, and many were thus driven into bankruptcy. The economic catastrophe temporarily transformed Skokie, Illinois, and Burbank, California, into ghost towns, grim reminders that land investment was not always profitable. Those who inaugurated their land-promotion schemes earlier generally had better luck, and some of their parklike communities—such as Coral Gables, Florida, and Roland Park in Baltimore—retain their prestige today. Qualitatively, the most successful American developer was Jesse Clyde Nichols, who began Kansas City's Country Club District in 1922. Quantitatively, the most important single developer was Harry Chandler of Los Angeles.[18]

Kansas City

"Few people in the world, or in America for that matter," wrote Andre Maurois, "realize that Kansas City is one of the prettiest cities on earth. . . . Why? Because one man wished it so, and insisted upon it." The man was Jesse Clyde Nichols who built perhaps 10 percent of the housing in the Kansas City area during a career as a commercial and residential developer that spanned the half-century between 1906 and 1953. His crown jewel was a planned suburb which sprawled over ten square miles on the southwest side of the metropolis. Influenced by a bicycle tour of garden communities in Europe and by the ideas of the City Beautiful Movement in the United States, Nichols purchased a garbage dump, a forlorn harness-racing track, and a brick kiln in 1908. Systematically adding to his holdings, he waited fourteen years before beginning the 6,000 homes and 160 apartment buildings that would ultimately house 35,000 residents. The Country Club District promised "spacious grounds for permanently protected homes, surrounded with ample space for air and sunshine," and it quickly won fame for its parkland, its setback lines, and its self-perpetuating deed restrictions. Nichols's homes also offered a full range of modern conveniences such as piped gas, electric service, household gadgets, and the nearby electric trolley, that enabled Kansas Citians to construct a congenial setting for self-contained family life.[19]

A graduate of Harvard who later founded the Urban Land Institute,

Nichols followed three precepts: no right angles or gridiron streets, no wanton destruction of trees, and no disregard of the natural contours of the land. His houses were expensive. Beginning with six dwellings per acre, Nichols progressively increased the minimum size of the lot and the minimum cost of the structure that could be built upon it. Exclusivity derived also from a laundry list of rigid restrictions. Land was sold subject to substantial setbacks from the street, to the provision for extensive foundation plantings, to the acceptance of racially restrictive covenants, and to the construction of driveways and garages. In addition, every purchaser had to join the Homeowners Association, the purpose of which was to supervise the quality of private lawn care, the cleaning of streets, and the collection of garbage. With an innovative shopping center (see Chapter 14) and exquisite landscaping, the Country Club District attracted planners and builders from around the world. By 1930 it was *the* place to live in Kansas City, and its new high school was among the most admired college-preparatory institutions in the United States.

Los Angeles

As Chicago had been the boom city of the nineteenth century, Los Angeles was the boom city of the twentieth. A community of only 11,200 in 1880 and of 102,000 in 1900, the City of Angels tripled in the 1900s, doubled in the 1910s, and doubled again in the 1920s. Land speculation was part of its heritage from the beginning of settlement. Real estate values rose 200 percent between 1865 and 1866 and another 500 percent in 1868. During the frenzied railroad boom of the 1880s, prices escalated wildly, only to collapse when investors realized that more lots had been subdivided than could possibly be sold. When growth resumed in the 1890s, street railway magnate Henry E. Huntington, investor H. J. Whitely, and entrepreneur Robert C. Gillis made fortunes transforming the barren countryside into housing tracts.

A major factor in the suburbanization of Los Angeles was the discovery of vast amounts of oil in the region in the mid-1890s. Employment opportunities at derricks, refineries, and tank farms in industrial suburbs like Whittier and Fullerton pulled an industrial working class to the edges of the vast metropolis, especially in southern Los Angeles County and northern Orange County. By World War I, a pattern had emerged in which industrial suburbs created a demand for their own residential suburbs, forming widely separated clusters. The discovery of additional oil fields in the 1920s created new suburbs between the existing clusters and made the boundaries separating the various communities indistinguishable to the eye. Thus was born the famed political fragmentation of the southern California metropolis.[20]

As Los Angeles surpassed Seattle and San Francisco to become the largest American metropolis west of the Mississippi River, it remained unusual in its extraordinary dispersal and in its dominance by detached houses. In 1930 about 94 percent of all dwellings in Los Angeles were single-family houses, a figure unmatched by any other city. In the 1920s alone, more than 3,200 subdivisions, encompassing about 250,000 houses, were begun, as native Americans from Iowa, Illinois, and Minnesota sought the end of the rainbow in the balmy climate of southern California.

Los Angeles suburbanization between the wars was unusual in two other ways as well. First, the enormous annexation of the San Fernando Valley in 1915 meant that professional and middle-class families seeking new housing typically moved to suburban-style neighborhoods that were actually within the legal city. Second, the many oil industry suburbs meant that working-class sections were often outside the legal city.[21]

The prototypical upper middle-class suburbs in the southern California region were Beverly Hills and Palos Verdes. At the turn of the century, the Amalgamated Oil Company, a subsidiary of the Huntington empire, purchased ranch land west of Los Angeles for exploratory drilling. When no oil was found the promoters turned to the other method of quick wealth in the West—real-estate speculation. In 1906 Burton E. Green formed the Rodeo Land and Water Company and subdivided thirty-two hundred acres north of Santa Monica Boulevard. In 1914 this development formed the basis for the incorporated city of Beverly Hills. The community featured curving streets instead of the ordinary checkerboard subdivision which up to that time had been the accepted method of subdividing in Los Angeles. Lots were wider than they were deep, which has remained a distinguishing characteristic of Beverly Hills. The village became fashionable when Mary Pickford and Douglas Fairbanks build "Pickfair" there, and it boomed in the 1920s, when its population rose by 2485 percent.

Palos Verdes was equally plush but well to the south of Los Angeles. After a number of planning proposals for the Palos Verdes Penninsula—including one by a New York banker to develop a "millionaires' colony"—the 16,000 acres of rolling hills overlooking the Pacific Ocean began to be developed in 1923, partially under the aegis of Frederick Law Olmsted, Jr. Spanish architecture was the only allowable residential type, and protective restrictions determined that half the land had to be reserved for parks and roads and that 90 percent of the remainder was to be reserved for private houses. Olmsted supervised the planting of hundreds of thousands of trees, and he built his own home on a hill overlooking the water.[22]

No one was more important in stimulating Los Angeles's suburban growth and no one benefitted from it more than New Hampshire-born Harry Chandler, who left Dartmouth College early because an undergraduate prank injured his lungs and forced him to seek a drier climate. Arriving in Southern California with few resources in 1883, he delivered newspapers so effectively that he gained control over circulation within large areas of the city. In 1894 he married the daughter of Harrison Gray Otis and soon thereafter became the owner of the *Los Angeles Times.*

After a brief interlude as a progressive reformer, Chandler turned his energies to real-estate development. Heading a syndicate of rich investors, Chandler purchased 47,500 acres of ranch land in the San Fernando Valley between 1903 and 1909. Because the property lacked any water sources and was in fact a desert, it went for only $2.5 million. But Chandler and his associates, who included transit magnate Henry E. Huntington and Hollywood developer Moses Sherman, were quietly arranging to have the City of Los Angeles buy the water rights of the fertile Owens Valley and pipe the precious substance 200 miles by aqueduct to the dry land of the San Fernando Valley. Chandler was a political conservative who usually opposed government enterprise, but in the case of the Owens Valley Aqueduct public improvements would advance his personal interests. Under pressure from the *Los Angeles Times* and the Chamber of Commerce, the city floated a $25 million bond issue, and the Metropolitan Water District of Southern California was born.

Meanwhile, the Pacific Electric Railway extended its service to Tract 1000, the first big subdivision of the syndicate. When the impact of these vast public works enterprises became known the real estate that Chandler's San Fernando Mission Company had bought for less than $3 million was worth about $120 million. Hardly pausing to count their profits, they raced still further afield to form another syndicate and to acquire 300,000 acres of the Tejon Ranch in Los Angeles and Kern Counties. In the process, Chandler became the largest and most influential land developer in the state and the most powerful single individual in the American West. Although his fame and prestige rested with the *Los Angeles Times,* the basis of his wealth (estimated at his death to range from $200 million to $500 million) was the suburban real estate that was incorporated into the city in the great annexation of 1915.[23]

The Los Angeles water project illustrates many of the basic ingredients of suburban land conversion in the United States. Politically powerful investors used public services and public transportation to raise land values and attract settlers. The influence of the developer made government largesse available. The consequence was the subdivision of large holdings, which increased the population, pumped up the economy, and encouraged suburbanization.[24]

Characteristics of Automotive Suburbanization Between the Wars

Whether in New York, Kansas City, or Los Angeles, the automobile suburbs that appeared in the 1920s differed in four major respects from their mass transit related predecessors: (1) the overall pattern of settlement, (2) the length and especially the direction of the journey-to-work, (3) the deconcentration of employment, and (4) new forms of low-density, residential architecture.

The pattern of settlement in the streetcar metropolis had been essentially finger-shaped. New homes were constructed and sold only within walking distance of the rail transportation corridors. In 1900 distance from a streetcar or elevated line usually produced an inferior neighborhood; the urban landscape was characterized by bands of high-priced real estate. A mile from the station or half a mile from the trolley tracks were the immutable limits. As Herbert Ladd Towle put it in 1913, ''The gates of paradise would not have tempted us further.''[25]

The purchase of an automobile, in contrast, released the potential home buyer from confining his choice of residence to one convenient to a bus or trolley line. The real significance of the motor vehicle lay in its ability to move laterally or perpendicular to the fixed tracks, and thus open up land for settlement previously regarded as too remote. This meant that vacant land between the transportation corridors could be platted and sold for home sites. Thus in Brooklyn, the southeastern neighborhood of Flatlands never received rapid transit and remained undeveloped until the 1920s, when the automobile enabled commuters to move away from the subway and elevated lines. In other cities as well, the interstitial land was developed between existing fingers of settlement, and the built-up area assumed a more symetrical shape.[26]

For the first two decades of automobility in America, or roughly until the outbreak of World War I, the motorcar was a means to adventure. As dependability and comfort improved, however, the automobile became less an instrument of sport and more a means of pleasure—to ''go riding'' on weekends or holidays was a common family experience into the 1950s. Offering more freedom and luxury than the older and more uncomfortable streetcars, the car replaced the trolley and bus for discretionary travel. As the basic mode of journey-to-work movement, acceptance was slower. A detailed study of sixty-eight cities in 1933 revealed that commutation by foot or by public carrier was more important. Joel A. Tarr has shown that working-class families did not own cars and were not auto-dependent. "In 1934," he writes, "at a time when 45 percent (less than half) of the chief wage earners in Pittsburgh owned automobiles, 28 per cent walked to work, 48.8 per cent rode the streetcar, 1.7 per cent used the commuter railroad, and (only) 20.3 per cent drove their automobiles." Even in the county outside of Pittsburgh, almost twice as many wage earners walked to work in 1934 as used automobiles. It should come as no surprise, therefore that master builder Robert Moses labeled his many New York area controlled-access thoroughfares "parkways" because he saw the car as a recreational rather than utilitarian vehicle.[27]

If it did not affect the majority of American bread-winners, the automobile nevertheless reoriented the lives of those who did use it for commutation. Not only did it increase their journey-to-work, but more importantly, it enabled wage-earners to work in peripheral areas outside their own neighborhood without penetrating the center or even the city at all. South Orange, New Jersey, a community that experienced its most dramatic growth in the 1920's, is illustrative of this process. Because it is located in the First Watchung Mountains and commands a dazzlng view of Newark, of the Orange Valley, and of the New York City skyline, South Orange attracted middle- and upper-class families between 1915 and 1940.[28]

An analysis of the workplaces of South Orange residents (see Appendix A-10) in 1914, 1934, and 1954 reveals the journey-to-work changes that took place in one community in the first half of the century. In 1914, as was typical of many similar suburbs, more than half of the heads of households made their living within the three square miles of the village. About a quarter of the population commuted to Manhattan on the railroad: almost no one worked in another suburb.[29]

By 1934 the proportion of commuters to New York and to Newark, to which good rail and streetcar lines were available, had increased sharply from about one in three to about two in three while the proportion that both lived and worked in South Orange plummeted. More significantly, the 1934 data reveal that about one family head in eight was already journeying to a workplace in another suburb. The trend continued into 1954 as the automobile, which was well suited to the flexible requirements of lateral movement to other suburbs, became the preferred mode of transportation. As Towle explained in 1913: "Perchance you have no car—as yet. But you have friends living five miles away by road. To visit them by rail you must go half a mile to the station ride ten miles to a junction wait an hour, and travel a dozen miles to a station half a mile from their home.''[30]

The automobile made it infinitely easier to commute in directions perpendicular to the trolley tracks. But of even greater significance in changing the way Americans live and work was the truck. Even in its primitive form before World War I, the truck could do four times the work of a horse-drawn wagon which took up the same street space, and as engines became more powerful the truck became a marvel of efficiency. Prior to 1910, when there was only one truck for every ten thousand Americans, factories clustered toward the centers and the rail junctions of large cities. Essentially it was easier and more economical to move people in and out of the core by public transit than it was to move freight by horses and wagons to and from scattered businesses. Thus, workers fanned out in the metropolis while factories crowded near the rail heads.[31]

Industry, which had not historically been associated with cities in the age of wind and sail, began moving from urban cores even at the height of the age of steam power. The high price and scarcity of land in central areas, coupled with municipal regulations and taxes were important encouragements to disperse. In 1859, for example, the Boudinot Mill in Paterson, New Jersey, was transplanted from the heavily congested center of the city to the open land of the periphery. And in 1873 the Singer Sewing Machine Company pulled out of Manhattan for more space in bucolic Elizabeth. (Elizabeth was not so bucolic in 1982, when Singer closed its local plant after 109 years of continuous operation, allegedly because of foreign competition.)[32]

With the twentieth century came the electrical revolution. The typical power unit was no longer a steam engine, but a thousand horse-power turbo-generator. Most importantly, electric power transmission and the introduction of small electric motors to replace large steam machines altered the technology of factory manufacture by making it more economical to build single-story rather than multi-story plants. In 1914 the National Electric Lamp Association abandoned its buildings in downtown Cleveland for forty acres of wooded, hilly land twelve miles east of the city center. The setting was so much like a campus that it was known as a ''university of industry.'' In 1915 Graham R. Taylor wrote the first book about industrial suburbs and he discussed the ''shifting of factories, one by one, to the edge of the city.'' East St. Louis, Pullman, Alton, and Granite City in Illinois; Fairfield in Alabama; Chester and Norristown in Pennsylvania; Norwood in Ohio; and Yonkers in New York were in fact already famous for their diversified products. A dozen years later Professor Robert Murray Haig of Columbia University documented not only the shift of certain manufactures to satellite areas in the New York region, but also the deleterious effects of this "decentralization" on older parts of the city.[33]

Trucks greatly stimulated a tendency that was already well under way. They were not generally available until 1909, but their impact was quick. K. H. Schaeffer and Elliott D. Sclar have calculated that manufacturing employment in Boston's inner ring (defined as communities from two to six miles from the Boston Common) grew much faster than those either closer in or farther out between 1909 and 1919. Between 1915 and 1930, when the number of American trucks jumped from 158,000 to 3.5 million and the proportion of trucks to private automobiles doubled, industrial deconcentration began to alter the basic spatial pattern of metropolitan areas. In conjunction with better highways and new methods of materials handling that emphasized one-story manufacture, the truck created a new efficiency for outlying industries that was not matched by similar economies in inner-city operations. Between 1920 and 1930, the proportion of factory employment located in central cities declined in every city of more than 100,000 residents in the United States.[34]

Warehousing and distribution activities followed the factories to the urban edges, where almost all new industrial construction took place after 1925. In 1921 the DuPont Company switched from trains to trucks in moving chemical products from its Philadelphia plant to the tanneries that were its major customers in Wilmington. In New York the food, meat, dairy, heavy machinery, wood, jewelry, and precious metal industries began migrating from Manhattan to the various suburbs and satellite cities of northern New Jersey. And because trucks could haul beer and liquor as easily as they could move machinery, even speakeasies, gambling dens, and houses of ill fame moved toward the roadhouses on the fringes. All this caused a drastic drop in the horse population, a process begun by the electric streetcar in 1887. Between 1910 and 1920, the number of horses in New York City decreased from 128,000 to 56,000; in Chicago from 68,000 to 30,000; in Baltimore from 15,000 to 7,000; and in Cleveland from 16,000 to 4,000.[35]

The most important characteristic of the automobile suburb was its lower density and larger average lot size as compared with anything ever previously experienced in an urban world. Because the motor vehicle opened up much more land than was possible with public transportation, the price of a square foot of real estate was lower in areas accessible only to cars than in neighborhoods served by good streetcar systems. With more developable land available at cheaper prices, the average size of a building lot rose from about three thousand square feet in streetcar suburbs to about five thousand square feet in automobile suburbs. Residential densities moved in the opposite direction from about twenty thousand per square mile in trolley-based areas to about half that in areas based solely on the motorcar. In fact, the residential density of a neighborhood today is largely a function of the type of transportation system that accompanied its early development. In older sections, as streets were widened, as houses and businesses gave way to parking lots and service stations and as upwardly mobile families moved up to newer structures, population density fell. In the Williamsburg/Greenpoint part of Brooklyn for example a 1920 density of 120 persons per acre (about 79,000 per square mile) had fallen to 90 persons per acre (58,000 per square mile) ten years later. In Manhattan, where new subway lines also encouraged centrifugal movement, population loss was especially dramatic. The Lower East Side (east of Broadway from the Battery to Houston Street), for example, contained 398,000 people in 1910, 303,000 in 1920, 182,000 in 1930, and 147,000 in 1940. To reformers who had long pressed for the depopulation of the slums, this leveling out of neighborhoods was a welcome and much celebrated relief.[36]

The greater space available for suburban living was accompanied by radical changes in the house itself. At the turn of the century, the parlor was a buffer zone between public and private space; it was the locus of formal entertaining and was always the front room in the house. Similarly, the front porch was a popular place for relaxation. There, one could observe the passing parade of life while remaining on private property.[37]

Economic catastrophes in the 1890s, the birth of the income tax in 1913, the leveling tendencies of World War I, and new architectural tastes and fashions gave houses a new look in the interwar years. Some observers argued that the dwelling itself had lost importance because the car had given so many people the freedom to move around. Even affluent families began to manage without servants, who had become, according to the prevailing wisdom, more expensive, more unreliable, and more difficult to find. Extravagant pretenses were curbed. Led by Frank Lloyd Wright, architects answered the need for simpler lifestyles and servantless domesticity by reviving Andrew Jackson Downing's nineteenth-century notion of the functional house. Wright's Usonian style of the 1930s emphasized one-story homes with low-slung roofs, carports, and generous amounts of glass. This model for the "ranch" houses would continue to characterize suburban development after World War II. The parlor and porch and the formal lifestyle they supported were among the first to go as the automobile restructured household patterns.[38]

The prototype of the new architecture was of course Wright's expansive prairie house. But Edward Bok, for thirty years the editor of the *Ladies' Home Journal,* also had a significant role in shaping a new image of the ideal home. Socially conservative, Bok wanted women to give up careers and outside activities for the joys of motherhood and domesticity, and he argued that the "closer we keep our children to the soil the healthier will they be physically. Particularly concerned about "the wretched architecture of small houses," he crusaded for dwellings free of "senseless ornamentation" and equipped with the latest in bathroom and kitchen facilities. In 1895 the *Ladies' Home Journal,* which was the most successful magazine not only in the United States but in the world during the first quarter of the twentieth century, began publishing the plans for a series of houses costing from $1,500 to $5,000. He offered his readers full building specifications, together with estimates from four builders in different parts of the United States for five dollars a set. A storm of criticism arose from architects who claimed that Bok was taking "the bread out of their mouths," but the magazine persevered, and entire colonies of "Ladies' Home Journal houses" sprung up. Shortly before architect Stanford White was murdered in 1906 in Madison Square Garden because of an affair he had with the wife of railroad millionaire Henry Kendall Thaw, he wrote: "I firmly believe that Edward Bok has more completely influenced American domestic architecture for the better than any man in this generation. When he began, I was shortsighted enough to discourage him, and refused to cooperate with him. If Bok came to me now, I would not only make plans for him, but I would waive my fee for them in retribution for my early mistake."[39]

By the time Bok retired in 1919, the *Ladies' Home Journal* had a circulation above two million and had helped popularize the bungalow, a style originally imported from India which by the 1920s had evolved into an unpretentious one-and-a-half-story dwelling with a horizontal appearance that reflected the increase in the average size of a lot. Advertisers offered mail-away bungalow-house plans for as little as one dollar; many of the structures depicted were a modest eight hundred square feet in size and were within the reach of the middle class. Variations appeared around the country, and former President Theodore Roosevelt noted approvingly; "Bok is the only man I ever heard of who changed, for the better, the architecture of an entire nation, and he did it so quickly and so effectively that we didn't know it was begun before it was finished."[40]

Suburbanization in the Great Depression

The prolonged and mammoth economic downturn that followed the stock-market crash in October 1929 brought the housing boom of the previous decade to an abrupt halt. Between 1928 and 1933, the construction of residential property fell by 95 percent, and expenditures on home repairs fell by 90 percent. Only aggressive sales campaigns, Federal Housing Administration mortgage programs (see the next chapter), and extensive advertising kept the vision of a home of one's own before the American people. The "model home" in particular became a popular marketing device, especially during a brief housing resurgence later in the decade. Every new suburban subdivision and home show seemed to include a full-scale ideal home, replete with hte latest appliances and equipment. In 1935 the General Electric Company sponsored an architectural competition whose subject was the small, single-family house. The 2,040 entrants were required to list the GE appliances incorporated into the designs; one architect used 76 such appliances. Two years later, with the unveiling of the *Ladies' Home Journal* "House of Tomorrow" exhibit, this form of mass advertising clearly became the province of the architectural profession. It glamorized a wide range of technological gadgets and hosted record crowds. Finally, in 1939, at the New York World's Fair, twenty-one single-family houses were organized into a "Town of Tomorrow."

Meanwhile, the automobile never retreated. In all but the three deepest depression years, motor-vehicle registrations continued to rise, and the 1940 total exceeded that of 1929 by 4.5 million. When a local transit company donated free rides to a WPA construction site, less than half the unemployed workers accepted the offer; the majority came by automobile. Similarly, when the Lynds returned to "Middletown" in 1933, they found that declines in retail trade since 1929 ranged from 38 to 85 percent. The single exception was the filling station, where sales fell by less than 5 percent, eloquent testimony to the reluctance of Muncie citizens to do without their personal transportation. "We'd rather do without clothes than give up the car," one working-class housewife said to the Lynds. "I'll go without food before I'll see us give up the car," insisted another.[41]

Early in 1933, as a final gasp of the Hoover Administration, the President's Committee on Recent Social Trends in the United States concluded that "the automobile has become a dominant influence in the life of the individual, and he, in a very real sense, has become dependent on it." In 1940 John Ford made a movie of *The Grapes of Wrath,* John

Steinbeck's novel of impoverished migratory workers struggling during the depression to get to California and find jobs. Automobiles brought the bankers and realtors to the farm to announce the evictions and tractors razed the shacks of the newly unemployed. The farmers' salvation was another motor vehicle, a "rolling junk," which was the family's only means of mobility. Communist authorities seized upon the film as a harrowing depiction of the distress of capitalism and showed it throughout the Soviet Union. But it was removed from the theaters after six weeks because the Russian people were less impressed by the poverty they saw on the screen than by the fact that even wandering peasants owned a car and wore shoes.[42]

The densely concentrated United States cities of the nineteenth century, with their industries, stores, and offices crowded together toward the middle, were a short-lived phenomenon brought on by the fact that interurban transportation—that between cities—was better than intraurban transportation—that within cities. The automobile so vastly changed the equation that cities began to "come apart" economically and functionally even as they had earlier begun to come apart legally with the breakdown of annexation. Indeed the automobile had a greater spatial and social impact on cities than any technological innovation since the development of the wheel. As early as 1938, William F. Ogburn wrote that "the invention of the automobile has had more influence on society than the combined exploits of Napoleon, Genghis Khan, and Julius Caeser," and in 1975 James J. Flink concluded that Henry Ford's major contributions—the moveable assembly line, the five-dollar day, and the Model T—had affected America in the twentieth century more than the Progressive Era and the New Deal combined. The liberation of our culture from the limitations of horse-drawn transport has yet to receive analytical treatment, however. Walter Prescott Webb showed the consequences of barbed wire, the revolver, and the windmill on the Great Plains; William F. Ogburn has analyzed the impact of the radio and the airplane; Lynn White has imaginatively written of the importance of the stirrup. But no one has yet gone beyond Flink's provocative *The Car Culture* and attempted to resolve the persistent paradox of the American love affair with the car and the less happy assessment of what happens to the environment when a society takes to the road.[43]

No other invention has altered urban form more than the internal-combustion engine. The automobile allowed its owner to leave and return when he wanted and along routes of his own choosing. Public transport, which had only recently broken the close spatial connection between work and residence no longer seemed so attractive and wonderful. Before 1920 developable real estate had to be located within walking distance of public transit. After 1920 suburbanization began to acquire a new character as residential developments multiplied, as cities expanded far beyond their old boundaries, and as the old distinctions between city and country began to erode. The enormity of the shift was apparent by 1941, when the Bureau of Public Roads surveyed commutation patterns. It found that 2,100 communities, with populations ranging from 2,500 to 50,000 did not have public transportation systems of any kind and were completely dependent upon the private automobile for personal travel. Such a situation would have been inconceivable twenty-five years earlier.[44]

11

Federal Subsidy and the Suburban Dream: How Washington Changed the American Housing Market

A nation of homeowners, of people who won a real share in their own land, is unconquerable.

—President Franklin D. Roosevelt

For at least the past two centuries, the easy availability of housing and land has distinguished the United States from other nations of the world. In 1920, when the Census Bureau announced that more than half the American population lived in urban areas, what was really unique about the United States was not the size of its huge cities, but the extent of their suburban sprawl; not the number of its workers, but the number of its commuters; not the height of its skyscrapers, but the proportion of its homeowners. Suburbanization had become a demographic phenomenon as important as the movement of eastern and southern Europeans to Ellis Island or the migration of American blacks to northern cities.[1] The appeal of low-density living over time and across regional, class, and ethnic lines was so powerful that some observers came to regard it as natural and inevitable, a trend "that no amount of government interference can reverse."[2] As a senior Federal Housing Administration (FHA) official told the 1939 convention of the American Institute of Planners: "Decentralization is taking place. It is not a policy, it is a reality—and it is as impossible for us to change this trend as it is to change the desire of birds to migrate to a more suitable location."[3]

Despite such protestations, there are many ways in which government largesse can affect where people live. For example, the federal tax code encourages businesses to abandon old structures before their useful life is at an end by permitting greater tax benefits for new construction than for the improvement of existing buildings. Thus, the government subsidizes an acceleration in the rate at which economic activity is dispersed to new locations.[4] Similarly, Roger Lotchin has recently begun important research on the significance of defense spending to the growth of Sunbelt cities since 1920. Military expenditures have meanwhile worked to the detriment of other areas. Estimates were common in the late 1970s that Washington was annually collecting between $6 billion and $11 billion more in the New York area than it was returning in expenditures, and the gap widened during the Reagan years as even larger proportions of the national budget was devoted to defense.[5]

On the urban-suburban level, the potential for federal influence is also enormous. For example, the Federal Highway Act of 1916 and the Interstate Highway Act of 1956 moved the government toward a transportation policy emphasizing and benefiting the road, the truck, and the private motorcar.[6] In conjunction with cheap fuel and mass-produced automobiles, the urban expressways led to lower marginal transport costs and greatly stimulated deconcentration. Equally important to most families is the incentive to detached-home living provided by the deduction of mortgage interest and real-estate taxes from their gross income. Even the reimbursement formulas for water-line and sewer construction have had an impact on the spatial patterns of metropolitan areas.[7]

The purpose of this chapter is to look at the impact of federal housing policies on how and where Americans live. More specifically, I seek to determine whether the results of such policies were foreseen by a government anxious to use its power and resources for the social control of ethnic and racial minorities. Has the American government been as benevolent—or at least as neutral—as its defenders have claimed?[8]

Government and Housing Before 1933

Although housing involves the largest capital costs of any human necessity, for the first three centuries of urban settlement in North America the provision of shelter was not regarded as an appropriate responsibility of government—whether that body was a colonial assembly or a state legislature, a town meeting or a city council, a Parliament in London or a Congress in Washington. Local governments occasionally outlawed wooden dwellings and thatched roofs in city centers in the seventeenth century,[9] and New York City passed restrictive housing laws as early as 1867, but the selection, construction, and purchase of a place to live was everywhere regarded as an essentially individual problem. Prior to the 1930s, federal involvement was limited to a survey of slum conditions in large cities in 1892, the creation of a Federal Land Bank System in 1916, and the construction of munitions and arms workers' housing during World War I.[10]

This last and potentially most important shift came in June 1918, a full year after the United States had entered the war, when Congress appropriated $110 million to begin two separate programs for housing war workers—the Emergency Fleet Corporation of the United States Shipping Board and the United States Housing Corporation. Although the two agencies operated differently, their purposes were similar: to provide residences for heads of households migrating to industrial areas in order to produce weapons for the European conflict. But because this war emergency effort began only five months before the Armistice, it resulted in only a few developments—Yorkship Village in Camden, New Jersey; Atlantic Heights in Portsmouth, New Hampshire; Union Park Gardens in Wilmington, Delaware; and several subdivisions in Bridgeport, Connecticut; Chester, Pennsylvania; and Kohler, Wisconsin—and most of them were not completed until after the war.[11]

The reason for the delay in beginning the programs was the general belief that homeownership promoted incentives to thrift and the lingering suspicion that subsidized rental units would be socialistic. Senator Albert Fall of New Mexico, for example, warned of "an insidious concerted effort . . . . to socialize this Government of ours, to overturn the entire Government of the United States." Senator Fall need not have worried. Less than 25,000 units were built in the entire nation under both programs. Although initially leased to their occupants, the houses were sold to private developers soon after the cessation of hostilities. By the early twenties, Washington was out of the housing business.

The first federal housing effort in the United States, therefore, was neither the result of a conscious effort to help the poor nor of an increased reform spirit. It was, as Charles Abrams wrote, "an exercise of the war power, not the disputed general welfare power." But it did demonstrate that Washington could intervene in a sacred sphere of private enterprise without falling victim to the dreaded Marxist demons. And the quality of the product was often quite good. Yorkship Village and Union Park Gardens, for example, were models of town planning. In Yorkship Village, winding and short straight streets led to an octagonal town square. The brick and stucco houses were placed in short rows and their broken roof lines avoided the monotony of the typical Philadelphia facade. In Union Park Gardens, the stylish row and duplex units featured steeply pitched roofs, staggered setbacks, and variation among house styles.[12]

As the United States returned to "normalcy" in the 1920s, the federal government adopted a hands-off policy with regard to housing. "Home, Sweet Home" remained a cherished ideal, of course, and the Department of Labor occasionally sponsored an "Own Your Own Home Week" to publicize the housing campaign of the National Association of Real Estate Boards. But the mechanics of construction and acquisition were left to the marketplace. As Senator William Calder of New York argued: "The Government is an organization to govern, not to build houses or operate mines or run railroads or banks."

Not until the advent of the Great Depression in 1929 did the American attitude toward government intervention shift in a fundamental way. The prolonged and mammoth economic catastrophe is too well known for elaboration here, but it inflicted crippling blows on both the housing industry and the homeowner. Between 1928 and 1933, the construction of residential property fell by 95 percent, and the expenditures on home repairs fell by 90 percent. In 1926, which may be taken as a typical year, about 68,000 homes were foreclosed in the United States. In 1930 about 150,000 non-farm households lost their property through foreclosure; in 1931, this increased to nearly 200,000; in 1932, to 250,000. In the spring of 1933, when fully half of all home mortgages in the United States were technically in default, and when foreclosures reached the astronomical rate of more than a thousand per day, the home-financing system was drifting toward complete collapse. Housing prices predictably declined—a typical $5,000 house in 1926 was worth about $3300 in 1932—virtually wiping out vast holdings in second and third mortgages as values fell below even the primary claim. Moreover, the victims were often middle-class families who were experiencing impoverishment for the first time.[13]

Theorizing that the predicament of the real-estate and construction industries was acting as a drag on the rest of the economy and believing that homeownership was "both the foundation of a sound economic and social system and a guarantee that our country will continue to develop rationally as changing conditions demand," Herbert Hoover convened the President's National Conference on Home Building and Home Ownership in 1931. More than four hundred specialists took part, including twenty-five fact-finding committees and six auxiliary groups. The avowed purpose of the meeting was to support homeownership for men "of sound character and industrious habits."[14] In an address at the opening session, President Hoover gave expression to this national preference for the private house:

> I am confident that the sentiment for home ownership is so embedded in the American heart that millions of people who dwell in tenements,

> apartments, and rented rooms . . . have the aspiration for wider opportunity in ownership of their own homes.[15]

To the Iowa farm boy whose road to wealth and the White House led through the corporate boardrooms of Manhattan, it was obvious that "Nothing makes for security and advancement more than devotion to the upbuilding of home life."

The conference made four recommendations that pointed to a new direction in federal housing policy and provided a boon to speculative builders, appliance manufacturers, and automobile companies: (1) the creation of long-term, amortized mortgages;[16] (2) the encouragement of low interest rates; (3) the institution of government aid to private efforts to house low-income families; and (4) the reduction of home construction costs. And the conference closed with a warning: "This committee is firmly of the opinion that private initiative taken by private capital is essential, at the present time, for the successful planning and operation of large scale projects. Still, if we do not accept this challenge, the alternative may have to be government housing."[17]

With support from the National Association of Home Builders, which insisted that contractors could not provide affordable houses at moderate cost without government assistance and—simultaneously—the freedom to build those houses as they wished, the Hoover administration tried to encourage homeownership in two ways. On July 22, 1932, the President affixed his signature to the Federal Home Loan Bank Act (Public Law 304) to establish a credit reserve for mortgage lenders and thus to increase the supply of capital in the housing market. But it was not designed to give help in cases of emergency distress and was able to give aid only where the risk was slight. The American public did not immediately perceive that, bureaucratic rhetoric aside, loans were only to go to families that did not need federal help, and within the first two years of the law's operation, 41,000 applications for direct loans were made to the banks by individual homeowners. Exactly *three* were approved. Although we should not minimize the satisfaction that those three families received from this evidence of federal compassion, their own good fortune was not sufficient to reverse the downhill slide of housing conditions. Public Law 304 was ineffective, and conditions became appreciably worse.[18]

A second measure, the Emergency Relief and Construction Act of 1932, also proved inconsequential. It empowered the Reconstruction Finance Commission to

> make loans to corporations formed wholly for the purpose of providing housing for families of low income, or for the reconstruction of slum areas, which are regulated by state or municipal law as to rents, capital structure,

rate of return, and areas and methods of operation, to aid in financing such projects undertaken by such corporations which are self-liquidating in character.[19]

Unfortunately, the legislation required the states to exempt such limited-dividend corporations from all taxes, and at the time only New York had such authority. As a result, Knickerbocker Village in New York City was the only project initiated under the legislation.

The Greenbelt Town Program

It remained for Franklin D. Roosevelt and his Democratic majority to develop successful new initiatives in housing. One of the freshest efforts of the New Deal was the Greenbelt Town Program. Inspired by Rexford G. Tugwell and administered by his Resettlement Administration (RA), the program was explicitly intended to foster deconcentration. Tugwell wanted to build ideal "greenbelt" communities based upon the planning theories of England's Ebenezer Howard, a turn-of-the-century visionary who was as appalled by traditional suburbs as much as he was by urban slums. Limited to 10,000 people, Tugwell's communities were to be characterized by decent housing and a high level of social and educational services and were to be surrounded by a belt of open land to prevent sprawl. As Tugwell explained it, "My idea was to go just outside centers of population, pick up cheap land, build a whole community, and entice people into them. Then go back into the cities and tear down whole slums and make parks of them."[20]

The Greenbelt Town Program came under vigorous conservative attack, however. A proposed New Jersey community never even made it off the drawing board, and the three garden communities that were built—Greenbelt in Maryland, Greenhills in Ohio, and Greendale in Wisconsin—were hurt by excessive construction costs and never served as models for future metropolitan development. The RA itself was scrapped by Congress in 1938.

Two other innovations of the New Deal—the Home Owners Loan Corporation and the Federal Housing Administration—were to have a more lasting and important impact upon the suburbanization of the United States.

The Home Owners Loan Corporation

On April 13, 1933, President Roosevelt urged the House and the Senate to pass a law that would protect the small homeowner from foreclosure, relieve him of part of the burden of excessive interest and principle payments incurred during a period of higher values and higher earning power,[21] and declare that it was national policy to protect homeownership. The measure received bipartisan support. As Republican Congressman Rich of Pennsylvania, a banker himself, remarked during the floor debate:

> I am opposed to the Government in business, but here is where I am going to do a little talking for the Government in business, because if aid is not going to be extended to these owners of small homes the Government will have to get into this business of trying to save their homes. The banker dares not loan for fear the depositor will draw out his deposit; then he must close his bank or the Comptroller of the Currency will close it for him.[22]

The resulting Home Owners Loan Corporation (HOLC), signed into law by FDR on June 13, 1933, was designed to serve urban needs; the Emergency Farm Mortgage Act, passed almost a month earlier, was intended to reduce rural foreclosures.[23]

The HOLC replaced the unworkable direct loan provisions of the Hoover administration's Federal Home Loan Bank Act and refinanced tens of thousands of mortgages in danger of default or foreclosure. It even granted loans at low-interest rates to permit owners to recover homes lost through forced sale. Between July 1933 and June 1935 alone, the HOLC supplied more than $3 billion for over one million mortgages, or loans for one-tenth of all owner-occupied, non-farm residences in the United States. Although applications varied widely by state—in Mississippi, 99 percent of the eligible owner-occupants applied for loans, while in Maine only 18 percent did so—nationally about 40 percent of eligible Americans sought HOLC assistance.[24]

The HOLC is important to history because it introduced, perfected, and proved in practice the feasibility of the long-term, self-amortizing mortgage with uniform payments spread over the whole life of the debt. In the nineteenth century, a stigma attached to the existence of a mortgage; well-established families were expected to purchase homes outright. After World War I, however, rising costs and increasing consumer debt made the mortgage a more typical instrument for the financing of a home. Indeed, housing became extraordinarily dependent on borrowed money, both to finance construction and to finance the final purchase. During the 1920s, a boom period in home building, the typical length of a mortgage was between five and ten years, and the loan itself was not fully paid off when the final settlement was due. Thus, the homeowner was periodically at the mercy of arbitrary and unpredictable forces in the money market. When money was easy, renewal every five or seven years was no problem. But if a mortgage expired at a time when money was tight, it might be impossible for the homeowner to secure a renewal, and foreclosure would ensue. Under the HOLC program, the loans were fully amortized, and the repayment period was extended to about twenty years.[25]

Aside from the larger number of mortgages that it helped to refinance on a long-term, low-interest basis, the HOLC systematized appraisal methods across the nation. Because it was dealing with problem mortgages—in some states over 40 percent of all HOLC loans were foreclosed even after refinancing—the HOLC had to make predictions and assumptions regarding the useful or productive life of housing it financed. Unlike refrigerators or shoes, dwellings were expected to be durable—how durable was the purpose of the investigation.

With care and extraordinary attention to detail, HOLC appraisers divided cities into neighborhoods and developed elaborate questionnaires relating to the occupation, income, and ethnicity of the inhabitants and the age, type of construction, price range, sales demand, and general state of repair of the housing stock. The element of novelty did not lie in the appraisal requirement itself—that had long been standard real-estate practice. Rather, it lay in the creation of a formal and uniform system of appraisal, reduced to writing, structured in defined procedures, and implemented by individuals only after intensive training. The ultimate aim was that one appraiser's judgment of value would have meaning to an investor located somewhere else. In evaluating such efforts, the distinguished economist C. Lowell Harriss has credited the HOLC training and evaluation procedures "with having helped raise the general level of American real estate appraisal methods." A less favorable judgement would be that the Home Owners Loan Corporation initiated the practice of "red lining."[26]

This occurred because HOLC devised a rating system that undervalued neighborhoods that were dense, mixed, or aging. Four categories of quality—imaginatively entitled First, Second, Third, and Fourth, with corresponding code letters of A, B, C, and D and colors of green, blue, yellow, and red—were established. The First grade (also A and green) areas were described as new, homogeneous, and "in demand as residential locations in good times and bad." Homogeneous meant "American business and professional men." Jewish neighborhoods, or even those with an "infiltration of Jews," could not be considered "best" any more than they could be considered "American."[27]

The Second security grade (blue) went to "still desirable" areas that had "reached their peak," but were expected to remain stable for many years. The Third grade (yellow or "C") neighborhoods were usually described as "definitely declining," while the Fourth grade (red) neighborhoods were defined as areas "in which the things taking place in C areas have already happened."[28]

HOLC assumptions about urban neighborhoods were based on both an ecological conception of change and a socioeconomic one. Adopting a dynamic view of the city and assuming that change was inevitable, its appraisers accepted as given the proposition that the natural tendency of any area was to decline—in part because of the increasing age and obsolescence of the physical structure and in part because of the filtering down of the housing stock to families of ever lower income. Thus physical deterioration was both a cause and an effect of population change, and HOLC officials made no attempt to sort them out. They were part and parcel of the same process. Thus, black neighborhoods were invariably rated as Fourth grade, but so also were any areas characterized by poor maintenance or vandalism. Similarly, those "definitely declining" sections that were marked Third grade or yellow received such a low rating in part because of age and in part because they were "within such a low price or rent range as to attract an undesirable element."[29]

The Home Owners Loan Corporation did not initiate the idea of considering race and ethnicity in real-estate appraisal. Bigotry has a long history in the United States, and the individuals who bought and sold houses were no better or worse than the rest of their countrymen. Realtors were well aware of the intense antagonisms which attended the attempts of middle-class black families to escape from ghetto areas, and their business practices reflected their observations. Indeed, so commonplace was the notion that race and ethnicity were important that Richard M. Hurd could write in the 1920s that the socioeconomic characteristics of a neighborhood determined the value of housing to a much greater extent than did structural characteristics. Prominent appraising texts, such as Frederick Babcock's *The Valuation of Real Estate* (1932) and *McMichael's Appraising Manual* (1931), echoed the same theme. Both advised appraisers to pay particular attention to "undesirable" or "least desirable" elements and suggested that the influx of certain ethnic groups was likely to precipitate price declines.[30]

This notion was codified and legitimized in the 1930s by Homer Hoyt and Robert Park at the University of Chicago. Developing a model of neighborhood change, Hoyt in particular showed that values declined as a function of the lowered status of residents and that the introduction of blacks into a neighborhood would first raise prices (the first black families had to pay a premium to break the color barrier) and then precipitate a drastic decline. In 1939 he systematized his theories in an influential study, *The Structure and Growth of Residential Neighborhoods in American Cities*.[31]

The HOLC simply applied these notions of ethnic and racial worth to real-estate appraising on an unprecedented scale. With the assistance of local realtors and banks, it assigned one of the four ratings to every block in every city. The resulting information was then translated into the appropriate color and duly recorded on secret "Residential Security Maps" in local HOLC offices. The maps themselves were placed in elaborate "City Survey Files," which consisted of reports, questionnaires, and workpapers relating to current and future values of real estate.

Because the Home Owners Loan Corporation and the Federal Housing Administration did not normally report data on anything other than a countywide basis, the St. Louis area was selected as a case study. There, the city and county were legally separated in 1876 (see Chapter 9), so the government had no alternative to city/suburb reporting. In addition, an even older industrial city, Newark, was selected because of the availability of a unique FHA study.

The Residential Security Map for the St. Louis area in 1937, as FIGURE 11-1 indicates, gave the highest ratings to the newer, affluent sub-

FIGURE 11-1. St. Louis Area Residential Security Map, 1937
SOURCE: Record Group 31, National Archives, Washington, D.C.

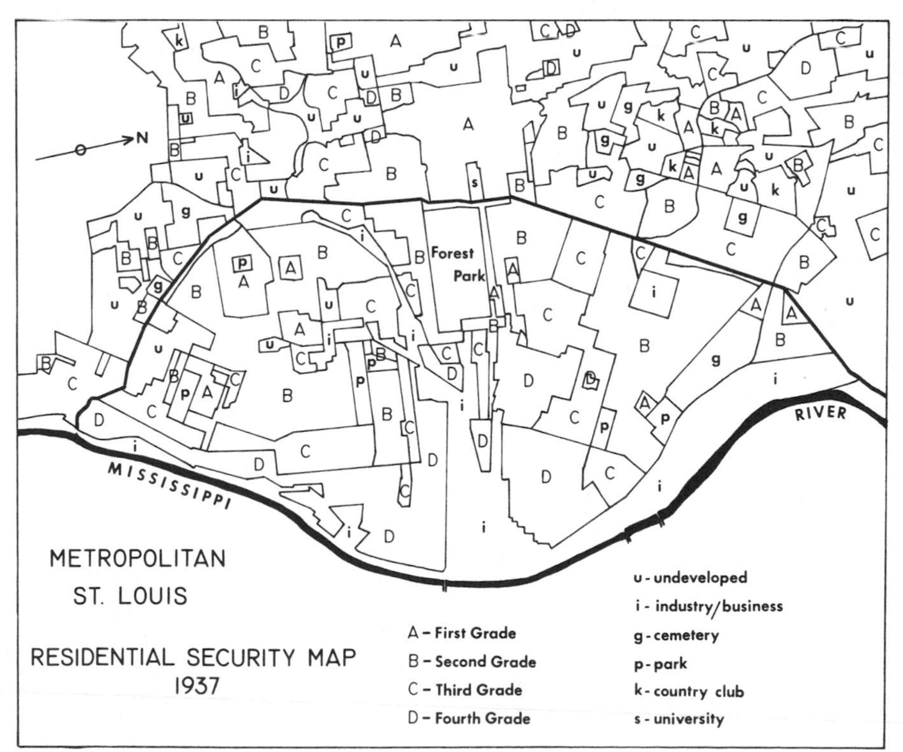

urbs that were strung out along curvilinear streets well away from the problems of the city. Three years later, in 1940, the advantage of the periphery over the center was even more marked. In both evaluations, the top of the scale was dominated by Ladue, a largely undeveloped section of high, rolling land, heavily wooded estates, and dozens of houses in the $20,000 to $50,000 range. Horses cantered through woodlands and glades on forty miles of bridle paths. In 1940 HOLC appraisers noted approvingly that the area's 4,535 acres, criss-crossed by streams, were "highly restricted" and occupied by "capitalists and other wealthy families." Reportedly not the home of "a single foreigner or negro," Ladue received a First grade (green) rating. In 1985, it remained private and preppy, *the* place in the St. Louis area for the rich, the powerful, and the socially elite to live.[32]

Other affluent St. Louis suburbs like Clayton, University City, and Webster Groves were also marked with green and blue on the 1937 and 1940 maps, indicating that they, too, were characterized by attractive homes on well-maintained plots, and that the appraisers felt confident about the safety of mortgages insured there. And well they might have been. In University City, almost 40 percent of the homes had been valued at more than $15,000 in 1930, while in Clayton the comparable figure was an astounding 72 percent (see Appendix A-11). Such statistics simply reflected the fact that for decades St. Louis's wealthier families had been forsaking the knolls and bluffs overlooking the Mississippi River and the mansions of the once fashionable Central West End for the elegant country style of the suburbs.

At the other end of the scale in St. Louis County were the rare Fourth grade areas. A few such neighborhoods were occupied by white laborers, such as "Ridgeview" in Kirkwood, where the garagelike shacks typically cost less than $1500. But the "D" regions in the county were usually black. One such place in 1937 was Lincoln Terrace, a small enclave of four- and five-room bungalows built in 1927. Originally intended for middle-class white families, the venture was unsuccessful, and the district quickly developed into a black neighborhood. But even though the homes were relatively new and of good quality, the HOLC gave the section (D-12 in 1937, D-8 in 1940) the lowest possible grade, asserting that the houses had "little or no value today, having suffered a tremendous decline in values due to the colored element now controlling the district."[33]

In contrast to the gently rolling terrain and sparse settlement of St. Louis County, the city had proportionately many more Third- and Fourth-grade neighborhoods, and more than twice as many renters as homeowners. As FIGURE 11-1 indicates, virtually all the residential sections along the Mississippi River or adjacent to the central business district received one of the two lowest ratings. This harsh judgement was in part a reflection of their badly deteriorated physical character. Just a few years earlier, the City Plan Commission of St. Louis had made a survey of 44 areas surrounding the business section. Only about 40 percent of the 8,447 living units had indoor toilets, and the tuberculosis morbidity rate was three times that of the city as a whole. As the St. Louis Regional Planning Report pessimistically concluded in 1936:

> The older residential districts which are depreciating in value and in character constitute one of the most serious problems in this region. They can never be absorbed by commercial and industrial uses. Even if owners wished to build new homes within them, it would be inadvisable because of the present character of the districts.[34]

Although HOLC appraisers marked down such neighborhoods because of true slum conditions, their negative attitudes toward city living in general also affected their judgements. The evaluation of a white, working-class neighborhood near St. Louis's Fairgrounds Park was typical. According to the description, "Lots are small, houses are only slightly set back from the sidewalks, and there is a general appearance of congestion." Although a city lover might have found this collection of cottages and abundant shade trees rather charming, the HOLC thought otherwise: "Age of properties, general mixture of type, proximity to industrial section on northeast and much less desirable areas to the south make this a good fourth grade area."[35]

As was the case in every city, any Afro-American presence was a source of substantial concern to the HOLC. In a confidential and generally pessimistic 1941 survey of the economic and real-estate prospects of the St. Louis metropolitan area, the Federal Home Loan Bank Board (the parent agency of the HOLC) repeatedly commented on the "rapidly increasing Negro population" and the resulting "problem in the maintenance of real estate values." The officials evinced a keen interest in the movement of black families and included maps of the density of black settlement with every analysis. Not surprisingly, even those neighborhoods with small proportions of black inhabitants were usually rated Fourth grade or "hazardous."[36]

Like St. Louis, Newark, New Jersey, has long symbolized the most extreme features of the urban crisis. In that troubled city, federal appraisers took note in the 1930s of the high tax rate, the heavy relief load, the per capita bonded debt, and the "strong tendency for years for people of larger incomes to move their homes outside the city." The 1939 Newark area Residential Security Map did not designate a single neighborhood in that city of more than 400,000 as worthy of an "A" rating. "High class Jewish" sections like Weequahic and Clinton Hill, as well as non-Jewish areas like Vailsburg and Forest Hill all received the Second grade, or "B." Typical Newark neighborhoods were rated even lower. The well-maintained and attractive working-class sections of Roseville, Woodside, and East Vailsburg were given Third-grade or "C" ratings; the remainder of the city, including immigrant Ironbound and every black neighborhood, was written off as Fourth grade or "hazardous."[37]

Immediately adjacent to Newark is New Jersey's Hudson County, which is among the half-dozen most densely settled and ethnically diverse political jurisdictions in the United States. Predictably, HOLC appraisers

TABLE 11-1

Distribution of HOLC Loans in Essex County (Newark), New Jersey and Shelby County (Memphis), Tennessee According to Neighborhood Classifications, 1935–1936

	Essex County		*Shelby County*	
Classifications	*Number*	*Percentage*	*Number*	*Percentage*
A—Best	685	10.2%	129	4.7%
B—Still Desirable	1,975	29.3	752	27.6
C—Definitely Declining	2,156	32.0	1,003	36.8
D—Hazardous	1,917	28.5	843	30.9

SOURCE: Compilations made from HOLC and FHA Reports in Record Group 195, National Archives.

had decided by 1940 that Hudson County was a lost cause. In the communities of Bayonne, Hoboken, Secaucus, Kearny, Union City, Weehawken, Harrison, and Jersey City, taken together, they designated only two very small Second grade areas and no First grade sections.[38]

The Home Owners Loan Corporation insisted "There is no implication that good mortgages do not exist or cannot be made in Third and Fourth grade areas." And, as TABLE 11-1 indicates, strong evidence indicates that the HOLC did in fact issue mortgage assistance impartially and make the majority of its obligations in "definitely declining" or "hazardous" neighborhoods. This seeming liberality was actually good business because the residents of poorer sections generally maintained a better pay-back record than did their more affluent cousins. As the Federal Home Loan Bank Board explained: "The rate of foreclosure per 1000 non-farm dwellings during 1939 was greater in St. Louis County than in St. Louis City by about 2 and one half to 1. A partial explanation or causation of this situation is the fact that County properties consist of a greater proportion of units in the higher priced brackets.''[39] The damage caused by the HOLC came not through its own actions, but through the influence of its appraisal system on the financial decisions of other institutions. During the late 1930s, the Federal Home Loan Bank Board circulated questionnaires to banks asking about their mortgage practices. Those returned by savings-and-loan associations and banks in Essex County (Newark), New Jersey indicated a clear relationship between public and private ''red lining'' practices. One specific question asked: ''What are the most desirable lending areas?'' The answers were often ''A and B'' or ''Blue'' or ''FHA only.'' Similarly, to the inquiry, ''Are there any areas in which loans will not be made?'' the responses included, ''Red and most yellow,'' ''C and D,'' ''Newark,'' ''Not in red,'' and ''D areas.'' Obviously, private banking institutions were privy to and influenced by the government's Residential Security Maps. And the pattern of discrimination was continued until at least 1970 by the Federal Home Loan Bank Board, whose examiners routinely ''red lined'' postal zip codes in which the symptoms of racial change and falling values were observed.[40]

Even more significantly, HOLC appraisal methods, and probably the maps themselves, were adopted by the Federal Housing Administration.

The Federal Housing Administration

No agency of the United States government has had a more pervasive and powerful impact on the American people over the past half-century than the Federal Housing Administration (FHA). It dates from the adoption of the National Housing Act on June 27, 1934. Designed by Winfield Riefler, Miles Lanier Colean, Frances Perkins, Marriner Eccles, Averell Harriman, and Henry Wallace to meet President Roosevelt's desire for at least one program that could stimulate building without government spending and that would rely instead on private enterprise, it was intended ''to encourage improvement in housing standards and conditions, to facilitate sound home financing on reasonable terms, and to exert a stabilizing influence on the mortgage market.'' The primary purpose of the legislation, however, was the alleviation of unemployment, which stood at about a quarter of the total work force in 1934 and which was particularly high in the construction industry. As the Federal Emergency Relief Administrator testified before the House Banking and Currency Committee on May 18, 1934:

> The building trades in America represent by all odds the largest single unit of our unemployment. Probably more than one-third of all the unemployed are identified, directly and indirectly, with the building trades. . . .
>
> Now, a purpose of this bill, a fundamental purpose of this bill, is an effort to get the people back to work.[41]

The FHA effort was later supplemented by the Servicemen's Readjustment Act of 1944 (more familiarly known as the GI Bill), which created a Veterans Administration (VA) program to help the sixteen million soldiers and sailors of World War II purchase a home after the defeat of Germany and Japan. Because the VA very largely followed FHA procedures and attitudes and was not itself on "the cutting edge of housing policy," the two programs can be considered as a single effort.

Between 1934 and 1968, and to a lesser extent until the present day, both the FHA and the VA (since 1944) have had a remarkable record of accomplishment. Essentially, they insure long-term mortgage loans made by private lenders for home construction and sale. To this end, they collect premiums, set up reserves for losses, and in the event of a default on a mortgage, indemnify the lender. They do not build houses or lend money. Instead, they induce lenders who have money to invest it in residential mortgages by insuring them against loss on such instruments, with the full weight of the United States Treasury behind the contract. And they have revolutionized the home finance industry in the following ways:

Before the FHA began operation, first mortgages were limited to one-half or two-thirds of the appraised value of the property. During the 1920s, for example, savings and loan associations held one-half of America's outstanding mortgage debt. Those mortgages averaged 58 per percent of estimated property value. Thus, prospective home buyers needed a down payment of at least 30 percent to close a deal. By contrast, the fraction of the collateral that the lender was able to lend for an FHA-secured loan was about 93 percent. Thus, down payments of more than 10 percent were unnecessary.[42]

Continuing a trend begun by the Home Owners Loan Corporation, FHA extended the repayment period for its guaranteed mortgages to twenty-five or thirty years and insisted that all loans be fully amortized. The effect was to reduce both the average monthly payment and the national rate of mortgage foreclosure. The latter declined from 250,000 non-farm units in 1932 to only 18,000 in 1951.

FHA established minimum standards for home construction that became almost standard in the industry. These regulations were not intended to make any particular structure fault-free, nor even to assure the owner's satisfaction with the purchase. But they were designed to insure with at least statistical accuracy that the dwelling would be free of gross structural or mechanical deficiencies. Although there was nothing innovative in considering the quality of a house in relation to the debt placed against it, two features of the system were new: first, that the standards were objective, uniform, and in writing; second, that they were to be enforced by actual on-site inspection—prior to insurance commitment in the case of an existing property, and at various fixed stages in the course of construction of new housing. Since World War II, the largest private contractors have built all their new houses to meet FHA standards, even though financing has often been arranged without FHA aid. This has occurred because many potential purchasers will not consider a house that cannot earn FHA approval.[43]

In the 1920s, the interest rate for first mortgages averaged between 6 and 8 percent. If a second mortgage were necessary, as it usually was for families of moderate means, the purchaser could obtain one by paying a discount to the lender, a higher interest rate on the loan, and perhaps a commission to a broker. Together, these charges added about 15 percent to the purchase price. Under the FHA (and later Veterans Administration) program, by contrast, there was very little risk to the banker if a loan turned sour. Reflecting this government guarantee, interest rates fell by two or three percentage points.[44]

These four changes substantially increased the number of American families who could reasonably expect to purchase homes. Builders went back to work, and housing starts and sales began to accelerate rapidly in 1936. They rose to 332,000 in 1937, to 399,000 in 1938, to 458,000 in 1939, to 530,000 in 1940, and to 619,000 in 1941. This was a startling lift from the 93,000 starts of 1933. After World War II, the numbers became even larger, and by the end of 1972, FHA had helped nearly eleven million families to own houses and another twenty-two million families to improve their properties. It had also insured 1.8 million dwellings in multi-unit projects. And in those same years between 1934 and 1972, the percentage of American families living in owner-occupied dwellings rose from 44 percent to 63 percent.[45]

Quite simply, it often became cheaper to buy than to rent. In 1939, for example, four hundred six-room houses were built just north of Wilmington, Delaware, in an FHA-backed development called Edgemoor Terrace. Using the tract techniques that would later be popularized by the Levitt organization after World War II—standardized models and lot sizes, routinized construction methods, and furnished models—the Wilmington Construction Company was able to offer the home for $5,150. The FHA mortgage guarantee meant that purchasers needed only $550 for a down payment and an incredible $29.61 monthly charge for twenty-five years to the bank. Advertisements for Edgemoor Terrace emphasized that it was cheaper to buy a new suburban home there than to rent a comparable structure in the city.[46]

Many developments in the United States could equal Edgemoor Terrace for value, and the new economic inducements to homeownership were essentially the same everywhere. Long Island builder Martin Winter recalled that in the early 1950s families living in the Kew Gardens section of Queens were paying about ninety dollars per month for small two-bedroom apartments. For less money, they could and often did, move to the new Levittown-type developments springing up along the highways from Manhattan. Even the working classes could aspire to homeownership. As one person who left New York City for suburban Dumont, New Jersey, remembered: "We had been paying $50 per month rent, and here we come up and live for $29 a month. That paid everything—taxes, principal, insurance on your mortgage, and interest." Not surprisingly, the middle-class suburban family with the new house and the long-term, fixed-rate, FHA-insured mortgage became a symbol, and perhaps a stereotype, of the American way of life.[47]

Unfortunately, the corollary to this achievement was the fact that FHA programs hastened the decay of inner-city neighborhoods by stripping them of much of their middle-class constituency. In practice, FHA insurance went to new residential developments on the edges of metropolitan areas, to the neglect of core cities. This occurred for three reasons. First, although the legislation nowhere mentioned an antiurban bias, it favored the construction of single-family projects and discouraged construction of multi-family projects through unpopular terms. Historically, single-family housing programs have been the heart of FHA's insured loan activities. Between 1941 and 1950, FHA-insured single family starts exceeded FHA multi-family starts by a ratio of almost four to one. In the next decade, the margin exceeded seven to one. Even in 1971, when FHA insured the largest number of multi-family units in its history, single-family houses were more numerous by 27 percent.[48]

Second, loans for the repair of existing structures were small and for short duration, which meant that a family could more easily purchase a new home than modernize an old one. The legislation required FHA to exercise more controls over rental than over sales housing, a circumstance that reflected the bias against non-owner-occupied structures. One part of the 1934 act was an embryonic authorization for mortgage insurance with respect to rental housing in regulated projects of public bodies or limited dividend corporations. Almost nothing was insured until 1938, and even thereafter the annual insurance for rental housing exceeded $1 billion only once between 1934 and 1962.[49]

The third and most important variety of suburban, middle-class favoritism had to do with the "unbiased professional estimate" that was a prerequisite to any loan guarantee. Required because maximum mortgage amounts were related to "appraised value," this mandatory judgement included a rating of the property itself, a rating of the mortgagor or borrower, and a rating of the neighborhood. The aim was to guarantee that at any time during the term of the mortgage the market value of the dwelling would exceed the outstanding debt.[50] The lower the valuation placed on properties, the less government risk and the less generous the aid to the potential buyers (and sellers). The purpose of the neighborhood evaluation was "to determine the degree of mortgage risk introduced in a mortgage insurance transaction because of the location of the property at a specific site." And unlike the Home Owners Loan Corporation, which used a similar procedure, the Federal Housing Administration allowed personal and agency bias in favor of all-white subdivisions in the suburbs to affect the kinds of loans it guaranteed—or, equally important, refused to guarantee. In this way, the bureacracy influenced the character of housing at least as much as the 1934 enabling legislation did.[51]

The Federal Housing Administration was quite precise in teaching its underwriters how to measure quality in residential areas. Eight criteria were established (the numbers in parentheses reflect the percentage weight given to each):

Relative economic stability (40 percent)
Protection from adverse influences (20 percent)
Freedom from special hazards (5 percent)
Adequacy of civic, social, and commercial centers (5 percent)
Adequacy of transportation (10 percent)
Sufficiency of utilities and conveniences (5 percent)
Level of taxes and special assessments (5 percent)
Appeal (10 percent)

Although FHA directives insisted that no project should be insured that involved a high degree of risk with regard to any of the eight categories, "economic stability" and "protection from adverse influences" together counted for more than the other six combined. Both were interpreted in ways that were prejudicial against heterogeneous environments. The 1939 *Underwriting Manual* taught that "crowded neighborhoods lessen desirability," and "older properties in a neighborhood have a tendency to accelerate the transition to lower class occupancy." Smoke and odor were considered "adverse influences," and appraisers were told to look carefully for any "inferior and non-productive characteristics of the areas surrounding the site." The agency endorsed restrictive zoning and insisted that any single-family residence it insured could not have facilities that allowed the dwelling to be used as a store, an office, or a rental unit.[52]

Obviously, prospective buyers could avoid many of these so-called undesirable features by locating in suburban sections. In 1939 FHA asked each of its fifty regional offices to send in plans for six "typical American houses." The photographs and dimensions were then used for a National Archives exhibit. Virtually all of the entries were bungalows or colonials on ample lots with driveways and garages.

In an attempt to standardize such ideal homes, the Federal Housing Administration set up minimum requirements for lot size, setback from the street, separation from adjacent structures, and even for the width of the house itself. While such requirements did provide light and air for new structures, they effectively eliminated whole categories of dwellings, such as the traditional 16-foot-wide row houses of Baltimore, from eligibility for loan guarantees. Even apartment-house owners were encouraged to look to suburbia: "Under the best of conditions a rental development under the FHA program is a project set in what amounts to a privately owned and privately controlled park area."[53]

Reflecting the racist tradition of the United States, the Federal Housing Administration was extraordinarily concerned with "inharmonious racial or nationality groups." It feared that an entire area could lose its investment value if rigid white-black separation was not maintained. Bluntly warning, "If a neighborhood is to retain stability, it is necessary that properties shall continue to be occupied by the same social and racial classes," the *Underwriting Manual* openly recommended "subdivision regulations and suitable restrictive covenants" that would be "superior to any mortgage."[54] Such covenants, which were legal provisions written into property deeds, were a common method of prohibiting black occupancy until the United States Supreme Court ruling in 1948 *(Shelley v. Kraemer)* that they were "unenforceable as law and contrary to public policy." Even then, it was not until 1949 that FHA announced that as of February 15, 1950, it would not insure mortgages on real estate subject to covenants. Although the press treated the FHA announcement as a major advancement in the field of racial justice, former housing administrator Nathan Straus noted that "the new policy in fact served only to warn speculative builders who had not filed covenants of their right to do so, and it gave them a convenient respite in which to file."[55]

In addition to recommending covenants, FHA compiled detailed reports and maps charting the present and most likely future residential locations of black families. In a March 1939 map of Brooklyn, for example, the presence of a single, non-white family on any block was sufficient to mark that entire block black. Similarly, very extensive maps of the District of Columbia depicted the spread of the black population and the percentage of dwelling units occupied by persons other than white.[56] As late as November 19, 1948, Assistant FHA Commissioner W. J. Lockwood could write that FHA "has never insured a housing project of mixed occupancy" because of the expectation that "such projects would probably in a short period of time become all-Negro or all-white."[57]

Occasionally, FHA decisions were particularly bizarre and capricious. In the late 1930s, for example, as Detroit grew outward, white families began to settle near a black enclave adjacent to Eight Mile Road. By 1940 the blacks were surrounded, but neither they nor the whites could get FHA insurance because of the proximity of an "inharmonious" racial group. So in 1941 an enterprising white developer built a concrete wall between the white and black areas. The FHA appraisers then took another look and approved mortgages on the white properties.[58]

The precise extent to which the agency discriminated against blacks and other minority groups is difficult to determine.[59] Although FHA has always collected reams of data regarding the price, floor area, lot size, number of bathrooms, type of roof, and structural characteristics of the single-family homes it has insured, it has been quite secretive about the location of these loans. For the period between 1942 and 1968, for example, when FHA had a vast influence on the suburbanization of the United States, the most detailed FHA statistics cannot be disaggregated below the county level.[60]

Such data as are available indicate that the neighborhood appraisals were very influential in determining "where it would be reasonably safe to insure mortgages." Indeed, the Preliminary Examiner was specifically instructed to refer to the Residential Security Maps—whether these were HOLC maps or new FHA maps with the same designations cannot be determined—in order "to segregate for rejection many of the applications involving locations not suitable for amortized mortgages." The result was a degree of suburban favoritism even greater than the documentary analysis would have suggested. Of a sample of 241 new homes insured by FHA throughout metropolitan St. Louis between 1935 and 1939, a full 220 or 91 percent were located in the suburbs. Moreover, more than half of these home buyers (135 of 241) had lived in the city immediately prior to their new home purchase. That FHA was helping to denude St. Louis of its middle-class residents is illustrated by an analysis of the HOLC Residential Security Map. As might be expected, the new suburbanites were not being drawn from the slums or from rural areas, but from the Second grade or "B" areas—generally sound but aging housing in middle-class neighborhoods of the central city.

A detailed analysis of two individual subdivisions in St. Louis County—Normandy and Affton—confirms the same point. Located just northwest of the city limits, Normandy is now a school district comprising twenty-five small neighborhoods. In 1937 it was made up of new five- and six-room houses costing between $4,000 and $7,500. Exactly 127 of these houses were sold under FHA-guaranteed mortgages in 1937 and 1938. One hundred of these purchasers (78 percent) moved out from the city, mostly from the solid, well-established blocks between West Florrissant and Easton Streets.

Never a wealthy area, Affton was on the opposite, or southwest, edge of St. Louis. Although it boomed after World War II as a middle-income alternative for returning veterans, it was the scene of considerable residential construction in 1938 and 1939. Of sixty-two families purchasing FHA-insured homes in Affton during these years, fifty-five were from the city of St. Louis. Most of them simply came out the four-lane Gravois Road from the southern part of the city to their new plots in the suburbs.

For the period since 1942, detailed analyses of FHA spatial patterns are difficult. But a reconstruction of FHA unpublished statistics for the St. Louis area over the course of a quarter of a century reveals the broad patterns of city-suburban activity. As TABLE 11-2 indicates, in the first twenty-seven years of FHA operation (through December 31, 1960), when tens of thousands of tract homes were built west of the city limits, the county of St. Louis was the beneficiary of more than five times as much mortgage insurance as the city of St. Louis, whether measured in number of mortgages, amount of mortgage insurance, or per capita assistance.

One possible explanation for the enormous city-county disparities in these figures is that the city had very little room for development, that the populace wanted to move to the suburbs, and that the periphery was where new housing could most easily be built. But in the 1930s, many more single-family homes were constructed in the city than in the county. Moreover, more than half of the FHA policies traditionally went to *existing* rather than *new* homes, and the city had a much larger inventory of existing housing than did the county in the period before 1960. Even in terms of home-improvement loans, a category in which the aging city was obviously more needy, only $44 million went to the city, while about three times that much, or $112 million, went to the county through 1960. In the late 1960s and early 1970s, in the wake of periodic urban rioting

TABLE 11-2

Cumulative Total of FHA Home Mortgage Activities and Per Capita Figures for Ten Selected United States Counties, 1934–1960

Jurisdiction	*Cumulative Number of Home Mortgages, 1934–1960*	*Cumulative Amount of Home Mortgages, 1934–1960*	*Per Capita Amount of Home Mortgages, as of January 1961*[a]
St. Louis County, Missouri	62,772	$558,913,633	$794
Fairfax County, Virginia	14,687	190,718,799	730
Nassau County, New York	87,183	781,378,559	601
Montgomery County, Maryland	14,702	159,246,550	467
Prince Georges County, Maryland	15,043	144,481,817	404
St. Louis City	12,166	94,173,422	126
District of Columbia	8,038	66,144,612	87
Kings County (Brooklyn), New York	15,438	140,330,137	53
Hudson County, New Jersey	1,056	7,263,320	12
Bronx County, New York	1,641	14,279,243	10

[a]The per capita amount was derived by dividing the cumulative amount of home mortgages by the 1960 population.

SOURCE: These calculations are based upon unpublished statistics available in the Single Family Insured Branch of the Management Information Systems Division of the Federal Housing Administration.

when the federal government attempted to redirect monies to the central cities, the previous inbalance was not corrected. Figures available through 1976 show a total of well over $1.1 billion for the county and only $314 million for the city. Thus, the suburbs continued their dominance.[61]

Although St. Louis County has done very well in terms of per-capita mortgage insurance in comparison with other areas of the nation, the Mississippi River city was not an isolated case of suburban favoritism. In Essex County, New Jersey, FHA committments went in overwhelming proportion to Newark's suburbs. And in neighboring Hudson County, residents received only twelve dollars of mortgage insurance per capita through 1960, the second lowest county total in the nation after the Bronx (TABLE 11-2).[62]

The New Jersey data reveal that the most favored areas for FHA mortgage insurance were not the wealthiest towns. Rather, the most likely areas for heavy FHA activity were those rated Second grade or "B" on the Residential Security Maps. In 1936 about 65 percent of new housing units in suburban Livingston were accepted for insurance; for Caldwell

FIGURE 11-2. New Home Mortgages Accepted for Insurance by the Federal Housing Administration in Essex County, New Jersey in 1936. Each dot represents one mortgage accepted for insurance.
SOURCE: Record Group 31, National Archives.

and Irvington, also solidly middle-class, the percentages were 59 and 42 respectively. In more elite districts, like South Orange, Glen Ridge, Milburn, and Maplewood, however, the FHA assistance rates were about the same as they were for Newark, or less than 25 percent. Presumably, this occurred because the allowable price limit for FHA mortgage insurance was originally $20,000, and also because persons who could afford to live in such posh neighborhoods did not require government financing.[63]

Even in the nation's capital, the outlying areas were considered more appropriate for federal assistance than older neighborhoods. FHA committments at the beginning of 1937 in the District of Columbia were heavily concentrated in two peripheral areas: between the United States Soldiers Home and Walter Reed Hospital in white and prosperous northwest Washington, and between Rock Creek Park and Connecticut Avenue, also in northwest Washington. Few mortgage guarantees were issued in the predominantly balck central and southeastern sections of the district. More importantly, at least two-thirds of FHA committments in the metropolitan area were located in the suburbs—especially in Arlington and Alexandria in Virginia and Silver Spring, Takoma Park, Chevy Chase, University Park, Westmoreland Hills, and West Haven in Maryland. Perhaps this was but a reflection of a dire FHA prediction in 1939 about the future of the capital city: "It should be noted in this connection that the "filtering-up" process, and the tendency of Negroes to congregate in the District, taken together, logically point to a situation where eventually the District will be populated by Negroes and the surrounding areas in Maryland and Virginia by white families."[64] Following a segregationist policy for at least the next twenty years, FHA did its part to see that the prophecy came true; through the end of 1960, as TABLE 11-2 indicates, the suburban counties received more than seven times as much mortgage insurance as the District.

For its part, the Federal Housing Administration usually responded that it was not created to help cities, but to revive home building, to stimulate homeownership, and to reduce unemployment. And it concentrated on convincing both the Congress and the public that it was, as its first Administrator, James Moffett, remarked, "a conservative business operation."[65] The agency emphasized its concern over sound loans no higher than the value of the assets and the repayment ability of the borrower would support. And FHA was unusual in the vast array of Washington programs because of its record of earning a small profit for the federal government.[66]

But FHA also helped to turn the building industry against the minority and inner-city housing market, and its policies supported the income and racial segregation of suburbia. For perhaps the first time, the federal government embraced the discriminatory attitudes of the marketplace. Previously, prejudices were personalized and individualized; FHA exhorted segregation and enshrined it as public policy. Whole areas of cities were declared ineligible for loan guarantees; as late as 1966, for example, FHA did not have a mortgage on a single home in Camden or Paterson, New Jersey, both declining industrial cities. This withdrawal of financing often resulted in an inability to sell houses in a neighborhood, so that vacant units often stood empty for months, producing a steep decline in value.[67]

Despite the fact that the government's leading housing agency openly exhorted segregation throughout the first thirty years of its operation, very few voices were raised against FHA red-lining practices. Between 1943 and 1945, Harland Bartholomew and Associates, the nation's leading urban planning firm, prepared a master plan for Dallas. Criticizing FHA for building "nearly all housing" in the suburbs, the company argued that "this policy has hastened the process of urban decentralization immeasurably." In 1955 Columbia Professor Charles Abrams pointed a much stronger accusatory finger at FHA for discriminatory practices. Writing in 1955, the famed urban planner said:

> A government offering such bounty to builders and lenders could have required compliance with a nondiscrimination policy. Or the agency could at least have pursued a course of evasion, or hidden behind the screen of local autonomy. Instead, FHA adopted a racial policy that could well have been culled from the Nuremberg laws. From its inception FHA set itself up as the protector of the all white neighborhood. It sent its agents into the field to keep Negroes and other minorities from buying houses in white neighborhoods.[68]

Not until the civil-rights movement of the 1960s did community groups realize that red lining and disinvestment were a major cause of community decline and that home-improvement loans were the "lifeblood of housing." In 1967 Martin Nolan summed up the indictment against FHA by asserting, "The imbalance against poor people and in favor of middle-income homeowners is so staggering that it makes all inquiries into the pathology of slums seem redundant." In the following year, Senator Paul Douglas of Illinois reported for the National Commission on Urban Problems on the role of the federal government in home finance:

> The poor and those on the fringes of poverty have been almost completely excluded. These and the lower middle class, together constituting the 40 percent of the population whose housing needs are greatest, received only 11 percent of the FHA mortgages. . . . Even middle-class residential districts in the central cities were suspect, since there was always the prospect that they, too, might turn as Negroes and poor whites continued to pour into the cities, and as middle and upper-middle-income whites continued to move out.[69]

Moreover, as urban analyst Jane Jacobs has said, "Credit blacklisting maps are accurate prophecies because they are self-fulfilling prophecies."

In 1966 FHA drastically shifted its policies with a view toward making much more mortgage insurance available for inner-city neighborhoods. Ironically, the primary effect of the change was to make it easier for white families to finance their escape from areas experiencing racial change. At the same time, the relaxed credit standards for black applicants meant that home improvement companies could buy properties at low cost, make cosmetic improvements, and sell the renovated home at inflated prices approved by FHA. Many of the minority purchasers could not afford the cost of maintenance, and FHA had to repossess thousands of homes. The final result was to increase the speed with which areas went through racial transformation and to victimize those it was designed to help. The only people to benefit were contractors and white, middle-class homeowners who were assisted in escaping from a distress position.[70]

In the 1930s, the Federal Home Loan Bank Board, the Home Owners Loan Corporation, and the Federal Housing Administration were churned out in rapid-fire succession by a government anxious to reduce unemployment and to provide a way for the home buyer to compete with large corporations for credit. The savings-and-loan industry's mandate was to encourage homeownership by taking the savings of small depositors and lending them out as mortgages. Washington, in turn, eased the risk to the system by insuring the mortgages through the Federal Housing Administration (and the deposits through the Federal Savings and Loan Insurance Corporation). When necessary, the government oiled the system by making additional low-cost funds available to lenders via the Federal Home Loan Bank Board.

In the course of accomplishing its mission, the HOLC developed real-estate appraisal methods that discriminated against racial and ethnic minorities and against older, industrial cities. But HOLC extended aid without regard for its own ratings and met the needs of a variety of families and neighborhoods. The Federal Housing Administration cooperated with HOLC and followed HOLC appraisal practices. But unlike the Home Owners Loan Corporation, FHA acted on the information in its files and clearly favored homogeneous subdivisions over industrial, aging, or heterogeneous neighborhoods.

From the perspective of the suburbs, but not most cities, the system worked remarkably well from 1933 until the late 1960s. As returning World War II veterans sought homes to raise their families, the government financed large tracts of houses on the periphery. Thus, the main beneficiary of the $119 billion in FHA mortgage insurance issued in the first four decades of FHA operation was suburbia, where almost half of all housing could claim FHA or VA financing in the 1950s and 1960s. And as the percentage of families who were homeowners increased from

44 percent in 1934 to 63 percent in 1972, the American suburb was transformed from an affluent preserve into the normal expectation of the middle class.

Not only did FHA help move mortgage funds from the cities to the suburbs, but two other housing innovations from Washington, the Federal National Mortgage Association (popularly known as Fannie Mae) and the Government National Mortgage Association (popularly known as Ginnie Mae), made possible the easy transfer of savings funds out of the cities of the Northeast and Middle West and toward the new developments of the South and West. Fannie Mae essentially created a standardized mortgage instrument that all states recognize, and on which banks and other institutions lend. "Mortgage funds can now move freely across the country to where needed," according to official doctrine. A typical result was that savings banks in the Bronx invested only about 10 percent of their funds in the 1970s in the borough and only about 30 percent in New York State. The rest went for investments elsewhere in the country, a result that would not have been possible except for Fannie Mae.[71]

Any serious indictment of federal lawmakers and federal officials for the miserable state of many American cities must take cognizance of two important points. First, and most obviously, it is hazardous to condemn a government for adopting policies in accord with the preference of a majority of its citizens. As novelist Anthony Trollope put it in 1867: "It is a very comfortable thing to stand on your own ground. Land is about the only thing that can't fly away." FHA helped to build houses, and where they were put was less important than that they were built. For more than a century, Americans have had a strong affinity for a detached home on a private lot. Obviously, some popular measures, such as gun control, are not adopted because of special-interest lobbies. But suburbanization was not willed on an innocent peasantry. Without a substantial amount of encouragement from the mainstream of public opinion, the bureaucrats would never have been able to push their projects as far as they did. The single-family house responded to the psychic value of privacy or castlehood. In fact, suburbanization was an ideal government policy because it met the needs of both citizens and business interests and because it earned the politicians' votes. It is a simple fact that homeownership introduced equity into the estates of over 35 million families between 1933 and 1978. The tract houses they often bought may have been dismissed as hopeless by highbrow architectural purists, but they were a lot less dreary to the people who raised families there and then sold to new families at a profit.

Federal housing policies were also not the *sine qua non* in the mushrooming of the suburbs. Mortgage insurance obviously made it easier for families to secure their dream houses, but the dominant residential drift in American cities had been toward the periphery for at least a century before the New Deal, and there is no reason to assume that the suburban trend would not have continued in the absence of direct federal assistance.

The lasting damage done by the national government was that it put its seal of approval on ethnic and racial discrimination and developed policies which had the result of the practical abandonment of large sections of older, industrial cities. More seriously, Washington actions were later picked up by private interests, so that banks and savings-and-loan institutions institutionalized the practice of denying mortgages "solely because of the geographical location of the property." The financial community saw blighted neighborhoods as physical evidence of the melting-pot mistake. To them, cities were risky because of their heterogeneity, because of their attempt to bring various people together harmoniously. Such mixing, they believed, had but two consequences—the decline of both the human race and of property values. As Mark Gelfand has observed, "Given the chance, bankers would do for their business what they had already done for themselves—leave the city."[72]

St. Louis illustrates the dilemma of many cities. Partly as a result of federal housing policies which have enabled the white, middle-class population to settle in the county, the city of St. Louis had become by 1984 a premier example of urban abandonment. Once the fourth largest city in America, the "Gateway to the West" is now twenty-seventh, a ghost of its former self. In 1940 it contained 816,000 inhabitants; in 1980 the census counted only 453,000. Many of its old neighborhoods have become dispiriting collections of burned-out buildings, eviscerated homes, and vacant lots. Although the drone of traffic on the nearby interstate highways is constant, there is an eerie remoteness to the pock-marked streets. The air is polluted, the sidewalks are filthy, the juvenile crime rate is horrendous, and the remaining industries are languishing. Grimy warehouses and aging loft factories are landscaped by weed-grown lots adjoining half-used rail yards. Like an elderly couple no longer sure of their purpose in life after their children have moved away, these neighborhoods face an undirected future.[73]

A particularly telling statistic is that, after Chicago, St. Louis is the nation's leading exporter of used bricks. Piled beside the railroad tracks that hug the Mississippi River, the great stacks of weathered bricks are destined to become parts of restoration projects in Atlanta or patios in

Houston. It is the supreme indignity. Having lost more than 300 factories in the 1970s to the Sunbelt, St. Louis itself is now being carted away.

The situation in the Mississippi River metropolis is more serious than that in most other cities, but the same broad patterns of downtown decline, inner-city deterioration, and exurban development so evident in St. Louis are actually typical of the large population centers of the United States. This same result might have been achieved in the absence of all federal intervention, but the simple fact is that the various government policies toward housing have had substantially the same result from Los Angeles to Boston. The poor in America have not shared in the postwar real-estate boom, in most of the major highway improvements, in property and income-tax write-offs, and in mortgage insurance programs. Public housing projects were intended to redress the imbalance. Unfortunately, as we shall see, it did not work out that way.

12

The Cost of Good Intentions: The Ghettoization of Public Housing in the United States

> If a healthy race is to be reared, it can be reared only in healthy homes; if infant mortality is to be reduced and tuberculosis is to be stamped out, the first essential is the improvement of housing conditions; if drink and crime are to be successfully combated, decent sanitary houses must be provided. If "unrest" is to be converted into contentment, the provision of good houses may prove to be one of the most potent agents in that conversion.
>
> —KING GEORGE V
> April 11, 1919

The long-term, low-interest mortgage was not the only federal housing innovation of the New Deal. More controversial was the attempt to meet the need of the poor for adequate shelter. The result, if not the intent, of the public housing program of the United States was to segregate the races, to concentrate the disadvantaged in inner cities, and to reinforce the image of suburbia as a place of refuge for the problems of race, crime, and poverty. By every measure, the Housing Act of 1937 was an important stimulus to deconcentration.

Prior to the 1930s, housing reform in the United States meant the improvement of slum conditions through the establishment of minimum standards of ventilation, sanitation, and density. New York City's pioneering codes of 1867, 1879, and 1901, each of which established progressively higher legal requirements for dwelling units, were illustrative of this trend, as were the ideas of the nation's preeminent nineteenth-century-housing reformers, Lawrence Veiller and E. R. L. Gould, both of whom opposed housing subsidies and the actual public construction of new housing. In their view, the role of government was simply to enforce the law. This was a common attitude in the nineteenth century, when private investors endlessly schemed to wring more profits from tenements. By the second decade of the twentieth century, however, entrepreneurs were shunning the low-income housing market, and most private investors displayed little interest in entering the field.[1]

At both the state and the national levels, governments in the United States remained absent from the housing field throughout the 1920s. Similarly, the American city-planning movement early in the century concerned itself only slightly with the housing problems of the poor. European lawmakers, however, were charting a different course. Government assistance in England dated back to an 1868 Artisans Dwelling Act to help people who could not afford to keep their homes in repair. The really massive boom came after World War I, however. The British Housing Act of 1919 started public housing, and during the next decade both Great Britain and Germany built more than one million publicly assisted "homes for heroes." In the Netherlands, the government rehoused one-fifth of the total population in the same fashion, while in the Soviet Union the transition to public responsibility was almost total. As the American housing reformer Edith Elmer Wood noted sadly in 1931: "Nearly all other European countries have developed some form of housing loan at low interest and some form of municipal housing or a thinly disguised substitute for it." England, according to her estimate, was half a century ahead of the United States in the field of shelter.[2]

Along with John Ihlder of Philadelphia and Representative George H. Tinkham of Massachusetts, Edith Elmer Wood (1870–1945) was one of the first Americans to champion "positive" rather than "negative" housing reform. The widely traveled daughter and later wife of naval officers, she argued that social behavior was conditioned by housing and that government action to replace the slums would improve citizenship, lower welfare costs, and reduce crime and delinquency. Having witnessed the initiatives European nations were taking to shelter their inhabitants, she subsequently devoted herself to the campaign for actual government construction of dwelling units. Her book, *The Housing of the Unskilled Wage Earner* (New York, 1919) became a classic and made her an international figure in housing reform. In it, she argued that private philanthropy was not the solution to the housing problem and that restrictive building codes simply raised the rent levels of tenements while doing nothing at all to increase the supply. In 1921, Mrs. Wood blasted Secretary of Commerce Herbert Hoover's plan to eliminate waste in the construction industry. "Efficiency is good," she remarked. "But we are still waiting to see it do for the poor man's home what Ford did for his car.''[3]

In later years Wood was joined in her efforts by Catherine Bauer, Mary Kingsbury Simkhovitch, Clarence Stein, Lewis Mumford, and Frederick Ackerman, most of whom were founders of the Regional Planning Association of America. And by the early 1930s, several lobbying groups, including the National Public Housing Conference, were working for ''public construction for those people who cannot be adequately housed at rents they can afford to pay.''[4] Prior to the New Deal, however, only the states of New York and North Dakota accepted the provision of housing as even a limited responsibility.[5]

In an important reversal of traditional federal policy, the administration of Franklin D. Roosevelt initiated its own construction program. The direct involvement of Uncle Sam began with the passage of the National Industrial Recovery Act during the First Hundred Days of 1933. The legislation had four purposes: to increase employment, to improve housing for the poor, to demonstrate to private industry the feasibility of large-scale community planning efforts, and to eradicate and rehabilitate slum areas in order ''to check the exodus to the outer limits of cities with consequent costly utility extensions and leaving the centrally located areas unable to pay their way.'' The first purpose was the most important; Congress wanted to create jobs, not housing.[6]

The 1933 housing law authorized the Public Works Administration (PWA) to accomplish these purposes through three mechanisms. First, the PWA Housing Division could lend money to private, limited-dividend corporations interested in slum clearance. Second, grants and loans could be made available to public authorities for the same purpose. Third, and most significant, the Housing Division was empowered to buy, condemn, sell, or lease property for developing new projects itself.[7]

Although PWA Administrator Harold Ickes bluntly complained that ''American cities cannot produce a single instance in which slums have been cleared and new dwellings built to rehouse the dispossessed occupants by private enterprise operating on a commercial basis,'' the PWA attempted to place the emphasis of the program on private development encouraged by federal loans. Only seven of five hundred limited-dividend applications were approved, however, partly because very few of the corporations had sufficient equity to qualify for the program, and partly because those with the equity seemed primarily anxious to sell land to the government at inflated prices. The 284-unit Carl Mackley project in Philadelphia was the first and most important of the few developments that resulted from this effort.[8]

Unable to rebuild the slums through the private enterprise provision of the law, the Housing Division terminated its limited-dividend program in mid-1934 and turned to its second alternative. The ability of the PWA to work with local authorities proved of limited usefulness, however. In 1933 no state or locality had the legal authority to engage in slum-clearance projects; as late as 1937, only New York, Ohio, Michigan, and South Carolina had passed the required enabling legislation, and only New York City had constructed public housing with local funds (the 120-unit First Houses on the Lower East Side).[9]

Finally, the Housing Division turned to its last option—the construction of its own low-income housing projects on land acquired for the government by condemnation or purchase. Between 1934 and 1937, when the Housing Division of PWA was replaced by the United States Housing Authority (USHA), the government began 21,000 units in forty-nine separate projects costing a total of $129 million. The most important of these efforts was a seven-building complex in Manhattan on the Harlem River Drive. Work began on the Harlem River Houses in 1936, and the project was dedicated by Mayor Fiorello LaGuardia on June 16, 1937. A short fireplug of a man, the feisty mayor thus hoped to "give the people of my city, in place of their tenements, decent, modern, cheerful housing, with a window in every room and a bit of sunshine in every window."

These early New Deal efforts at subsidized shelter were soon stymied, however. In a landmark decision handed down in January 1935, Federal Judge Charles I. Dawson of Kentucky ruled that acquiring land for public housing in Louisville by condemnation (eminent domain) was not constitutional and that the Public Works Administration could not therefore exercise this power. In the words of Judge Dawson:

> [Low cost housing] is certainly not a public use, in the sense that the property is proposed to be used by the federal government for performing any of the legitimate functions of the government itself. Surely it is not a governmental function to construct buildings in a state for the purpose of selling or leasing them to private citizens for occupancy as houses.[10]

Judge Dawson's ruling was upheld by the Court of Appeals for the Sixth Circuit, and lawyers for the Public Works Administration decided at the last moment to withdraw their petition before the United States Supreme Court. The PWA was thus forced to purchase sites at private market prices, increasing the costs and reducing the number of projects initiated. Costs also rose because this first New Deal housing effort was a hastily designed program to put men back to work. Thus, by the time PWA erected apartments or houses the minimum rents that had to be charged precluded occupancy by the urban poor. Elements within the Roosevelt Administration suggested that the projects charge low rents and be operated at a loss, but the proposal was shelved when the Comptroller General ruled that there was no law to authorize such a subsidy.[11]

Well aware that adverse judicial decisions and escalating costs would effectively cripple the PWA housing program, Senator Robert F. Wagner of New York and Representative Henry Ellenbogen of Pennsylvania introduced new legislation in 1934 to create a permanent public housing agency. Initially, President Roosevelt gave the Wagner-Ellenbogen measure only lukewarm support, and it died in the House Banking and Currency Committee.[12]

The outlook for change remained bleak the following year. The 1936 Democratic Party Platform contained only a weak plank endorsing federal involvement in housing, and the issue was of minor significance in the election. A hint of change came in October, however, when Roosevelt promised a large New York City audience:

> We have too long neglected the housing problem for our lower income groups. . . . We have not yet begun adequately to spend money in order to help the families in the overcrowded sections of our cities live as American citizens have a right to live. You and I will not be content until city, state, and federal governments have joined with private capital in helping every American live that way. . . . I am confident that the next Congress will start us on our way with a sound housing policy.[13]

The unprecedented magnitude of his re-election landslide made the President feel safe in his advocacy of a stronger Washington role in housing. Moreover, Catherine Bauer had systematically built labor support, collecting endorsements from 525 local unions and almost every state federation of labor. In his second inaugural address, Roosevelt came out dramatically on the issue:

> But here is the challenge to our democracy. In this nation I see tens of millions of its citizens—a substantial part of its whole population—who at this very moment are denied the greater part of what the very lowest standards of today call the necessities of life. . . . I see one-third of a nation ill-housed, ill-clad, ill-nourished.[14]

Although the President was personally more interested in back-to-the-land movements than in public housing, and although like many during the 1930s he confused slum clearance with improving conditions for the poor, Roosevelt did give his personal and powerful support to public housing in the spring of 1937. Within a matter of months, the United States Housing Act (also known as the Wagner-Steagall Act) had passed the Senate by 64-16 and the House by 275-86. Signed into law on September 1, 1937, it marked the first time the federal government accepted permanent responsibility for the construction of decent, low-cost homes. Long-time reformer Catherine Bauer called it a ''radical piece of legislation,'' and the *New York Times* added: ''With the President's signature the Wagner-Steagall bill becomes law and at last America makes a real start toward wiping out its city slums.''[15]

The legislation empowered the United States Housing Authority (USHA) to develop public projects by funding duly constituted local housing agencies. The USHA was to funnel this money to municipalities through two mechanisms: first, by lending up to 90 percent of the capital costs of a project to local officials (for loans of up to sixty years, which meant lower rentals), and second, by subsidizing construction and maintenance costs. So enthusiastic was President Roosevelt that when work began on the first five projects under the new procedures on March 17, 1938, he wrote to Nathan Straus, his chief housing official: ''Today marks the beginning of a new era in the economic and social life of America. Today, we are launching an attack on the slums of this country which must go forward until every American family has a decent home.''[16]

On one level public housing was a resounding success. By the end of 1938, thirty-three states had passed enabling legislation, and 221 local authorities had been established. By 1941 the USHA had sponsored 130,000 new units in 300 projects scattered throughout the nation. And by the end of 1962, more than two million people lived in the half-million units built under various public housing programs. If the quality and design of the projects frequently invited derision, they were nevertheless superior to the delapidated structures they replaced.[17]

On another level, however, public housing did not fulfill the expectations of its supporters. There never was enough of it, in part because conservatives found frequent opportunities to reduce its funds. In 1949, for example, Congress authorized 810,000 units over six years; only 322,000 new starts were actually funded over the next eleven years. Thus, in 1980, while publicly owned housing accounted for only about one percent of the United States housing market, it comprised 46 percent of the market in England and Wales, and 37 percent of the French housing market. And the problem was a shortage of funding, not a shortage of need. The real purpose of the 1937 law was to alleviate ''present and recurring unemployment''; it emphasized economic stimulus over social and architectural objectives. As soon as the nation entered wartime and postwar housing booms, the public sector was relegated to a low priority.[18]

Of particular importance to the spatial distribution of the limited amount of public housing that was built was the decentralized nature of the program. In view of Judge Dawson's ruling and of widespread opinion that federal use of the power of eminent domain for housing was unconstitutional, Senator Wagner's bill created the USHA as a "low-rent housing and slum clearance measure . . . drawing its strength from *local initiative and responsibility* [italics mine]." It required that any city desiring public housing had to provide tax exemptions for the project and had to create a municipal housing agency. The 1937 requirement of local participation was strengthened and broadened in the Housing Act of 1949. Thus, every community had to make its own decision as to whether or not a need existed; the resulting application for federally subsidized housing had to be a *voluntary* action.[19]

The distinction was critical. Because municipalities had discretion on where and when to build public housing, the projects invariably reinforced racial segregation. A suburb that did not wish to tarnish its exclusive image by having public housing within its precincts could simply refuse to create a housing agency. No local housing authority from another jurisdiction and no national official could force it to do otherwise. By contrast, in Britain the municipality itself is the "housing authority," and in Japan the national government buys inexpensive land in very distant areas as the only practical means of acquiring space for public-housing projects.[20]

Needless to say, hundreds of suburbs throughout the United States did not create housing agencies and did not apply for federal funding. As a result, low-income housing did not go up on the cheaper, vacant land of the suburbs, but in the heart of cities. Parma, Ohio, for example, a suburb of 100,000 people just a few miles southeast of Cleveland, had no low-income projects in 1981, largely because of a local law requiring approval by referendum of any proposed subsidized housing. Other communities used slightly different tactics with the same result. At the other end of the spectrum, some communities went after public housing funds with extraordinary vigor. Newark, New Jersey, for example, built more units of public housing per capita than any other city in the nation. Because its neighbors were more protective of their image, however, Newark attracted an even larger percentage of the very poor, and by 1970 it was the most troubled metropolis in the United States by any of a half-dozen measures of urban pathology.[21]

A second feature of the legislation that tended to concentrate public housing in the center rather than in the suburbs was the fact that housing authorities were typically made up of prominent citizens who were more anxious to clear slums and to protect real-estate values than they were to rehouse the poor. As John F. Bauman has clearly demonstrated in a study of Philadelphia, the public housing authority was especially anxious to boost the sagging tax structure, to halt the spread of blight, and to raise property values.[22]

Finally, there was a requirement that one slum unit be eliminated for every unit of public housing erected. Thus, only localities with significant numbers of inadequate dwellings could receive assistance. The following exchange between an inquisitive Congressman and the Commissioner of the United States Housing Authority underscores the point:

> Congressman Kunkel (Pennsylvania): "Under this program, no area in which there is no substandard housing would be eligible for any public housing. Is that correct?"
>
> Commissioner Egan: "That is correct. If there were no slums in that locality, regardless of how acute the housing shortage was, and if we knew we could not get the equivalent elimination required by the act, we could not go in there."[23]

Even the most progressive Congressional leaders accepted such limitations out of concern that to do otherwise would imperil passage. In 1949, when a major new public housing bill was being debated, Republican opponents proposed an amendment that forbade any form of racial or ethnic discrimination in any public housing project. This placed many of the bill's supporters in a dilemma.If the amendment succeeded, the Southern senators would be certain to vote against the entire bill, thus ensuring its defeat. Yet, Northern liberals were not inclined to vote against an amendment promising racial justice.[24]

Senator Paul Douglas, as decent a man as ever served on Capitol Hill, urged his liberal colleagues to put aside their principles temporarily in order to give the measure a chance. He told the chamber that the amendment "necessarily creates a sharp conflict within the hearts of all of us who want on one hand to clear the slums and to provide housing for the slum dwellers and who, at the same time, feel very keenly that we should not treat any race as second-class citizens." He went on to say:

> I am ready to appeal to history and to time that it is in the best interests of the Negro race that we carry through the housing program as planned, rather than put in the bill an amendment which will inevitably defeat it, and defeat all hopes for rehousing 4 million persons.[25]

Senator Douglas was successful. When the amendment came to a vote it was defeated 49 to 31. And when the bill was voted on the next day it passed easily by 57 to 13. Public housing remained a federal commitment. But the fundamental problem remained. The new dwellings were not built at low density on the less costly land outside the central cities, as was the case for example in Great Britain, where two-thirds of the publicly owned units constructed between 1920 and 1980 were houses and only one-third were apartments. Instead, because determination of need and site selection were left up to localities, public housing was confined to existing slums. It further concentrated the poor in the central cities and reinforced the image of suburbia as a place of refuge from the social pathologies of the disadvantaged.[26]

Title I of the 1949 Housing Act not only did nothing to correct this problem, but it went so far as to encourage existing inequities. One section of the act required that ''there be a feasible method for the temporary relocation of families displaced from the project area'' into ''decent, safe, and sanitary dwellings.'' Quite often, such housing was not available. And even when it was, there was often malfeasance on the part of public officials. Robert A. Caro, for example, has documented the ways in which New York City's powerful construction czar, Robert Moses, managed to move out the poor with scant attention to their resettlement.

The destruction of deteriorating buildings to make room for public housing often created problems in adjacent neighborhoods. An excellent example of this phenomenon can be found in Brooklyn. When blocks of slums in the Brownsville district were cleared to make room for public housing in the 1950s, thousands of displaced families moved into the neighboring district of East New York, which at that time was a vibrant, predominantly white, middle-class area with a stable economy. The sudden influx of large, lower-income, black and Hispanic families from Brownsville strained the physical and social services of the community. A mass exodus of the white population began. Within six years, a healthy community became one of the most decayed and dangerous neighborhoods in the United States. If the government had invested its funds in maintenance programs for the older housing of the inner city, the poor might have inherited stable neighborhoods, and the cities might have avoided the sorry spectacle of abandoned areas.[27]

The original concept of public housing was that it was for the ''working poor,'' the ''deserving poor,'' and the honest man temporarily down on his luck. Long-term welfare families, loafers, and unwed mothers were not welcomed. By the 1960s, however, this concept had been discarded, and admissions policies were changed to allow welfare recipients into the structures. Thereafter, public housing came to be seen as the shelter of last resort, as a permanent home for the underclass rather than a temporary refuge for ''respectable'' families. By 1980, not only was public housing segregated and isolated, but black youths were defining the ghetto as ''the projects.'' They saw the concentration of the poor in public housing, and they knew that "the projects" were themselves concentrated in particular parts of cities.[28] In Chicago, for example, 150,000 persons lived in low-income public housing in 1978. A few scattered developments were in marginal white neighborhoods, such as Mayor Richard Daley's beloved Bridgeport. Their occupants were predominantly white and aged. The other 95 percent of Chicago's public housing, however, was dumped into the most poverty-impacted black ghettoes in the city.

By the late 1960s, public housing was being criticized on sociological, economic, and architectural grounds. The most famous social criticism was Lee Rainwater's study of the Pruitt-Igoe housing development in St. Louis. The huge highrise project typified what was wrong with public housing across the nation. Riddled by crime and vandalism, it was unable to fill vacancies when they opened up, and it was torn down in 1976.

The economic attack on public housing was led by Martin Anderson, then a Columbia Business School professor and later an important functionary in the Nixon administration. Anderson's book, *The Federal Bulldozer,* maintained that Washington had actually destroyed more low-income residences than it had created and that government could best serve the dwelling needs of the poor by abandoning urban renewal and public housing.

The architectural criticism of highrise, superblock housing was led by Oscar Newman, whose *Defensible Space* won international acclaim. Calling for more entrances in buildings, more low-rise structures instead of apartment towers, and more observable, usable space where people could congregate, Newman contended that there were four attributes that contribute to the safety or lack of safety of any housing project: the capacity of an area to create perceived territorial influence, the capacity of a building to provide surveillance opportunities over the immediate area, the capacity of design to decrease the occupant's perception of isolation and stigma, and the availability of adjacent commercial activities.[29]

Sadly, most projects lacked the amenities Newman identified, and public housing nationally continued on a downward spiral into the 1980s. Poorly maintained, segregated, cheaply constructed, and often physically dangerous, the projects had become "the dumping ground for the poor." Skeptics even charged that the only group that really gained from public housing was the investor in the tax-exempt, federally guaranteed bonds that were issued by most local housing authorities. Successive presidential administrations gave it a low priority, and real-estate interests were always quick to protest any government competition in the housing market. As one "heavy taxpayer and supporter of the party in power" wrote on November 9, 1933: "We have too many cheap houses now and rent values are very low and there are too many subdivisions upon which all people who made purchases have lost practically all their investments." A half-century later, the objections were essentially the same.[30]

In the mind of the average citizen, the failures of public housing were due to cultural characteristics of the poor themselves, who were seen to be resisting improvement. If government-assisted shelter—a free ride in the view of many citizens—failed to alleviate the characteristics of poverty, then the poor had only themselves to blame. In this sense, public housing was similar to other "poverty programs," which assumed that poverty could be eliminated simply by altering one aspect of the life of the poor.

Actually, the fault was not with public housing or with the tenants, but with the expectation that any one solution could so vastly reduce poverty and social pathology. The public housing concept remains valid, however, and the success of "the projects" in some cities suggests that the fault may rest with management or with funding levels rather than with the idea. In New York City, for example, the 1984 waiting list for public housing included more than 175,000 names, and the annual turnover rate was less than 4 percent. While public housing might still be described as "federally built and supported slums," the need for adequate, clean, and inexpensive shelter for the poor obviously remains.

The proliferation of public housing is a twentieth-century phenomenon that has changed the appearance and character of cities around the world. In countries as diverse as the Soviet Union and South Africa, state-constructed projects have become an integral part of national ideologies (socialism in the U.S.S.R. and apartheid in South Africa). In many industrial countries, such as Great Britain and West Germany, public housing has become a standard extension of the welfare state. In impoverished Third World nations, public housing is an important tool with which to address massive social problems. Whether inspired by ideology or necessity, public housing has become an important institution in most countries.

The major exception is the United States, where government assistance has served mainly to create invidious distinctions between city and suburban life. Beginning in the 1930s, the American government began two major thrusts in the broad field of housing. On the one hand, it encouraged homeownership through the long-term, low-interest mortgage. The beneficiaries of these programs were typically white and middle-class and their destinations were usually suburban.

The second major housing initiative of the federal government involved the actual construction of dwellings by public tax monies for the benefit of those who could not pay market rates for shelter. This program slowly evolved from one offering temporary relief for the working poor to one offering a kind of permanent and dismal protection for the most disadvantaged members of the society. The locations of these structures were almost always in the poorest parts of central cities.

Federal policies are often at cross-purposes with one another, or, as is sometimes said, the right hand does not know what the left hand is doing. Perhaps in an organization as immense as the United States government it would be impossible to have a single, coherent, consistent housing policy. But however confused the situation appears, however much government officials argue that Washington programs have been consistently motivated by the desire to produce social benefit for all income groups, the basic direction of federal policies toward housing has been the concentration of the poor in the central city and the dispersal of the affluent to the suburbs. American housing policy was not only devoid of social objectives, but instead helped establish the basis for social inequities. Uncle Sam was not impartial, but instead contributed to the general disbenefit of the cities and to the general prosperity of the suburbs.[31]

13

The Baby Boom and the Age of the Subdivision

What the Blandings wanted . . . was simple enough: a two-story house in quiet, modern good taste, . . . a good-sized living room with a fire place, a dining room, pantry, and kitchen, a small lavatory, four bedrooms and accompanying baths . . . a roomy cellar . . . plenty of closets.

—ERIC HODGINS,
Mr. Blandings Builds His Dream House (1939)

No man who owns his own house and lot can be a Communist. He has too much to do.

—WILLIAM J. LEVITT, 1948

At 7 P.M. (Eastern time) on August 14, 1945, radio stations across the nation interrupted normal programming for President Harry S. Truman's announcement of the surrender of Japan. It was a moment in time that those who experienced it will never forget. World War II was over. Across the nation, Americans gathered to celebrate their victory. In New York City two million people converged on Times Square as though it were New Year's Eve. In smaller cities and towns, the response was no less tumultuous, as spontaneous cheers, horns, sirens, and church bells telegraphed the news to every household and hamlet, convincing even small children that it was a very special day. To the average person, the most important consequence of victory was not the end of shortages, not the restructuring of international boundaries or reparations payments or big power politics, but the survival of husbands and sons. Some women regretted that their first decent-paying, responsible jobs would be taken away by returning veterans. Most, however, felt a collective sigh of relief. Normal family life could resume. The long vigil was over. Their men would be coming home.[1]

In truth, the United States was no better prepared for peace than it had been for war when the German *Wehrmacht* crossed the Polish frontier in the predawn hours of September 1, 1939. For more than five years military necessity had taken priority over consumer goods, and by 1945 almost everyone had a long list of unfilled material wants.

Housing was the area of most pressing need. Through sixteen years of depression and war, the residential construction industry had been dormant, with new home starts averaging less than 100,000 per year. Almost one million people had migrated to defense areas in the early 1940s, but new housing for them was designated as "temporary," in part as an economy move and in part because the real-estate lobby did not want emergency housing converted to permanent use after the war. Meanwhile, the marriage rate, after a decade of decline, had begun a steep rise in 1940, as war became increasingly likely and the possibility of separation added a spur to decision-making. In addition, married servicemen received an additional fifty dollars per month allotment, which went directly to the wives. Soon thereafter, the birth rate began to climb, reaching 22 per 1,000 in 1943, the highest in two decades. Many of the newcomers were "good-bye babies," conceived just before the husbands shipped out, partly because of an absence of birth control, partly because the wife's allotment check would be increased with each child, and partly as a tangible reminder of a father who could not know when, or if, he would return. During the war, government and industry both played up the suburban house to the families of absent servicemen, and between 1941 and 1946 some of the nation's most promising architects published their "dream houses" in a series in the *Ladies' Home Journal*.[2]

After the war, both the marriage and the birth rates continued at a high level. In individual terms, this rise in family formation coupled with the decline in housing starts meant that there were virtually no homes for sale or apartments for rent at war's end. Continuing a trend begun during the Great Depression, six million families were doubling up with relatives or friends by 1947, and another 500,000 were occupying quonset huts or temporary quarters. Neither figure included families living in substandard dwellings or those in desperate need of more room. In Chicago, 250 former trolley cars were sold as homes. In New York City a newly wed couple set up housekeeping for two days in a department store window in hopes that the publicity would help them find an apartment. In Omaha a newspaper advertisement proposed: "Big Ice Box, 7 × 17 feet, could be fixed up to live in." In Atlanta the city bought 100 trailers for veterans. In North Dakota surplus grain bins were turned into apartments. In brief, the demand for housing was unprecedented.[3]

The federal government responded to an immediate need for five million new homes by underwriting a vast new construction program. In the decade after the war Congress regularly approved billions of dollars worth of additional mortgage insurance for the Federal Housing Administration. Even more important was the Servicemen's Readjustment Act of 1944, which created a Veterans Administration mortgage program similar to that of FHA. This law gave official endorsement and support to the view that the 16 million GI's of World War II should return to civilian life with a home of their own. Also, it accepted the builders' contention that they needed an end to government controls but not to government insurance on their investments in residential construction. According to novelist John Keats, "The real estate boys read the Bill, looked at one another in happy amazement, and the dry, rasping noise they made rubbing their hands together could have been heard as far away as Tawi Tawi."[4]

It is not recorded how far the noise carried, but anyone in the residential construction business had ample reason to rub their hands. The assurance of federal mortgage guarantees—at whatever price the builder set—stimulated an unprecedented building boom. Single-family housing starts spurted from only 114,000 in 1944, to 937,000 in 1946, to 1,183,000 in 1948, and to 1,692,000 in 1950, an all-time high. However, as Barry Checkoway has noted, what distinguished the period was an increase in the number, importance, and size of large builders. Residential construction in the United States had always been highly fragmented in comparison with other industries, and dominated by small and poorly organized house builders who had to subcontract much of the work because their low volume did not justify the hiring of all the craftsmen needed to put up a dwelling. In housing, as in other areas of the economy, World War II was beneficial to large businesses. Whereas before 1945, the typical contractor had put up fewer than five houses per year, by 1959, the median single-family builder put up twenty-two structures. As early as 1949, fully 70 percent of new homes were constructed by only 10 percent of the firms (a percentage that would remain roughly stable for the next three decades), and by 1955 subdivisions accounted for more than three-quarters of all new housing in metropolitan areas.[5]

Viewed from an international perspective, however, the building of homes in the United States remained a small-scale enterprise. In 1969, for example, the percentage of all new units built by builders of more than 500 units per year was only 8.1 percent in the United States, compared with 24 percent in Great Britain and 33 percent in France. World War II, therefore, did not transform the American housing industry as radically as it did that of Europe.[6]

Levittown

The family that had the greatest impact on postwar housing in the United States was Abraham Levitt and his sons, William and Alfred, who ultimately built more than 140,000 houses and turned a cottage industry into a major manufacturing process. They began on a small scale on Long Island in 1929 and concentrated for years on substantial houses in Rockville Center. Increasing their pace in 1934 with a 200-unit subdivision called "Strathmore" in Manhasset, the Levitts continued to focus on the upper-middle class and marketed their tudor-style houses at between $9,100 and $18,500. Private commissions and smaller subdivisions carried the firm through the remainder of the prewar period.[7]

In 1941 Levitt and Sons received a government contract for 1,600 (later increased to 2,350) war worker's homes in Norfolk, Virginia. The effort was a nightmare, but the brothers learned how to lay dozens of concrete foundations in a single day and to preassemble uniform walls and roofs. Additional contracts for more federal housing in Portsmouth, Virginia, and for barracks for shipyard workers at Pearl Harbor provided supplemental experience, as did William's service with the Navy Seabees from 1943 to 1945. Thus, the Levitts were among the nation's largest home builders even before construction of the first Levittown.[8]

Returning to Long Island after the war, the Levitts built 2,250 houses in Roslyn in 1946 in the $17,500 to $23,500 price range, well beyond the means of the average veteran. In that same year, however, they began the acquisition of 4,000 acres of potato farms in the Town of Hempstead, where they planned the biggest private housing project in American history.[9]

The formula for Island Trees, soon renamed Levittown, was simple. After bulldozing the land and removing the trees, trucks carefully dropped off building materials at precise 60-foot intervals. Each house was built on a concrete slab (no cellar); the floors were of asphalt and the walls of composition rock-board. Plywood replaced ¾-inch strip lap, ¾-inch double lap was changed to ⅜-inch for roofing, and the horse and scoop were replaced by the bulldozer. New power hand tools like saws, routers, and nailers helped increase worker productivity. Freight cars loaded with lumber went directly into a cutting yard where one man cut parts for ten houses in one day.

The construction process itself was divided into twenty-seven distinct steps—beginning with laying the foundation and ending with a clean sweep of the new home. Crews were trained to do one job—one day the white-paint men, then the red-paint men, then the tile layers. Every possible part, and especially the most difficult ones, were preassembled in central shops, whereas most builders did it on site. Thus, the Levitts reduced the skilled component to 20–40 percent. The five-day work week was standard, but they were the five days during which building was possible; Saturday and Sunday were considered to be the days when it rained. In the process, the Levitts defied unions and union work rules (against spray painting, for example) and insisted that subcontractors work only for them. Vertical integration also meant that the firm made its own concrete, grew its own timber, and cut its own lumber. It also bought all appliances from wholly owned subsidiaries. More than thirty houses went up each day at the peak of production.[10]

Initially limited to veterans, this first "Levittown" was twenty-five miles east of Manhattan and particularly attractive to new families that had been formed during and just after the war. Squashed in with their in-laws or in tiny apartments where landlords frowned on children, the GI's looked upon Levittown as the answer to their most pressing need. Months before the first three hundred Levitt houses were occupied in October 1947, customers stood in line for the four-room Cape Cod box renting at sixty dollars per month. The first eighteen hundred houses were initially available only for rental, with an option to buy after a year's residence. Because the total for mortgage, interest, principal, and taxes was *less* than the rent, almost everyone bought; after 1949 all units were for sale only. So many of the purchasers were young families that the first issue of *Island Trees,* the community newspaper, opined that "our lives are held closely together because most of us are within the same age bracket, in similar income groups, live in almost identical houses and have common problems."[11] And so many babies were born to them that the suburb came to be known as "Fertility Valley" and "The Rabbit Hutch."

Ultimately encompassing more than 17,400 separate houses and 82,000 residents, Levittown was the largest housing development ever put up by a single builder, and it served the American dream-house market at close to the lowest prices the industry could attain. The typical Cape Cod was down-to-earth and unpretentious; the intention was not to stir the imagination, but to provide the best shelter at the least price. Each dwelling included a twelve-by-sixteen-foot living-room with a fireplace, one bath, and two bedrooms (about 750 square feet), with easy expansion possibilities upstairs in the unfinished attic or outward into the yard. Most importantly, the floor plan was practical and well-designed, with the kitchen moved to the front of the house near the entrance so that mothers could watch their children from kitchen windows and do their washing and cooking with a minimum of movement. Similarly, the living room was placed in the rear and given a picture window overlooking the back yard. This early Levitt house was as basic to post World War II suburban development as the Model T had been to the automobile. In each case, the actual design features were less important than the fact that they were mass-produced and thus priced within the reach of the middle class.[12]

William Jaird Levitt, who assumed primary operating responsibility for the firm soon after the war, disposed of houses as quickly as other men disposed of cars. Pricing his Cape Cods at $7,990 (the earliest models went for $6,990) and his ranches at $9,500, he promised no down payment, no closing costs, and "no hidden extras." With FHA and VA "production advances," Levitt boasted the largest line of credit ever offered a private home builder. He simplified the paperwork required for purchase and reduced the entire financing and titling transaction to two half-hour steps. His full-page advertisements offered a sweetener to eliminate lingering resistance—a Bendix washer was included in the purchase price. Other inducements included an eight-inch television set (for which the family would pay for the next thirty years). So efficient was the operation that *Harper's Magazine* reported in 1948 that Levitt undersold his nearest competition by $1,500 and still made a $1,000 profit on each house. As *New York Times'* architecture critic Paul Goldberger has noted, "Levittown houses were social creations more than architectural ones—they turned the detached, single-family house from a distant dream to a real possibility for thousands of middle-class American families."[13]

Buyers received more than shelter for their money. When the initial families arrived with their baby strollers and play pens, there were no trees, schools, churches, or private telephones. Grocery shopping was a planned adventure, and picking up the mail required sloshing through the mud to Hicksville. The Levitts planted apple, cherry, and evergreen trees on each plot, however, and the development ultimately assumed a more parklike appearance. To facilitate development as a garden community, streets were curvilinear (and invariably called "roads" or "lanes"), and through traffic was shunted to peripheral thoroughfares. Nine swimming pools, sixty playgrounds, ten baseball diamonds, and seven "village greens" provided open space and recreational opportunities. The Levitts forbade fences (a practice later ignored) and permitted outdoor clothes drying only on specially designed, collapsible racks. They even supervised lawn-cutting for the first few years—doing the jobs themselves if necessary and sending the laggard families the bill.[14]

Architectural critics, many of whom were unaccustomed to the tastes or resources of moderate-income people, were generally unimpressed by the repetitious houses on 60-by-100-foot "cookie cutter lots" and referred to Levittown as "degraded in conception and impoverished in form." From the Wantagh Parkway, the town stretched away to the east as far as the eye could see, house after identical house, a horizon broken only by telephone poles. Paul Goldberger, who admired the individual designs, thought that the whole was "an urban planning disaster," while Lewis Mumford complained that Levittown's narrow range of house type and income range resulted in a one-class community and a backward design. He noted that the Levitts used "new-fashioned methods to compound old-fashioned mistakes."[15]

But Levittown was a huge popular success where it counted—in the marketplace. On a single day in March 1949, fourteen hundred contracts were drawn, some with families that had been in line for four days. "I truly loved it," recalled one early resident. "When they built the Village Green, our big event was walking down there for ice cream."[16]

In the 1950s the Levitts shifted their attention from Long Island to an equally large project near Philadelphia. Located on former broccoli and spinach farms in lower Bucks County, Pennsylvania, this new Levittown was built within a few miles of the new Fairless Works of the United States Steel Corporation, where the largest percentage of the community's residents were employed. It was composed on eight master blocks, each of about one square mile and focusing on its own recreational facilities. Totaling about 16,000 homes when completed late in the decade, the town included light industry and a big, 55-acre shopping center. According to Levitt, "We planned every foot of it—every store, filling station, school, house, apartment, church, color, tree, and shrub."[17]

In the 1960s, the Levitt forces shifted once again, this time to Willingboro, New Jersey, where a third Levittown was constructed within distant commuting range of Philadelphia. This last town was the focus of Herbert Gans's well-known account of *The Levittowners*. The Cape Cod remained the basic style, but Levitt improved the older models to resemble more closely the pseudo-colonial design that was so popular in the Northeast.[18]

If imitation is the sincerest form of flattery, then William Levitt has been much honored in the past forty years.[19] His replacement of basement foundations with the radiantly heated concrete slab was being widely copied as early as 1950. Levitt did not actually pioneer many of the mass-production techniques—the use of plywood, particle board, and gypsum board, as well as power hand tools like saws, routers, and nailers, for example—but his developments were so widely publicized that in every large metropolitan area, large builders appeared who adopted similar methods—Joseph Kelly in Boston, Frank White in Portland, Louis H. Boyar and Fritz B. Burns in Los Angeles, Del Webb in Phoenix, William G. Farrington in Houston, Franklin L. Burns in Denver, Wallace E. Johnson in Memphis, Ray Ellison in San Antonio, Maurice Fishman in Cleveland, Waverly Taylor in Washington, Irving Blietz and Phillip Klutznick in Chicago, John Mowbray in Baltimore, and Carl Gellert and Ellie Stoneman in San Francisco, to name just the more well-known builders.[20]

FHA and VA programs made possible the financing of their immense developments. Title VI of the National Housing Act of 1934 allowed a builder to insure 90 percent of the mortgage of a house costing up to nine thousand dollars. Most importantly, an ambitious entrepreneur could get an FHA "commitment" to insure the mortgage, and then use that "commitment" to sign himself up as a temporary mortgagor. The mortgage lender (a bank of savings and loan institution) would then make "production advances" to the contractor as the work progressed, so that the builder needed to invest very little of his own hard cash. Previously, even the largest builders could not bring together the capital to undertake thousand-house developments. FHA alone insured three thousand houses in Henry J. Kaiser's Panorama City, California; five thousand in Frank Sharp's Oak Forest; and eight thousand in Klutznick's Park Forest project.[21]

Characteristics of Postwar Suburbs

However financed and by whomever built, the new subdivisions that were typical of American urban development between 1945 and 1973 tended to share five common characteristics. The first was peripheral location. A Bureau of Labor Statistics survey of home building in 1946–1947 in six metropolitan regions determined that the suburbs accounted for at least 62 percent of construction. By 1950 the national suburban growth rate was ten times that of central cities, and in 1954 the editors of *Fortune* estimated that 9 million people had moved to the suburbs in the previous decade. The inner cities did have some empty lots—serviced by sewers, electrical connections, gas lines, and streets—available for development. But the filling-in process was not amenable to mass production techniques, and it satisfied neither the economic nor the psychological temper of the times.[22]

The few new neighborhoods that were located within the boundaries of major cities tended also to be on the open land at the edges of the built-up sections. In New York City, the only area in the 1946–1947 study where city construction was greater than that of the suburbs, the big growth was on the outer edges of Queens, a borough that had been largely undeveloped in 1945. In Memphis new development moved east out Summer, Poplar, Walnut Grove, and Park Avenues, where FHA and VA subdivisions advertised "No Down Payment" or "One Dollar Down" on giant billboards. In Los Angeles the fastest-growing American city in the immediate postwar period, the area of rapid building focused on the San Fernando Valley, a vast space that had remained largely vacant since its annexation to the city in 1915. In Philadelphia thousands of new houses were put up in farming areas that had legally been part of the city since 1854, but which in fact had functioned as agricultural settlements for generations.

The second major characteristic of the postwar suburbs was their relatively low density. In all except the most isolated instances, the row house completely lost favor; between 1946 and 1956, about 97 percent of all new single-family dwellings were completely detached, surrounded on every side by their own plots. Typical lot sizes were relatively uniform around the country, averaging between 1/5 (80 by 100 feet) and 1/10 (40 by 100 feet) of an acre and varying more with distance from the center than by region. Moreover, the new subdivisions alloted a higher proportion of their land area to streets and open spaces. Levittown, Long Island, for example, was settled at a density of 10,500 per square mile, which was about average for postwar suburbs but less than half as dense as the streetcar suburbs of a half-century earlier. This design of new neighborhoods on the assumption that residents would have automobiles meant that those without cars faced severe handicaps in access to jobs and shopping facilities.[23]

This low-density pattern was in marked contrast with Europe. In war-ravaged countries east of the Rhine River, the concentration upon apartment buildings can be explained by the overriding necessity to provide shelter quickly for masses of displaced and homeless people. But in comparatively unscathed France, Denmark, and Spain, the single-family house was also a rarity. In Sweden, Stockholm committed itself to a suburban pattern along subway lines, a decision that implied a high-density residential pattern. Nowhere in Europe was there the land, the money, or the tradition for single-family home construction.[24]

The third major characteristic of the postwar suburbs was their architectural similarity. A few custom homes were built for the rich, and mobile homes gained popularity with the poor and the transient, but for most American families in search of a new place to live some form of tract house was the most likely option. In order to simplify their production methods and reduce design fees, most of the larger developers offered no more than a half-dozen basic house plans, and some offered half that number. The result was a monotony and repetition that was especially stark in the early years of the subdivision, before the individual owners had transformed their homes and yards according to personal taste.

But the architectural similarity extended beyond the particular tract to the nation as a whole. Historically, each region of the country had developed an indigenous residential style—the colonial-style homes of New England, the row houses of Atlantic coastal cities, the famous Charleston town houses with their ends to the street, the raised plantation homes of the damp bayou country of Louisiana, and the encircled patios and massive walls of the Southwest. This regionalism of design extended to relatively small areas; early in the twentieth century a house on the South Carolina coast looked quite different from a house in the Piedmont a few hundred miles away.

This tradition began eroding after World War I, when the American dream house became, as already noted, the Cape Cod cottage, a quaint one-and-a-half-story dwelling. This design remained popular into the post World War II years, when Levittown featured it as a bargain for veterans. In subsequent years, one fad after another became the rage. First, it was the split-level, then the ranch, then the modified colonial. In each case, the style tended to find support throughout the continent, so that by the 1960s the casual suburban visitor would have a difficult time deciphering whether she was in the environs of Boston or Dallas.

The ranch style, in particular, was evocative of the expansive mood of the post-World War II suburbs and of the disappearing regionality of style. It was almost as popular in Westchester County as in Los Angeles County. Remotely derived from the adobe dwellings of the Spanish colonial tradition and more directly derived from the famed prairie houses of Frank Lloyd Wright, with their low-pitched roofs, deep eaves, and pronounced horizontal lines, the typical ranch style houses of the 1950s were no larger than the average home a generation earlier. But the one-level ranch house suggested spacious living and an easy relationship with the outdoors. Mothers with small children did not have to contend with stairs. Most importantly, the postwar ranch home represented newness. In 1945 the publisher of the *Saturday Evening Post* reported that only 14 percent of the population wanted to live in an apartment or a ''used'' house. Whatever the style, the post-World War II house, in contrast to its turn-of-the-century predecessor, had no hall, no parlor, no stairs, and no porch. And the portion of the structure that projected farthest toward the street was the garage.[25]

The fourth characteristic of post-World War II housing was its easy availability and thus its reduced suggestion of wealth. To be sure, upper-income suburbs and developments sprouted across the land, and some set high standards of style and design. Typically, they offered expansive lots, spacious and individualized designs, and affluent neighbors. But the most important income development of the period was the lowering of the threshhold of purchase. At every previous time in American history, and indeed for the 1980s as well, the successful acquisition of a family home required savings and effort of a major order. After World War II, however, because of mass-production techniques, government financing, high wages, and low interest rates, it was quite simply cheaper to buy new housing in the suburbs than it was to reinvest in central city properties or to rent at the market price.[26]

The fifth and perhaps most important characteristic of the postwar suburb was economic and racial homogeneity. The sorting out of families by income and color began even before the Civil war and was stimulated by the growth of the factory system. This pattern was noticeable in both the exclusive Main Line suburbs of Philadelphia and New York and in the more bourgeois streetcar developments which were part of every city. The automobile accentuated this discriminatory "Jim Crow" pattern. In Atlanta where large numbers of whites flocked to the fast-growing and wealthy suburbs north of the city in the 1920s, Howard L. Preston has reported that: "By 1930, if racism could be measured in miles and minutes, blacks and whites were more segregated in the city of Atlanta than ever before." But many pre-1930 suburbs—places like Greenwich, Connecticut; Englewood, New Jersey; Evanston, Illinois; and Chestnut Hill, Massachusetts—maintained an exclusive image despite the presence of low-income or minority groups living in slums near or within the community.[27]

The post-1945 developments took place against a background of the decline of factory-dominated cities. What was unusual in the new circumstances was not the presence of discrimination—Jews and Catholics as well as blacks had been excluded from certain neighborhoods for generations—but the thoroughness of the physical separation which it entailed. The Levitt organization, which was no more culpable in this regard than any other urban or suburban firm, publically and officially refused to sell to blacks for two decades after the war. Nor did resellers deal with minorities. As William Levitt explained, "We can solve a housing problem, or we can try to solve a racial problem. But we cannot combine the two." Not surprisingly, in 1960 not a single one of the Long Island Levittown's 82,000 residents was black.[28]

The economic and age homogeneity of large subdivisions and sometimes entire suburbs was almost as complete as the racial distinction. Although this tendency had been present even in the nineteenth century, the introduction of zoning—beginning with a New York City ordinance in 1916—served the general purpose of preserving residential class segregation and property values. In theory zoning was designed to protect the interests of all citizens by limiting land speculation and congestion. And it was popular. Although it represented an extraordinary growth of municipal power, nearly everyone supported zoning. By 1926 seventy-six cities had adopted ordinances similar to that of New York. By 1936, 1,322 cities (85 percent of the total) had them, and zoning laws were affecting more property than all national laws relating to business.

In actuality zoning was a device to keep poor people and obnoxious industries out of affluent areas. And in time, it also became a cudgel used by suburban areas to whack the central city. Advocates of land-use restrictions in overwhelming proportion were residents of the fringe. They sought through minimum lot and set-back requirements to insure that only members of acceptable social classes could settle in their privileged sanctuaries. Southern cities even used zoning to enforce racial segregation. And in suburbs everywhere, North and South, zoning was used by the people who already lived within the arbitrary boundaries of a community as a method of keeping everyone else out. Apartments, factories, and ''blight,'' euphemisms for blacks and people of limited means, were rigidly excluded.

While zoning provided a way for suburban areas to become secure enclaves for the well-to-do, it forced the city to provide economic facilities for the whole area and homes for people the suburbs refused to admit. Simply put, land-use restrictions tended to protect residential interests in the suburbs and commercial interests in the cities because the residents of the core usually lived on land owned by absentee landlords who were more interested in financial returns than neighborhood preferences. For the man who owned land but did not live on it, the ideal situation was to have his parcel of earth zoned for commercial or industrial use. With more options, the property often gained in value. In Chicago, for example, three times as much land was zoned for commercial use as could ever have been profitably employed for such purposes. This overzoning prevented inner-city residents from receiving the same protection from commercial incursions as was afforded suburbanites. Instead of becoming a useful tool for the rational ordering of land in metropolitan areas, zoning became a way for suburbs to pirate from the city only its desirable functions and residents. Suburban governments became like so many residential hotels, fighting for the upper-income trade while trying to force the deadbeats to go elsewhere.

Because zoning restrictions typically excluded all apartments and houses and lots of less than a certain number of square feet, new home purchasers were often from a similar income and social group. In this regard, the postwar suburbs were no different from many nineteenth-century neighborhoods when they were first built. Moreover, Levittown was originally a mix of young professionals and lower-middle-class blue-collar workers.

As the aspiring professionals moved out, however, Levittowns became a community of the most class-stratifying sort possible.[29] This phenomenon was the subject of one of the most important books of the 1950s. Focusing on a 2,400-acre project put up by the former Public Housing Administrator Phillip Klutznick, William H. Whyte's *The Organization Man* sent shudders through armchair sociologists. Although Whyte found that Park Forest, Illinois, offered its residents "leadership training" and an "ability to chew on real problems," the basic portrait was unflattering. Reporting excessive conformity and a mindless conservatism, he showed Park Foresters to be almost interchangeable as they fought their way up the corporate ladder, and his "organization man" stereotype unfortunately became the norm for judging similar communities throughout the nation.

By 1961, when President John F. Kennedy proclaimed his New Frontier and challenged Americans to send a man to the moon within the decade, his countrymen had already remade the nation's metropolitan areas in the short space of sixteen years. From Boston to Los Angeles, vast new subdivisions and virtually new towns sprawled where a generation earlier nature had held sway. In an era of low inflation, plentiful energy, federal subsidies, and expansive optimism, Americans showed the way to a more abundant and more perfect lifestyle. Almost every contractor-built, post-World War II home had central heating, indoor plumbing, telephones, automatic stoves, refrigerators, and washing machines.

There was a darker side to the outward movement. By making it possible for young couples to have separate households of their own, abundance further weakened the extended family in America and ordained that most children would grow up in intimate contact only with their parents and siblings. The housing arrangements of the new prosperity were evident as early as 1950. In that year there were 45,983,000 dwelling units to accommodate the 38,310,000, families in the United States and 84 percent of American households reported less than one person per room.

Critics regarded the peripheral environment as devastating particularly to women and children. The suburban world was a female world, especially during the day. Betty Friedan's 1968 classic *The Feminine Mystique* challenged the notion that the American dream home was emotionally fulfilling for women. As Gwendolyn Wright has observed, their isolation from work opportunities and from contact with employed adults led to stifled frustration and deep psychological problems. Similarly, Sidonie M. Gruenberg warned in the *New York Times Magazine* that "Mass produced, standardized housing breeds standardized individuals, too—especially among youngsters." Offering neither the urbanity and sophistication of the city nor the tranquility and repose of the farm, the suburb came to be regarded less as an intelligent compromise than a cultural, economic, and emotional wasteland. No observer was more critical than Lewis Mumford, however. In his 1961 analysis of *The City in History,* which covered the entire sweep of civilization, the famed author reiterated sentiments he had first expressed more than four decades earlier and scorned the new developments which were surrounding every American city:

> In the mass movement into suburban areas a new kind of community was produced, which caricatured both the historic city and the archetypal suburban refuge: a multitude of uniform, unidentifiable houses, lined up inflexibly, at uniform distances, on uniform roads, in a treeless communal waste, inhabited by people of the same class, the same income, the same age group, witnessing the same television performances, eating the same tasteless pre-fabricated foods, from the same freezers, conforming in every outward and inward respect to a common mold, manufactured in the central metropolis. thus, the ultimate effect of the suburban escape in our own time is, ironically, a low-grade uniform environment from which escape is impossible.[30]

Secondly, because the federally supported home-building boom was of such enormous proportions, the new houses of the suburbs were a major cause of the decline of central cities. Because FHA and VA terms for new construction were so favorable as to make the suburbs accessible to almost all white, middle-income families, the inner-city housing market was deprived of the purchasers who could perhaps have supplied an appropriate demand for the evacuated neighborhoods.[31]

The young families who joyously moved into the new homes of the suburbs were not terribly concerned about the problems of the inner-city housing market or the snobbish views of Lewis Mumford and other social critics. They were concerned about their hopes and their dreams. There were looking for good schools, private space, and personal safety, and places like Levittown could provide those amenities on a scale and at a price that crowded city neighborhoods, both in the Old World and in the new, could not match. The single-family tract house—post-World War II style—whatever its aesthetic failings, offered growing families a private haven in a heartless world. If the dream did not include minorities or the elderly, if it was accompanied by the isolation of nuclear families, by the decline of public transportation, and by the deterioration of urban neighborhoods, the creation of good, inexpensive suburban housing on an unprecedented scale was a unique achievement in the world.

14

The Drive-in Culture of Contemporary America

> The human animal has two profound and conflicting impulses; he wants to be safe and warm, snug, enclosed, 'at home.' And he wants to roam the wide world, to see what is out there beyond the horizon. The automobile is a kind of house on wheels, but it will take you anywhere you want to go. You can conduct your sex life in it, you can eat and drink in it, go to the movies, listen to Vivaldi or the Stones, and you can dominate others, if you have more power and are adept with the gearshift lever. It is a whole existence. Or it is till the gas runs out.
>
> —McDonald Harris
> *New York Times,* May 16, 1979

The postwar years brought unprecedented prosperity to the United States, as color televisions, stereo systems, frost-free freezers, electric blenders, and automatic garbage disposals became basic equipment in the middle-class American home. But the best symbol of individual success and identity was a sleek, air-conditioned, high-powered, personal statement on wheels. Between 1950 and 1980, when the American population increased by 50 percent, the number of their automobiles increased by 200 percent. In high school the most important rite of passage came to be the earning of a driver's license and the freedom to press an accelerator to the floor. Educational administrators across the country had to make parking space for hundreds of student vehicles. A car became one's identity, and the important question was: "What does he drive?" Not only teenagers, but also millions of older persons literally defined themselves in terms of the number, cost, style, and horse-power of their vehicles. "Escape," thinks a character in a novel by Joyce Carol Oates. "As long as he had his own car he was an American and could not die."

Unfortunately, Americans did die, often behind the wheel. On September 9, 1899, as he was stepping off a streetcar at 74th Street and

Central Park West in New York, Henry H. Bliss was struck and killed by a motor vehicle, thus becoming the first fatality in the long war between flesh and steel. Thereafter, the carnage increased almost annually until Americans were sustaining about 50,000 traffic deaths and about 2 million nonfatal injuries per year. Automobility proved to be far more deadly than war for the United States. It was as if a Pearl Harbor attack took place on the highways every two weeks, with crashes becoming so commonplace that an entire industry sprang up to provide medical, legal, and insurance services for the victims.

The environmental cost was almost as high as the human toll. In 1984 the 159 million cars, trucks, and buses on the nation's roads were guzzling millions of barrels of oil every day, causing traffic jams that shattered nerves and clogged the cities they were supposed to open up and turning much of the countryside to pavement. Not surprisingly, when gasoline shortages created long lines at the pumps in 1974 and 1979, behavioral scientists noted that many people experienced anger, depression, frustration, and insecurity, as well as a formidable sense of loss.[1]

Such reactions were possible because the automobile and the suburb have combined to create a drive-in culture that is part of the daily experience of most Americans. Because of unemployment and war, per capita motor-vehicle ownership was stable (at about 30 million vehicles) between 1930 and 1948, and as late as 1950 (when registrations had jumped to 49 million) an astonishing 41 percent of all American families and a majority of working-class families still did not own a car. Postwar prosperity and rising real wages, however, made possible vastly higher market penetration, and by 1984 there were about seventy motor vehicles for every one hundred citizens, and more cars than either households or workers. Schaeffer and Sclar have argued that high auto ownership is the result of real economic needs rather than some "love affair" with private transportation. Moreover, the American people have proven to be no more prone to motor vehicle purchases than the citizens of other lands. After World War II, the Europeans and the Japanese began to catch up, and by 1980 both had achieved the same level of automobile ownership that the United States had reached in 1950. In automotive technology, American dominance slipped away in the postwar years as German, Swedish, and Japanese engineers pioneered the development of diesel engines, front-wheel drives, disc brakes, fuel-injection, and rotary engines.[2]

Although it is not accurate to speak of a uniquely American love affair with the automobile, and although John B. Rae claimed too much when he wrote in 1971 that "modern suburbia is a creature of the automobile and could not exist without it," the motor vehicle has fundamentally restructured the pattern of everyday life in the United States. As a young man, Lewis Mumford advised his countrymen to "forget the damned motor car and build cities for lovers and friends." As it was, of course, the nation followed a different pattern. Writing in the *American Builder* in 1929, the critic Willard Morgan noted that the building of drive-in structures to serve a motor-driven population had ushered in "a completely new architectural form."[3]

The Interstate Highway

The most popular exhibit at the New York World's Fair in 1939 was General Motors' "Futurama." Looking twenty-five years ahead, it offered a "magic Aladdin-like flight through time and space." Fair-goers stood in hour-long lines, waiting to travel on a moving sidewalk above a huge model created by designer Norman Bel Geddes. Miniature superhighways with 50,000 automated cars wove past model farms en route to model cities. Five million persons peered eventually at such novelties as elevated freeways, expressway traffic moving at 100 miles per hour, and "modern and efficient city planning—breath-taking architecture—each city block a complete unit in itself (with) broad, one-way thoroughfares—space, sunshine, light, and air." The message of "Futurama" was as impressive as its millions of model parts: "The job of building the future is one which will demand our best energies, our most fruitful imagination; and that with it will come greater opportunities for all."[4]

The promise of a national system of impressive roadways attracted a diverse group of lobbyists, including the Automobile Manufacturers Association, state-highway administrators, motor-bus operators, the American Trucking Association, and even the American Parking Association—for the more cars on the road, the more cars would be parked at the end of the journey. Truck companies, for example, promoted legislation to spend state gasoline taxes on highways, rather than on schools, hospitals, welfare, or public transit. In 1943 these groups came together as the American Road Builders Association, with General Motors as the largest contributor, to form a lobbying enterprise second only to that of the munitions industry. By the mid-1950s, it had become one of the most broad-based of all pressure groups, consisting of the oil, rubber, asphalt, and construction industries; the car dealers and renters; the trucking and bus concerns; the banks and advertising agencies that depended upon the companies involved; and the labor unions. On the local level, professional real-estate groups and home-builders associations joined the movement in the hope that highways would cause a spurt in housing turnover and a jump in prices. They envisaged no mere widening of existing roads, but the creation of an entirely new superhighway system and the initiation of the largest peacetime construction project in history.[5]

The highway lobby inaugurated a comprehensive public relations program in 1953 by sponsoring a national essay contest on the need for better roads. The winner of the $25,000 grand prize was Robert Moses, the greatest builder the world has yet known and a passionate advocate of the urban expressway. The title of his work was "How to Plan and Pay for Better Highways." As his biographer Robert A. Caro has noted, Moses was "the world's most vocal, effective and prestigious apologist for the automobile," and he did more than any other single urban official to encourage more hesitant officials to launch major road-building efforts in their cities.[6]

The Cold War provided an additional stimulus to the campaign for more elaborate expressways. In 1951 the *Bulletin of the Atomic Scientists* devoted an entire issue to "Defense through Decentralization." Their argument was simple. To avoid national destruction in a nuclear attack, the United States should disperse existing large cities into smaller settlements. The ideal model was a depopulated urban core surrounded by satellite cities and low-density suburbs.

Sensitive to mounting political pressure, President Dwight Eisenhower appointed a committee in 1954 to "study" the nation's highway requirements. Its conclusions were foregone, in part because the chairman was Lucius D. Clay, a member of the board of directors of General Motors. The committee considered no alternative to a massive highway system, and it suggested a major redirection of national policy to benefit the car and the truck. The Interstate Highway Act became law in 1956, when the Congress provided for a 41,000-mile (eventually expanded to a 42,500-mile) system, with the federal government paying 90 percent of the cost. President Eisenhower gave four reasons for signing the measure: current highways were unsafe; cars too often became snarled in traffic jams; poor roads saddled business with high costs for transportation; and modern highways were needed because "in case of atomic attack on our key cities, the road net must permit quick evacuation of target areas." Not a single word was said about the impact of highways on cities and suburbs, although the concrete thoroughfares and the thirty-five-ton tractor-trailers which used them encouraged the continued outward movement of industries toward the beltways and interchanges. Moreover, the interstate system helped continue the downward spiral of public transportation and virtually guaranteed that future urban growth would perpetuate a centerless sprawl. Soon after the bill was passed by the Senate, Lewis Mumford wrote sadly: ''When the American people, through their Congress, voted a little while ago for a $26 billion highway program, the most charitable thing to assume is that they hadn't the faintest notion of what they were doing.''

Once begun, the Interstate Highway System of the United States became a concrete colossus that grew bigger with every passing year. The secret of its success lay in the principle of non-divertibility of highway revenues collected from gasoline taxes. The Highway Trust Fund, as it was called, was to be held separately from general taxes. Although no less a personage than Winston Churchill called the idea of a non-divertible road fund ''nonsense,'' ''absurd,'' and ''an outrage upon . . . common sense,'' the trust fund had powerful friends in the United States, and it easily swept all opposition before it. Unlike European governments, Washington used taxes to support the highway infrastructure while refusing assistance to railroads. According to Senator Gaylord Nelson of Wisconsin, 75 percent of government expenditures for transportation in the United States in the postwar generation went for highways as opposed to 1 percent for urban mass transit.[7]

The inevitable result of the bias in American transport funding, a bias that existed for a generation before the Interstate Highway program was initiated, is that the United States now has the world's best road system and very nearly its worst public-transit offerings. Los Angeles, in particular, provides the nation's most dramatic example of urban sprawl tailored to the mobility of the automobile. Its vast, amorphous conglomeration of housing tracts, shpping centers, industrial parks, freeways, and independent towns blend into each other in a seamless fabric of concrete and asphalt, and nothing over the years has succeeded in gluing this automobile-oriented civilization into any kind of cohesion—save that of individual routine. Los Angles's basic shape comes from three factors, all of which long preceded the freeway system. The first was cheap land (in the 1920s rather than 1970s) and the desire for single-family houses. In 1950, for example, nearly two-thirds of all the dwelling units in the Los Angeles area were fully detached, a much higher percentage than in Chicago (28 percent), New York City (20 percent), or Philadelphia (15 percent), and its residential density was the lowest of major cities. The second was the dispersed location of its oil fields and refineries, which led to the creation of industrial suburbs like Whittier and Fullerton and of residential suburbs like La Habra, which housed oil workers and their families. The third was its once excellent mass-transit system, which at its peak included more than1,100 miles of track and constituted the largest electric interurban railway in the world.[8]

The Pacific Electric Company collapsed in the 1920s, however, and

The close connection between road building and suburban growth is apparent in this 1922 photograph of Westwood outside of Los Angeles. The sign in the foreground announces the construction of a new 50-foot boulevard while the sign in the background proclaims the availability of residential sites. *Courtesy Los Angeles Public Library.*

One of the most striking characteristics of American cities is their broad streets and avenues, especially in comparison with the rather narrow thoroughfares which exist elsewhere in the world. This photograph of 24th Street in Brooklyn on March 24, 1925, shows the ample roadways that were typical in even the most populous of America's great cities. The two lonely automobiles in this picture make it clear that the width of the street was not necessitated by heavy vehicular traffic, but rather by the ideal of spaciousness itself. *Courtesy of the Municipal Archives, Department or Records and Information Services, City of New York.*

The colonial-style American house, which traced its origins to the pre-Revolutionary period; became immensely popular after World War II, especially in the East, Middle West, and South. Its appeal derived partly from its spaciousness, partly from its suggestion of affluence, and partly from its symbolic connection to an earlier period. This photograph was taken on November 26, 1937, in the Cannon Hill Development of Huntington, Long Island, and it depicts a style that would dominate many suburban areas in the post-war period. *Courtesy Samuel H. Gottscho Collection, Avery Architectural Library, Columbia University.*

(Opposite) During the years between the two World Wars, the Country Club District in Kansas City was the nation's most famous example of planned residential development. This view of Pembroke Lane just south of 56th Street was in the Mission Hills portion of the affluent community. *Courtesy of J. C. Nichols Company.*

Because it could carry many times the load of a horse-drawn wagon, the truck completely restructured the shape of the American metropolis by greatly stimulating the deconcentration of industry. This 1920 photograph shows how the D. L. Knight Moving Company of Louisville, Kentucky incorporated the new horseless technology into a business that required yearround operation. *Courtesy of the Caufield and Shook Collection, University of Louisville Photographic Archives.*

The first Holiday Inn opened on Summer Avenue in Memphis and featured rooms that were clean, affordable, and respectable. This 1952 view of the chain's first "hotel court" shows the automobile orientation of the company. The large signs were intended to be easily visible from the highway and the rooms themselves were designed for easy access from the family car. *Courtesy of Kemmons Wilson.*

(Opposite) The garage has gone through many incarnations in its 75-year history, but only in California does the garage itself dominate the front facade of the typical home. This 1984 view of a house in Irvine shows how the driveway dominates the small front yard and how the two-car garage occupies two-thirds of the width of the entire residence.

The American service station began as an appendage to a general store or a livery stable and only gradually evolved into an all-purpose emporium where one could purchase gasoline, arrange for automotive repairs, and buy groceries. This New York City station at First Avenue and 124th Street in East Harlem is an example of the first generation of free-standing service stations, replete with a repair area and off-street pumps. The date is July 10, 1934. *Courtesy of the Municipal Archives, Department of Records and Information Services, City of New York.*

Country Club Plaza in Kansas City was one of the first planned shopping centers in the world. Designed by J. C. Nichols in the 1920s, it offered a mix of retailing and office functions. These photographs were taken in the earliest years of the plaza, but it continues to thrive in 1985. *Courtesy of Bill Mott photography and the J. C. Nichols Company.*

The Poplar Plaza Shopping Center in Memphis, Tennessee, which opened in 1949, was one of the first off-street retailing complexes to be built after World War II. This 1952 aerial photograph reveals an important difference between the first generation of American shopping centers and those which were built after the mid 1950s. Notice, for example, that the stores are set back less than one hundred feet from the street, even though there was ample room for expansion at the top of the picture. This type of shopping center was rare after 1960, when the typical pattern was to center the stores in the midst of a vast parking lot. In fact, the Poplar Plaza Shopping Center itself later expanded toward the top left of the picture, and the number of parking spaces was expanded by several times. *Courtesy T. David Goodwin.*

(Opposite) Few cities anywhere could match Chicago's combination of subway, elevated, bus, and commuter railroad lines when this photograph was taken of Grant Park in the early 1950s. But the automobile presence was obviously important as central business district workers abandoned public transit for the convenience and privacy of the motorcar. *Courtesy of the National Archives.*

The main streets of most American communities, suburban and otherwise, are no longer along the familiar sidewalks of downtown areas. Instead, they are within the artificial environments of the indoor malls, which dot the American landscape in the 1980s and which are the new hangouts of the adolescent generation. This photograph of the interior of the Paramus Mall in New Jersey shows a typical sequence of retail shops.

The mobile home park concept was pioneered in the 1920s and 1930s, but the largest experiment with manufactured or prefabricated houses came in World War II. This 1945 aerial photograph was taken of a part of Oak Ridge, Tennessee, a community that had not even existed in 1940. In the next few years, as atomic workers poured into the area, more than 5,000 trailers supplemented 9,600 prefabricated houses and 16,000 barracks to provide temporary dwellings in this top-secret facility. *Courtesy of the Atomic Energy Commission.*

This 1984 photograph of the Crystal Cathedral in Garden Grove, California, depicts the extraordinary success of the Reverend Robert Schuller's drive-in church concept. Although most worshippers can be accommodated within the glass-walled structure itself, many people continue to prefer the privacy of their personal automobiles. During the Reverend Schuller's sermons, a great glass door swings open so that the drive-in communicants can have a direct view of their pastor.

(Opposite) The house on wheels has come a long way in appearance and in acceptability since trailer homes first developed in the years after World War I. Perhaps the most significant single advance has been an innovation called a "double wide," which makes it possible to link two—or sometimes even three—mobile homes to create a structure that in its exterior dimensions resembles an ordinary "stick-built" house. This photograph of a double wide in Indiana in 1981 illustrates the desire of the "manufactured housing" industry to have its products accepted as normal, stationary, and permanent dwellings. *Courtesy of Camilo J. Vergara.*

McDonald's
HAMBURGERS
OVER 19 BILLION SERVED
McDonald

The General Foods Corporation, which moved from Manhattan to White Plains, New York in 1954, was one of the first large American companies to move from a central city location to a suburban office park. This aerial view of the General Foods headquarters complex illustrates the basic ingredients of corporate flight—ample parking, abundant open space, and easily accessible highways. *Courtesy of the General Foods Corporation.*

(Opposite) No single symbol of America's drive-in culture is more ubiquitous than the golden arches of McDonald's hamburger stands. The formula of standardized menus, low prices, dependable quality, fast service, and easy road access was ultimately copied by dozens of imitators, but none were as successful or as widely recognized as the trademarks of the enormous chain.

The suburban corporate office park is already into its second generation, as this view of General Foods' new headquarters building attests. Opened in 1982 only a few miles from its 1954 complex (which General Foods continues to occupy), this architecturally distinguished edifice takes the bucolic image one step further than its predecessor, with a large reflecting lake along one side of the main building. *Courtesy of the General Foods Corporation.*

since that time Los Angeles has been more dependent upon the private automobile than other large American cities. Beginning in 1942, the Los Angeles Chamber of Commerce, the automobile club, and elected officials met regularly to plan for a region-wide expressway network. They succeeded, and southern California's fabled 715 miles of freeways now constitute a grid that channels virtually all traffic and sets many communal boundaries. They are the primary form of transportation for most residents, who seem to regard time spent in their cars as more pleasurable than time walking to, waiting for, or riding on the bus. More than a third of the Los Angeles area is consumed by highways, parking lots, and interchanges, and in the downtown section this proportion rises to two-thirds. Not surprisingly, efforts to restore the region's public transportation to excellence have thus far failed. In 1976, for example, the state of California attempted to discourage single-passenger automobiles by reserving one lane in each direction on the Santa Monica Freeway for express buses and car pools. An emotional explosion ensued that dominated radio talk shows and television news, and Los Angeles' so-called "diamond lanes" were soon abolished.[9]

More recently, southern California has followed the growing national enthusiasm for rail transit, and Los Angeles broke ground in 1984 for an 18-mile, $3.3 billion subway that will cut underneath the densely built, heavily trafficked Wilshire Boulevard corridor, cut through Hollywood, and end up in the residential San Fernando Valley. The underground will hopefully be the centerpiece of an eventual 160-mile network, second in size in the United States only to New York City's.

The Garage

The drive-in structure that is closest to the hearts, bodies, and cars of the American family is the garage. It is the link between the home and the outside world. The word is French, meaning storage space, but its transformation into a multi-purpose enclosure internally integrated with the dwelling is distinctively American.

In the streetcar era, curbs had been unbroken and driveways were almost unknown. A family wealthy enough to have a horse and carriage would have stored such possessions either in a public livery stable or in a private structure at the rear of the property. The owners of the first automobiles were usually sufficiently affluent to maintain a private stable. The first cars, therefore, which were open to the elements, often found lodging in a corner of the stable, side by side with the carriages they were soon to replace. These early accommodations for the automobile were often provided with gasoline tanks, for filling stations at the time were few and far between. This and the fact that cars often caught fire were good and sufficient reasons to keep the motor vehicles away from the family.[10]

After World War I, house plans of the expensive variety began to include garages, and by the mid-1920s driveways were commonplace and garages had become important selling points. The popular 1928 *Home Builders* pattern book offered designs for fifty garages in wood, Tudor, and brick varieties. In affluent sections, such large and efficiently planned structures included housing above for the family chauffeur. In less pretentious neighborhoods, the small, single-purpose garages were scarcely larger than the vehicles themselves, and they were simply portable and prefabricated structures, similar to those in Quebec today, that were camouflaged with greenery and trellises. As one architect complained in 1924: "The majority of owners are really ashamed of their garages and really endeavor to keep them from view," and he implored his readers to build a garage "that may be worthy of standing alongside your house." Although there was a tendency to move garages closer to the house, they typically remained at the rear of the property before 1925, often with access via an alley which ran parallel to the street. The car was still thought of as something similar to a horse—dependable and important, but not something that one needed to be close to in the evening.[11]

By 1935, however, the garage was beginning to merge into the house itself, and in 1937 the *Architectural Record* noted that "the garage has become a very essential part of the residence." The tendency accelerated after World War II, as alleys went the way of the horse-drawn wagon, as property widths more often exceeded fifty feet, and as the car became not only a status symbol, but almost a member of the family, to be cared for and sheltered. The introduction of a canopied and unenclosed structure called a "car port" represented an inexpensive solution to the problem, particularly in mild climates, but in the 1950s the enclosed garage was back in favor and a necessity even in a tract house. Easy access to the automobile became a key aspect of residential design, and not only for the well-to-do. By the 1960s garages often occupied about 400 square feet (about one-third that of the house itself) and usually contained space for two automobiles and a variety of lawn and woodworking tools. Offering direct access to the house (a conveniently placed door usually led directly into the kitchen), the garage had become an integrated part of the dwelling, and it dominated the front facades of new houses. In California garages and driveways were often so prominent that the house could almost be described as accessory to the garage. Few people, however, went to the extremes common in England, where the automobile was often so precious that living rooms were often converted to garages.[12]

The Motel

As the United States became a rubber-tire civilization, a new kind of roadside architecture was created to convey an instantly recognizable image to the fast-moving traveler. Criticized as tasteless, cheap, forgettable, and flimsy by most commentators, drive-in structures did attract the attention of some talent architects, most notably Los Angeles's Richard Neutra. For him, the automobile symbolized modernity, and its design paralleled his own ideals of precision and efficiency. This correlation between the structure and the car began to be celebrated in the late 1960s and 1970s when architects Robert Venturi, Denise Scott Brown, and Steven Izenour developed such concepts as "architecture as symbol" and the "architecture of communication." Their book, *Learning From Las Vegas,* was instrumental in encouraging a shift in taste from general condemnation to appreciation of the commercial strip and especially of the huge and garish signs which were easily recognized by passing motorists.[13]

A ubiquitous example of the drive-in culture is the motel. In the middle of the nineteenth century, every city, every county seat, every aspiring mining town, every wide place in the road with aspirations to larger size, had to have a hotel. Whether such structures were grand palaces on the order of Boston's Tremont House or New York's Fifth Avenue Hotel, or whether they were jerry-built shacks, they were typically located at the center of the business district, at the focal point of community activities. To a considerable extent, the hotel was the place for informal social interaction and business, and the very heart and soul of the city.[14]

Between 1910 and 1920, however, increasing numbers of traveling motorists created a market for overnight accommodation along the highways. The first tourists simply camped wherever they chose along the road. By 1924, several thousand municipal campgrounds were opened which offered cold water spigots and outdoor privies. Next came the "cabin camps," which consisted of tiny, white clapboard cottages arranged in a semicircle and often set in a grove of trees. Initially called "tourist courts," these establishments were cheap, convenient, and informal, and by 1926 there were an estimated two thousand of them, mostly in the West and in Florida.

Soon after clean linens and comfortable rooms became available along the nation's highways, it became apparent that overnight travelers were not the only, or even the largest, pool of customers. Convenience and privacy were especially appealing to couples seeking a romantic retreat. A well-publicized Southern Methodist University study in 1935 reported that 75 percent of Dallas area motel business consisted of one man and one woman remaining for only a short stay. Whatever the motivation of patrons, the success of the new-style hotels prompted Sinclair Lewis to predict in 1920:

> Somewhere in these states there is a young man who is going to become rich. He is going to start a chain of small, clean, pleasant hotels, standardized and nationally advertised, along every important motor route in the country. He is not going to waste money on glit and onyx, but he is going to have agreeable clerks, good coffee, endurable mattresses and good lighting.[15]

It was not until 1952 that Kemmons Wilson and Wallace E. Johnson opened their first "Holiday Inn" on Summer Avenue in Memphis. But long before that, in 1926, a San Luis Obispo, California, proprietor had coined a new word, "motel," to describe an establishment that allowed a guest to park his car just outside his room. New terminology did not immediately erase the unsavory image of the roadside establishments, however. In 1940 FBI Director J. Edgar Hoover declared that most motels were assignation camps and hideouts for criminals. Perhaps he was thinking of Bonnie and Clyde, who had a brief encounter with the law at the Red Crown Cabin Camp near Platte City, Missouri, one evening in July of 1933. Many of Hoover's "dens of vice" were once decent places that, unable to keep up, turned to the "hot pillow trade." Some Texas cabins, said the FBI director, were rented as many as sixteen times a night, while establishments elsewhere did business by the hour, with "a knock on the door when the hour was up."[16]

Motels began to thrive after World War II, when the typical establishment was larger and more expensive than the earlier cabins. Major chains set standards for prices, services, and respectability that the traveling public could depend on. As early as 1948, there were 26,000 self-styled motels in the United States. Hard-won respectability attracted more middle-class families, and by 1960 there were 60,000 such places, a figure that doubled again by 1972. By that time an old hotel was closing somewhere in downtown America every thirty hours. And somewhere in suburban America, a plastic and glass Shangri La was rising to take its place.[17]

Typical of the inner-city hotels was the Heritage in Detroit. The big bands once played on its roof, and aspiring socialites enjoyed crepe-thin pancakes. In 1975 a disillusioned former employee gestured futilely, "It's dying; the whole place is dying," as the famed hotel closed its doors. By 1984 about fifty historic establishments in downtown areas, such as the Peabody in Memphis, the Mayflower in Washington, the Galvez in Houston, the Menger in San Antonio, and the Biltmore in Providence were reopening with antique-filled rooms and oak-paneled bars. But the trend remained with the standard, two-story motel.[18]

The Drive-in Theater

The downtown movie theaters and old vaudeville houses faced a similar challenge from the automobile. In 1933 Richard M. Hollinshead set up a 16-mm projector in front of his garage in Riverton, New Jersey, and then settled down to watch a movie. Recognizing a nation addicted to the motorcar when he saw one, Hollinshead and Willis Smith opened the world's first drive-in movie in a forty-car parking lot in Camden on June 6, 1933. Hollinshead profited only slightly from his brainchild, however, because in 1938 the United States Supreme Court refused to hear his appeal against Loew's Theaters, thus accepting the argument that the drive-in movie was not a patentable item. The idea never caught on in Europe, but by 1958 more than four thousand outdoor screens dotted the American landscape. Because drive-ins offered bargain-basement prices and double or triple bills, the theaters tended to favor movies that were either second-run or second-rate. Horror films and teenage romance were the order of the night, as *Beach Blanket Bingo* or *Invasion of the Body Snatchers* typified the offerings. Pundits often commented that there was a better show in the cars than on the screen.[19]

In the 1960s and 1970s the drive-in movie began to slip in popularity. Rising fuel costs and a season that lasted only six months contributed to the problem, but skyrocketing land values were the main factor. When drive-ins were originally opened, they were typically out in the hinterlands. When subdivisions and shopping malls came closer, the drive-ins could not match the potential returns from other forms of investments. According to the National Association of Theater Owners, only 2,935 open-air theaters still operated in the United States in 1983, even though the total number of commercial movie screens in the nation, 18,772, was at a 35-year high. The increase was picked up not by the downtown and the neighborhood theaters, but by new multi-screen cinemas in shopping centers. Realizing that the large parking lots of indoor malls were relatively empty in the evening, shopping center moguls came to regard theaters as an important part of a successful retailing mix.[20]

The Gasoline Service Station

The purchase of gasoline in the United States has thus far passed through five distinct epochs. The first stage was clearly the worst for the motorist, who had to buy fuel by the bucketful at a livery stable, repair shop, or dry goods store. Occasionally, vendors sold gasoline from small tank cars which they pushed up and down the streets. In any event, the automobile owner had to pour gasoline from a bucket through a funnel into his tank. The entire procedure was inefficient, smelly, wasteful, and occasionally dangerous.[21]

The second stage began about 1905, when C. H. Laessig of St. Louis equipped a hot-water heater with a glass gauge and a garden hose and turned the whole thing on its end. With this simple maneuver, he invented an easy way to transfer gasoline from a storage tank to an automobile without using a bucket. Later in the same year, Sylvanus F. Bowser invented a gasoline pump which automatically measured the outflow. The entire assembly was labeled a "filling station." At this stage, which lasted until about 1920, such an apparatus consisted of a single pump outside a retail store which was primarily engaged in other businesses and which provided precious few services for the motorist. Many were located on the edge of town for safety and to be near the bulk stations; those few stations in the heart of the city did not even afford the luxury of off-street parking.

Between 1920 and 1950, service stations entered into a third phase and became, as a group, one of the most widespread kinds of commercial buildings in the United States. Providing under one roof all the functions of gasoline distribution and normal automotive maintenance, these full-service structures were often built in the form of little colonial houses, Greek temples, Chinese pagodas, and Art Deco palaces. Many were local landmarks and a source of community pride. One cartoonist in the 1920s mocked such structures with a drawing in which a newcomer to town confused the gas station with the state capitol. Grandiose at the time, many of them molder today—deserted, forlorn structures with weeds growing in the concrete where gasoline pumps once stood. Their bays stand empty and silent, rendered that way by changing economics, changing styles, and changing consumer preferences.

After 1935 the gasoline station evolved again, this time into a more homogeneous entity that was standardized across the entire country and that reflected the mass-marketing techniques of billion-dollar oil companies. Some of the more familiar designs were innovative or memorable, such as the drumlike Mobil station by New York architect Frederick Frost, which featured a dramatically curving facade while conveying the corporate identity. Another popular service station style was the Texaco design of Walter Dorwin Teague—a smooth white exterior with elegant trim and the familiar red star and bold red lettering. Whatever the product or design, the stations tended to be operated by a single entrepreneur and represented an important part of small business in American life.

The fifth stage of gasoline-station development began in the 1970s, with the slow demise of the traditional service-station businessman. New gasoline outlets were of two types. The first was the super station, often owned and operated by the oil companies themselves. Most featured a combination of self-service and full-service pumping consoles, as well as fully equipped "car care centers." Service areas were separated from the pumping sections so that the two functions would not interfere with each other. Mechanics never broke off work to sell gas.

The more pervasive second type might be termed the "mini-mart station." The operators of such establishments have now gone full circle since the early twentieth century. Typically, they know nothing about automobiles and expect the customers themselves to pump the gasoline. Thus, "the man who wears the star" has given way to the teenager who sells six-packs, bags of ice, and pre-prepared sandwiches.[22]

The Shopping Center

Large-scale retailing, long associated with central business districts, began moving away from the urban cores between the world wars. The first experiments to capture the growing suburban retail markets were made by major department stores in New York and Chicago in the 1920s, with Robert E. Wood, Sears's vice president in charge of factories and retail stores, as the leader of the movement. A student of population trends, Wood decided in 1925 that motor-vehicle registrations had outstripped the parking space available in metropolitan cores, and he insisted that Sears's new "A" stores (their other retail outlets were much smaller) be located in low-density areas which would offer the advantages of lower rentals and yet, because of the automobile, be within reach of potential customers. With the exception of Sears's flagship store on State Street in Chicago (which was itself closed in 1983), Woods's dictum of ample free parking was rigorously followed throughout the United States. Early examples of the formula were the Pico Boulevard store in

Los Angeles and the Crosstown store in Memphis. A revolution in retailing followed. Writing in the *American Builder* in 1929, the critic Willard Morgan found it natural that traffic congestion at the center would drive thousands of prospective customers to turn instead to suburban marketing centers.[23]

Another threat to the primacy of the central business district was the "string street" or "shopping strip," which emerged in the 1920s and which were designed to serve vehicular rather than pedestrian traffic. These bypass roads encouraged city dwellers with cars to patronize businesses on the outskirts of town. Short parades of shops could already have been found near the streetcar and rapid transit stops, but, as has been noted, these new retailing thoroughfares generally radiated out from the city business district toward low-density, residential areas, functionally dominating the urban street system. They were the prototypes for the familiar highway strips of the 1980s which stretch far into the countryside.[24]

Sears's big stores were initially isolated from other stores, while the retail establishments of highway strips were rarely unified into a coordinated whole. The multiple-store shopping center with free, off-street parking represented the ultimate retail adaptation to the requirements of automobility. Although the *Guinness Book of World Records* lists the Roland Park Shopping Center (1896) as the world's first shopping center, the first of the modern variety was Country Club Plaza in Kansas City. It was the effort of a single entrepreneur, Jesse Clyde Nichols, who put together a concentration of retail stores, and used leasing policy to determine the composition of stores in the concentration. By doing that, Nichols created the idea of the planned regional shopping center.

Begun in 1923 in a Spanish-Moorish style with red tile roofs and little towers—its Giralda Tower is actually a replica of the original in Seville—Country Club Plaza featured waterfalls, fountains, flowers, tree-lined walks, and expensive landscaping. As the first automobile-oriented shopping center, it offered extensive parking lots behind ornamented brick walls. Most buildings were two stories high, with the second-floor offices typically occupied by physicians, dentists, and attorneys, whose presence would help stimulate a constant flow of well-heeled visitors. An enormous commercial success, Country Club Plaza stood in organic harmony with the prairie surroundings, and it soon became the hub of Kansas City's business and cultural activities.[25]

Nichols's Country Club Plaza generated considerable favorable publicity after it became fully operational in 1925, and by the mid-1930s the concept of the planned shopping center, as a concentration of a number of businesses under one management and with convenient parking facilities, was well known and was recognized as the best method of serving the growing market of drive-in customers. But the Great Depression and World War II had a chilling effect on private construction, and as late as 1946 there were only eight shopping centers in the entire United States. They included Upper Darby Center in West Philadelphia (1927); Suburban Square in Ardmore, Pennsylvania (1928); Highland Park Shopping Village outside Dallas (1931); River Oaks in Houston (1937); Hampton Village in St. Louis (1941); Colony in Toledo (1944); Shirlington in Arlington, Virginia (1944); and Belleview Square in Seattle (1946). Importantly, however, they provided many of the amenities that shoppers would take for granted half a century later. In 1931, for example, Highland Park Village outside Dallas offered department, drug, and food stores, as well as banks, a theater, beauty and barber shops, offices, studios, and parking for seven hundred cars. The Spanish architecture was uniform throughout, and the rental charge include a maintenance fee to insure that the property was adequately cared for during the term of the lease.[26]

The first major planned retail shopping center in the world went up in Raleigh, North Carolina in 1949, the brainchild of Homer Hoyt, a well-known author and demographer best known for his sector model of urban growth. Thereafter, the shopping-center idea caught on rapidly in the United States and less rapidly in Canada, where the first shopping center—Dixie Plaza near Toronto—did not open until 1954. The most successful early examples, such as Poplar Plaza in Memphis, offered at least thirty retail small retailers, one large department store, and parking for five hundred or more cars. By 1984 the nation's 20,000 large shopping centers accounted for almost two-thirds of all retail trade, and even in relatively centralized cities like New York, Boston, and San Francisco downtown merchants adapted to the suburban shift. Easy facilities for parking gave such collections of stores decisive advantages over central city establishments.[27]

The concept of the enclosed, climate-controlled mall, first introduced at the Southdale Shopping Center near Minneapolis in 1956, added to the suburban advantage. A few of the indoor malls, such as the mammoth Midtown Plaza in Rochester, New York, were located downtown, but more typical were Paramus Park and Bergen Mall in New Jersey; Woodfield Mall in Schaumburg outside Chicago; King's Plaza and Cross County outside Gotham; and Raleigh Mall in Memphis—all of which were located on outlying highways and all of which attracted shoppers from trading areas of a hundred square miles and more. Edward J. Bartolo, Sr., a self-made millionaire and workaholic, operated from a base in Youngstown, Ohio, to become the most prominent mall developer in the United States, but large insurance companies, especially the Equitable Life Assurance Society, increasingly sought high yields as shopping-center landlords.

During the 1970s, a new phenomenon—the super regional mall—added a more elaborate twist to suburban shopping. Prototypical of the new breed was Tyson's Corner, on the Washington Beltway in Fairfax County, Virginia. Anchored by Bloomingdale's, it did over $165 million in business in 1983 and provided employment to more than 14,000 persons. Even larger was Long Island's Roosevelt Field, a 180-store, 2.2 million square foot mega-mall that attracted 275,000 visitors a week and did $230 million in business in 1980. Most elaborate of all was Houston's Galleria, a world-famed setting for 240 prestigious boutiques, a quartet of cinemas, 26 restaurants, an olympic-sized ice-skating pavilion, and two luxury hotels. There were few windows in these mausoleums of merchandising, and clocks were rarely seen—just as in gambling casinos.[28]

Boosters of such mega-malls argue that they are taking the place of the old central business districts and becoming the identifiable collecting points for the rootless families of the newer areas. As weekend and afternoon attractions, they have a special lure for teenagers, who often go there on shopping dates or to see the opposite sex. As one official noted in 1971: "These malls are now their street corners. The new shopping centers have killed the little merchant, closed most movies, and are now supplanting the older shopping centers in the suburbs." They are also especially attractive to mothers with young children and to the elderly, many of whom visit regularly to get out of the house without having to worry about crime or inclement weather.[29]

In reality, even the largest malls are almost the opposite of downtown areas because they are self-contained and because they impose a uniformity of tastes and interests. They cater exclusively to middle-class tastes and contain no unsavory bars or pornography shops, no threatening-looking characters, no litter, no rain, and no excessive heat or cold. As Anthony Zube-Jackson has noted, their emphasis on cleanliness and safety is symptomatic of a very lopsided view of urban culture.

Despite their blandness, the shopping malls and the drive-in culture of which they are a part have clearly eclipsed the traditional central business districts, and in many medium-sized cities the last of the downtown department stores has already closed. The drive-in blight that killed them, like the Dutch Elm disease that ravaged Eastern towns in years past, has played hopscotch from one town to another, bringing down institutions that had once appeared invincible. The targets of this scourge, however, were not trees, but businesses, specifically the once-mighty department stores that anchored many a Main Street.

The most famous retailing victim of the drive-in culture thus far has been the stately J. L. Hudson Company of Detroit. It was a simple fact that all roads in the Motor City led to Hudson's. Featuring tall chandeliers, wood-paneled corridors, and brass-buttoned doormen, the 25-story, full-square-block emporium at its height ranked with Macy's in New York and Marshall Field in Chicago as one of the country's three largest stores. After 1950, however, the once-proud store was choked by its own branches, all of them in outlying shopping centers. As soon as Hudson's opened Northland, its biggest suburban outlet and one of the earliest in the nation, sales downtown began to fall. They declined from a peak in 1953 of $153 million to $45 million in 1981. Finally, in 1981, the downtown landmark closed its doors for good.[30] Hudson's was a victim of the product that made Detroit: the car.

In a Christmastime obituary for Detroit's most famous retailer, a WWJ radio commentator maintained that white flight to the suburbs, hastened by the Motor City's 1967 race riot, helped deal Hudson's a mortal blow. Actually, the 91-year-old store was killed by the free parking, easy accessibility, and controlled environment of the mega-malls.

By the 1960s, the primary rival to the shopping center as the locus of brief, informal communication and interaction had become the highway strip, with its flashing neon signs and tacky automobile showrooms. Especially in medium-sized cities, the vitality after dark is concentrated in the shopping malls or along the highway, not along Main Street.

The House Trailer and Mobile Home

The phenomenon of a nation on wheels is perhaps best symbolized by the uniquely American development of the mobile home. "Trailers are here to stay," predicted the writer Howard O'Brien in 1936. Although in its infancy at that time, the mobile-home industry has flourished in the United States. The house trailer itself came into existence in the teens of this century as an individually designed variation on a truck or a car, and it began to be produced commercially in the 1920s. Originally, trailers were designed to travel, and they were used primarily for vacation purposes. During the Great Depression of the 1930s, however, many people, especially salesmen, entertainers, construction workers, and farm laborers, were forced into a nomadic way of life as they searched for work, any work. They found that these temporary trailers on rubber tires provided the necessary shelter while also meeting their economic and migratory requirements. Meanwhile, Wally Byam and other designers were streamlining the mobile home into the classic tear-drop form made famous by Airstream.[31]

During World War II, the United States government got into the act by purchasing tens of thousands of trailers for war workers and by forbidding their sale to the general public. By 1943 the National Housing Agency alone owned 35,000 of the aluminum boxes, and more than 60 percent of the nation's 200,000 mobile homes were in defense areas. The government also built prefabricated homes without wheels near weapons factories. The ticky-tacky quality of these prefabricated shanty towns gave prefabs a lingering bad image, which remained after the war, when trailers found a growing market among migratory farm workers and military personnel, both of whom had to move frequently.

Not until the mid-1950s did the term "mobile home" begin to refer to a place where respectable people could marry, mature, and die. By then it was less a "mobile" than a "manufactured" home. No longer a trailer, it became a modern industrialized residence with almost all the accoutrements of a normal house. By the late 1950s, widths were increased to ten feet, the Federal Housing Administration (FHA) began to recognize the mobile home as a type of housing suitable for mortgage insurance, and the maturities on sales contracts were increased from three to five years.

In the 1960s, twelve-foot widths were introduced, and then fourteen, and manufacturers began to add fireplaces, skylights, and cathedral ceilings. In 1967 two trailers were attached side by side to form the first "double wide." These new dimensions allowed for a greater variety of room arrangement and became particularly attractive to retired persons with fixed incomes. They also made the homes less mobile. By 1979 even the single-width "trailer" could be seventeen feet wide (by about sixty feet long), and according to the Manufactured Housing Institute, fewer than 2 percent were ever being moved from their original site. Partly as a result of this increasing permanence, individual communities and the courts began to define the structures as real property and thus subject to real-estate taxes rather than as motor vehicles subject only to license fees.[32]

Although it continued to be popularly perceived as a shabby substitute for "stick" housing (a derogatory word used to describe the ordinary American balloon-frame dwelling), the residence on wheels reflected American values and industrial practices. Built with easily machined and processed materials, such as sheet metal and plastic, it represented a total consumer package, complete with interior furnishings, carpets, and appliances. More importantly, it provided a suburban type alternative to the inner-city housing that would otherwise have been available to blue-collar workers, newly married couples, and retired persons. After 1965 the production of factory-made housing (the term preferred by the industry) rarely fell below 200,000 per year, and in Florida, Wyoming, and Montana they typically accounted for more than a quarter of all new housing units. By 1979 manufactured housing was a $3.1 billion industry, and the nation counted more than ten million mobile-home dwellers. These figures exclude the "motor homes" made popular by Winnebago in the 1970s, the modular homes that are built on a floor system like a conventional house, and the prefabricated houses for which parts are built in a factory and shipped in sections to be assembled on the site.[33]

A Drive-in Society

Drive-in motels, drive-in movies, and drive-in shopping facilities were only a few of the many new institutions that followed in the exhaust of the internal-combustion engine. By 1984 mom-and-pop grocery stores had given way almost everywhere to supermarkets, most banks had drive-in windows, and a few funeral homes were making it possible for mourners to view the deceased, sign the register, and pay their respects without emerging from their cars. Odessa Community College in Texas even opened a drive-through registration window.

Particularly pervasive were fast-food franchises, which not only decimated the family-style restaurants but cut deeply into grocery store sales. In 1915 James G. Huneker, a raconteur whose tales of early twentieth-century American life were compiled as *New Cosmopolis,* complained of the infusion of cheap, quick-fire "food hells," and of the replacement of relaxed dining with "canned music and automatic lunch taverns." With the automobile came the notion of "grabbing" something to eat. The first drive-in restaurant, Royce Hailey's Pig Stand, opened in Dallas in 1921, and later in the decade, the first fast-food franchise, "White Tower," decided that families touring in motorcars needed convenient meals along the way. The places had to look clean, so they were painted white. They had to be familiar, so a minimal menu was standardized at every outlet. To catch the eye, they were built like little castles, replete with fake ramparts and turrets. And to forestall any problem with a land lease, the little white castles were built to be moveable.

The biggest restaurant operation of all began in 1954, when Ray A. Kroc, a Chicago area milkshake-machine salesman, joined forces with Richard and Maurice McDonald, the owners of a fast-food emporium in San Bernardino, California. In 1955 the first of Mr. Kroc's "McDonald's" outlets was opened in Des Plaines, a Chicago suburb long famous as the site of an annual Methodist encampment. The second and third, both in

California, opened later in 1955. Within five years, there were 228 golden arches drive-ins selling hamburgers for 15 cents, french fries for 10 cents, and milkshakes for 20 cents. In 1961 Kroc bought out the McDonald brothers, and in the next twenty years this son of an unsuccessful realtor whose family came from Bohemia built an empire of 7,500 outlets and amassed a family fortune in excess of $500 million. Appropriately headquartered in suburban Oak Brook, Illinois, the McDonald's enterprise is based on free parking and drive-in access, and its methods have been copied by dozens of imitators. Late in 1984, on an interstate highway north of Minneapolis, McDonald's began construction of the most complete drive-in complex in the world. To be called McStop, it will feature a motel, gas station, convenience store, and, of course, a McDonald's restaurant.[34]

Even church pews occasionally were replaced by the automobile. In early 1955, in suburban Garden Grove, California, the Reverend Robert Schuller, a member of the Reformed Church in America, began his ministry on a shoestring. With no sanctuary and virtually no money, he rented the Orange Drive-In movie theater on Sunday mornings and delivered his sermons while standing on top of the concession stand. The parishioners listened through speakers available at each parking space. What began as a necessity became a virtue when Schuller began attracting communicants who were more comfortable and receptive in their vehicles than in a pew. Word of the experiment—"Worship as you are . . . In the family car"—spread, the congregation grew, and in 1956 Schuller constructed a modest edifice for indoor services and administrative needs. But the Drive-in Church, as it was then called, continued to offer religious inspiration for automobile-bound parishoners, and in succeeding sanctuaries facilities were always included for those who did not want a "walk-in" church. By 1969 he had six thousand members in his church, and architect Richard Neutra had designed a huge, star-shaped "Tower of Power," situated appropriately on twenty-two acres just past Disneyland on the Santa Ana Freeway. It looked like and was called "a shopping center for Jesus Christ."[35]

In 1980 a "Crystal Cathedral" was dedicated on the grounds. Designed by Philip Johnson, the $26 million structure is one of the most impressive and gargantuan religious buildings on earth. More than 125 feet high and 415 feet wide, its interior is a stunning cavern without columns, clad in over 10,000 panes of transparent glass. Yet the drive-in feature remains. Instead of separate services for his indoor and outdoor followers, Schuller broadcasts his message over the radio from an indoor/outdoor pulpit. At the beginning of each session, two 90-foot glass walls swing open so that the minister can be seen by drive-in worshippers. Traditionalists come inside the 3,000-seat "Crystal Cathedral," while those who remain in the "pews from Detroit" are directed to the announcement: "If you have a car radio, please turn to 540 on your dial for this service. If you do not have a radio, please park by the amplifiers in the back row." The appeal has been enormously successful. By 1984 Schuller's Garden Grove Community Church claimed to be the largest walk-in, drive-in church in the world. Its Sunday broadcasts were viewed by an estimated one million Californians and commanded the nation's highest ratings for religious programming.

The Centerless City

More than anyplace else, California became the symbol of the postwar suburban culture. It pioneered the booms in sports cars, foreign cars, vans, and motor homes, and by 1984 its 26 million citizens owned almost 19 million motor vehicles and had access to the world's most extensive freeway system. The result has been a new type of centerless city, best exemplified by once sleepy and out-of-the-way Orange County, just south and east of Los Angeles. After Walt Disney came down from Hollywood, bought out the ranchers, and opened Disneyland in 1955, Orange County began to evolve from a rural backwater into a suburb and then into a collection of medium and small towns. It had never had a true urban focus, in large part because its oil-producing sections each spawned independent suburban centers, none of which was particularly dominant over the others. The tradition continued when the area became a subdivider's dream in the 1960s and 1970s. By 1980 there were 26 Orange County cities, none with more than 225,000 residents. Like the begats of the Book of Genesis, they merged and multiplied into a huge agglomeration of two million people with its own Census Bureau metropolitan area designation—Anaheim, Santa Ana, Garden Grove. Unlike the traditional American metropolitan region, however, Orange County lacked a commutation focus, a place that could obviously be accepted as the center of local life. Instead, the experience of a local resident was typical: "I live in Garden Grove, work in Irvine, shop in Santa Ana, go to the dentist in Anaheim, my husband works in Long Beach, and I used to be the president of the League of Women Voters in Fullerton."[36]

A centerless city also developed in Santa Clara County, which lies forty-five miles south of San Francisco and which is best known as the home of "Silicon Valley." Stretching from Palo Alto on the north to the garlic and lettuce fields of Gilroy to the south, Santa Clara County has the world's most extensive concentration of electronics concerns. In 1940, however, it was best known for prunes and apricots, and it was not until after World War II that its largest city, San Jose, also became the nation's largest suburb. With fewer than 70,000 residents in 1940, San Jose exploded to 636,000 by 1980, superseding San Francisco as the region's largest municipality. As the automobile-based circulation system matured, the county's spacious orchards were easily developed, and bulldozers uprooted fruit trees for shopping centers and streets. Home builders, encouraged by a San Jose city government that annexed new territory at a rapid pace and borrowed heavily to build new utilities and schools on the fringes of town, moved farther and farther into the rural outskirts. Dozens of semiconductor and aerospace companies expanded and built plants there. In time, this brought twice-daily ordeals of bumper-to-bumper traffic on congested freeways. The driving time of some six-mile commutes lengthened to forty-five minutes, and the hills grew hazy behind the smog. As Santa Clara County became a national symbol of the excesses of uncontrolled growth, its residents began to fear that the high-technology superstars were generating jobs and taxes, but that the jobs attracted more people, and the taxes failed to cover the costs of new roads, schools, sewers, and expanded police and fire departments.[37]

The numbers were larger in California, but the pattern was the same on the edges of every American city, from Buffalo Grove and Schaumburg near Chicago, to Germantown and Collierville near Memphis, to Creve Couer and Ladue near St. Louis. And perhaps more important than the growing number of people living outside of city boundaries was the sheer physical sprawl of metropolitan areas. Between 1950 and 1970, the urbanized area of Washington, D.C., grew from 181 to 523 square miles, of Miami from 116 to 429, while in the larger megalopolises of New York, Chicago, and Los Angeles, the region of settlement was measured in the thousands of square miles.

The Decentralization of Factories and Offices

The deconcentration of post-World War II American cities was not simply a matter of split-level homes and neighborhood schools. It involved almost every facet of national life, from manufacturing to shopping to professional services. Most importantly, it involved the location of the workplace, and the erosion of the concept of suburb as a place from which wage-earners commuted daily to jobs in the center. So far had the trend progressed by 1970 that in nine of the fifteen largest metropolitan areas suburbs were the principle sources of employment, and in some cities, like San Francisco, almost three-fourths of all work trips were by people who neither lived nor worked in the core city. In Wilmington, Delaware, 66 percent of area jobs in 1940 were in the core city; by 1970, the figure had fallen below one quarter. And despite the fact that Manhattan contained the world's highest concentration of office space and business activity, in 1970, about 78 percent of the residents in the New York suburbs also worked in the suburbs. Many outlying communities thus achieved a kind of autonomy from the older downtown areas. A new "Americanism" even entered the language—"beltway"—to describe the broad expressways that encircled every important city by 1975 and that attracted employers of every description.[38]

Manufacturing is now among the most dispersed of non-residential activities. As the proportion of industrial jobs in the United States work force fell from 29 percent to 23 percent of the total in the 1970s, those manufacturing enterprises that survived often relocated either to the suburbs or to the lower-cost South and West. Even tertiary industries, which do not utilize assembly-line processes and which require less flat space than larger factories, have adapted to the internal-combustion engine with peripheral sites. As early as 1963, industrial employment in the United States was more than half suburban based, and by 1981, about two-thirds of all manufacturing activity took place in the "industrial parks" and new physical plants of the suburbs. The transition has been especially hard on older workshop cities, where venerable factories are abandoned as employers are lured outward by the promise of open land, easy access to interstate highways, and federal investment tax credits. Between 1970 and 1980, for example, Philadelphia lost 140,000 jobs, many of them with the closing down or moving away of such Quaker City mainstays as Philco-Ford, Cuneo Eastern Press, Midvale Heppenstall Steel, Bayuk Cigar, Eaton and Cooper Industries' Plumb Tool Division, and the Container Corporation.[39]

Office functions, once thought to be securely anchored to the streets of big cities, have followed the suburban trend. In the nineteenth century, businesses tried to keep all their operations under one centralized roof. It was the most efficient way to run a company when the mails were slow and uncertain and communication among employees was limited to the distance that a human voice could carry. More recently, the economics of real estate and a revolution in communications have changed these circumstances, and many companies are now balkanizing their accounting departments, data-processing divisions, and billing departments. Just as insurance companies, branch banks, regional sales staffs, and doctors offices have reduced their costs and presumably increased their accessibility by moving to suburban locations, so also have back-office functions been splitting away from front offices and moving away from central business districts.

Corporate headquarters relocations have been particularly well-publicized. Although the publishing firm of Doubleday and Company moved to quiet Garden City on Long Island in 1910 and Reader's Digest shifted to Pleasantville, New York, in Westchester County in 1936, the overall trend of corporate movement was toward central business districts until about 1950. The outward trend began in earnest in 1954, when the General Foods Corporation moved its home office from midtown Manhattan to a spacious, low-slung campus surrounded by acres of trees and free parking in suburban White Plains. The exodus reached a peak between 1955 and 1980, when, arguing, "It's an altogether more pleasant way of life for all," more than fifty corporations, including such giants as International Business Machines, Gulf Oil, Texaco, Union Carbide, General Telephone, American Cyanamid, Xerox, Pepsico, U. S. Tobacco, Cheeseborough Ponds, Nestlé, American Can, Singer, Champion International, and Olin, abandoned their headquarters in New York City.[40]

Because Manhattan remained the dominant center of the nation's corporate and financial life, most companies simply moved within the region to more bucolic surroundings, principally in one of three small areas: a strip of central Westchester County from the Hudson River past White Plains to the Connecticut border, the downtown of Stamford and adjacent Greenwich in Fairfield County, Connecticut, and a narrow slice through the heartland of Morris and Somerset Counties in New Jersey. All three areas built more than 16 million square feet of office space between 1972 and 1985, or more than exists in all but a handful of American cities.

The trend was particularly strong toward Connecticut, where executives could have the benefit of Gotham's business and cultural advantages without the bother of New York State's income taxes. In 1960 when the first urban renewal plans were drawn up for downtown Stamford, no consideration was given to building any commercial office space there. In the next three decades, however, while the original proposals were delayed by community resistance, Stamford's urban-renewal plans were redrawn to reflect changes in corporate attitudes toward relocating out of Gotham and into more comfortable suburban locations. For Stamford the delay was beneficial. When companies began their Manhattan exodus, Stamford had available space downtown. By 1984 Fairfield County was the third leading corporate headquarters site in the United States, after only New York City and Chicago.

Several studies have pointed out that the most important variable in determining the direction of a corporate shift was the location of the home and country club of the chief executive officer of the particular company. In fact, top officers were often the only ones to benefit from the suburban shifts. When A & W Beverages made the move from Manhattan to White Plains early in 1984, the company lost its entire support staff in the transition and had to spend a small fortune on severance costs. "Some of these people had been with us for many years, so we had to ask ourselves what we should do with loyal and good workers who will no longer have a job," said Craig Honeycutt, director of personnel for A & W, about the employees who quit rather than commute from Manhattan, Brooklyn, or New Jersey to White Plains.[41]

Because the construction of suburban office headquarters tends to be expensive, the purpose of most such moves is to improve employee morale and productivity as much as to reduce costs. To this end, a company typically hires a well-known architect to design a rustic complex on the model of a college campus or a self-contained village. Free parking and easy access to interstate highways presumably make possible a longer work day, while stone piazzas, landscaped gardens, impressive sculpture, and splashing water fountains, as well as gymnasiums, showers, and saunas presumably make possible a more relaxed one. Company-owned cafeterias replace the downtown restaurants, shopping districts, and even noontime concerts of the city centers. To some employees the result is "close to perfect." Others find the campus environment boring and bemoan that "the main thing of interest out here is what's new in the gift shop."

Corporate relocation in the postwar period has been overwhelmingly a city-to-suburb phenomenon rather than a regional shift. The move of Gulf Oil to Houston and of American Airlines to Dallas, both from New York, were exceptions to this general rule. Only occasionally have large firms shifted both from a city to a suburb and from one region to another. The Johns-Manville Company, which transferred in the 1970s from a Manhattan office tower to a sleek and gleeming spaceship-style structure in the midst of a 10,000-acre ranch in the foothills of the Rocky Mountains, is a clear exception. Perhaps coincidentally the Johns-Manville Corporation was saved from bankruptcy in 1982 only by the intervention of a court.

Since World War II, the American people have experienced a transformation of the man-made environment around them. Commercial, residential, and industrial structures have been redesigned to fit the needs of the motorist rather than the pedestrian. Garish signs, large parking lots, one-way streets, drive-in windows, and throw-away fast-food buildings—all associated with the world of suburbia—have replaced the slower-paced, neighborhood-oriented institutions of an earlier generation. Some observers of the automobile revolution have argued that the car has created a new and better urban environment and that the change in spatial scale, based upon swift transportation, has formed a new kind of organic entity, speeding up personal communication and rendering obsolete the older urban settings. Lewis Mumford, writing from his small-town retreat in Amenia, New York, has emphatically disagreed. His prize-winning book, *The City in History,* was a celebration of the medieval community and an excoriation of "the formless urban exudation" that he saw American cities becoming. He noted that the automobile megalopolis was not a final stage in city development but an anticity which "annihilates the city whenever it collides with it."[42]

The most damning indictment of private transportation remains, however, the 1958 work of the acid-tongued John Keats, *The Insolent Chariots*. He forcefully argued, as have others since that time, that highway engineers were wrong in constantly calling for more lanes of concrete to accommodate yet more lines of automobiles. Instead, Keats's position was that motorcars actually created the demand for more highways, which in turn increased the need for more vehicles, and so on ad infinitum. More ominously, he surmised, public expenditures for the automobile culture diverted funds from mass transit and needed social services.[43]

The automobile lobby swept everything and everybody before it, however, and it was not until the first oil boycott of 1973 that Americans would seriously ponder the full implications of their drive-in culture. Especially in the 1950s, expressways represented progress and modernity, and mayors and public officials stumbled over themselves in seeking federal largesse for more and wider roads. Only a few people realized that high-speed roads accelerated deconcentration, displaced inner-city residents, contributed to the decay of central business districts, and hastened the deterioration of existing transportation systems. As Raymond Tucker, mayor of St. Louis and former president of the American Municipal Association, put it, "The plain fact of the matter is that we just cannot build enough lanes of highways to move all of our people by private automobile and create enough parking space to store the cars without completely paving over our cities and removing all of the . . . economic, social, and cultural establishments that the people were trying to reach in the first place."

Because structures built to accommodate the demands of the automobile are likely to have an ephemeral life, it is a mistake for cities to duplicate suburban conditions. In 1973 a RAND study of St. Louis suggested as an alternative strategy that the city become "one of many large suburban centers of economic and residential life" rather than try to revive traditional central city functions. Such advice is for those who study statistics rather than cities. Too late, municipal leaders will realize than a slavish duplication of suburbia destroys the urban fabric that makes cities interesting. Memphis's Union Avenue, once a grand boulevard lined with the homes of the well-to-do, has recently fallen victim to the drive-in culture. In 1979 one of the last surviving landmarks, an elegant stone mansion, was leveled to make room for yet another fast-food outlet. Within three years, the plastic-and-glass hamburger emporium was bankrupt, but the scar on Union Avenue remained.

There are some signs that the halcyon days of the drive-in culture and automobile are behind us. More than one hundred thousand gasoline stations, or about one-third of the American total, have been eliminated in the last decade. Empty tourist courts and boarded-up motels are reminders that the fast pace of change can make commercial structures obsolete within a quarter-century of their erection. Even that suburban bellweather, the shopping center, which revolutionized merchandising after World War II, has come to seem small and out-of-date as newer covered malls attract both the trendy and the family trade. Some older centers have been recycled as bowling alleys or industrial buildings, and some have been remodeled to appeal to larger tenants and better-heeled customers. But others stand forlorn and boarded up. Similarly, the characteristic fast-food emporiums of the 1950s, with uniformed "car hops" who took orders at the automobile window, are now relics of the past. One of the survivors, Delores Drive-In, which opened in Beverly Hills in 1946, was recently proposed as an historic landmark, a sure sign that the species is in danger.[44]

15

The Loss of Community in Metropolitan America

A major casualty of America's drive-in culture is the weakened ''sense of community'' which prevails in most metropolitan areas. I refer to a tendency for social life to become ''privatized,'' and to a reduced feeling of concern and responsibility among families for their neighbors and among suburbanites in general for residents of the inner city. The term ''community'' implies cooperation. If, for example, the ''sense of community'' in metropolitan Chicago were strong, then most citizens of Lake Forest, Barrington Hills, Flossmoor, Harvey, South Holland, and two hundred other suburbs would have a clear and positive identification with the Windy City. They would believe that they are united in a way that other citizens of Illinois are not.[1]

Citizen identification with the city is now less than it was a century or more ago. To be sure, nineteenth-century communities were bothered by crime, class rigidity, social unrest, racial prejudice, epidemics, alcohol abuse, and fires. But they possessed a significant sense of local pride and spirit as a result of their struggles with other cities for canals, railroads, factories, and state institutions. In our own time, most observers have noted that alienation and *anomie* are more characteristic of urban life than a sense of participation and belonging.[2]

This is reflected in a general shift in the meaning of the word ''suburban.'' Whereas it once implied a relationship with the city, the term today is more likely to represent a distinction from the city. The nomenclature of peripheral communities provides many clues to this new circumstance. In the nineteenth-century suburbs, names were often adopted that suggested their direction from the central city, such as North Chicago, South Chicago, or West Chicago. No longer. Rather, the pronounced trend in the twentieth century has been to choose something more suggestive of the countryside than of downtown. In the Chicago area, for example, twenty-four separate communities have taken either "Park" or "Forest" (including both a Park Forest and a Forest Park) in their names; other popular terms are Rolling Meadows, Highland Hills, Sleepy Hollow, River Grove, and Lake Villa. Accurate description is rather less important than bucolic imagery. A good example of the semantic shift was provided in 1973 by East Paterson, New Jersey, which changed its name to Elmwood Park. The former title designated a close spatial relationship with a seedy, industrial city; the second was more suggestive of a quiet residential setting. Similarly, East Detroit became Erin Heights in 1984.

The observant traveler can witness a similar phenomenon in the naming of streets and subdivisions. In the older sections of almost every American city—large and small, east or west—the streets of central areas are numbered. City fathers did this because of their belief that numbered streets were a sign of prominence and promise—were not Philadelphia, New York, Chicago, and Cincinnati famous for their numbered gridiron systems? To live on a Fourteenth Street, therefore conveyed several messages. It meant that your residence was fourteen blocks away from the central business district, and it also meant that you lived in an urban place. Not every nineteenth-century thoroughfare was numbered, obviously, but given names did have a certain logic to them. Many took the title of the city to which they ultimately led, as in Bedford Road, or after an important function or institution that was located there, as in Schoolhouse Road, Dock Street, Market Street, or Railroad Street. Thus, the street layout, the street name, and even the use of the word "street" itself all conveyed an image of urbanity.

Contemporary suburbs, of course, seek to suggest quiet repose rather than commercial importance. Beginning in the middle of the nineteenth century in places like Llewellyn Park, New Jersey, but becoming important only after the 1920s, residential developers began to name streets after the bucolic and the peaceful. They abandoned the grid plan wherever possible and began to name rights-of-way with utter disregard for topography, function, or history. The result is familiar to us all—the enterprising entrepreneur simply combines acceptable word choices (rolling, fields, tall, lake, view, hills, timber, roaring, brook, green, farms, forest) into a three- or four-word combination. The new concoction is never followed by the word "street," but rather by lane, cove, road, way, fairway, or terrace. In California and other parts of the Southwest, Spanish names are substituted for the English, but the intent is the same. History, circumstance, and geography are discarded in a conscious attempt to market houses according to the suburban ideal.

Finally, professional sports nomenclature offers a clue to the demise of community. The designation of a place or a team by a name—a specific name under which fans or residents can unite—is one piece of evidence that a community exists. Until about 1960, professional athletic teams were almost always known by the names of the central cities they represented: the New York Yankees, the Montreal Canadiens, the Boston Celtics, or the Pittsburgh Steelers. In recent years, however, there has been a trend away from naming teams for cities, as if an association with the core city would limit box office appeal. Thus we have the New Jersey Nets, the Minnesota Twins, the Texas Rangers, the Golden State Warriors, the California Angels, and the New England Patriots. At Chavez Ravine, the home of the Los Angeles Dodgers, there are two parking spaces for every seven seats, a telling index of a way of life made possible by the car. In the team's former home, Brooklyn, the "trolley dodgers" had derived their very name from the streetcar system.

Changing town, street, and athletic nomenclature is symptomatic of the deeper and more fundamental fragmentation of metropolitan America. As we have seen, the private automobile has been the most important catalyst for this shift. But there are also three related reasons why this circumstance has come about: the polarization of urban neighborhoods by function, by income, and by race; the failure of municipalities to extend their boundaries through annexation and consolidation; and the changing nature of modern entertainment.

The Polarization of the Metropolis

As we noted earlier, America's large cities underwent a startling spatial transformation between 1815 and 1875. By the 1920s, the exodus of the middle and upper classes from the urban centers had proceeded so far that sociologists at the University of Chicago constructed a concentric-zone model to describe the way in which residential neighborhoods improved in quality with increasing distance from the core. This Park-Burgess model has been attacked by two generations of academic urbanologists, but the gist of the disagreement has been about detail, not substance. After 1920, no one could deny that the inner cities were poor and that the suburbs were, relatively speaking, rich. Children learned this at an early age. In 1971 a suburban rabbi confessed that when he was growing up in Brooklyn, the posh Five Towns area of Long Island, even more than Israel, represented "the promised land."[3]

As the suburbs drew off the wealthy, central cities became identified with social problems. Newark offers a good example of this trend. In the nineteenth century, the New Jersey metropolis was one of the nation's leading industrial centers. Its heavy industries, its whirring factories, its prosperous building trades, and its noted public works made it a confident and optimistic community. As late as 1927, a prominent businessman could boast:

> Great is Newark's vitality. It is the red blood in its veins—this basic strength that is going to carry it over whatever hurdles it may encounter, enable it to recover from whatever losses it may suffer and battle its way to still higher achievement industrially and financially, making it eventually perhaps the greatest industrial center in the world.[4]

Yet the suburban trend was already draining away Newark's most successful and prosperous citizens. In 1925 more than 40 percent of all attorneys whose offices were in Newark were already living in the suburbs; by 1947 the figure had jumped to 63 percent; and by 1965, 78 percent. Members of the city's leading booster association abandoned their home community in even greater numbers. As early as 1932, more than 86 percent of the officers and board members of the Newark Chamber of Commerce lived in the suburbs[5] (TABLES A-4, A-7, A-8).

Patterns in other old cities were less startling but similar. In staid Boston, more than half of the Hub's lawyers were living outside the city in 1911, two years before the first car rolled off a moving assembly line. Sixty years of automobility increased the percentage only to three out of four. In New York City 38 percent of the attorneys with offices in Manhattan lived outside the borough in 1908, the year Henry Ford introduced the Model T. On the eve of World War I, the percentage was up to 47, but half a century later it had not risen beyond two-thirds. A profession in the law became more attainable to ordinary people in the course of the twentieth century, as the prestige of an address in one of Manhattan's towering office towers remained. During the day, it was important to operate in the reflected glow of the world's corporate, financial, and communications center, but at night the attorneys traveled to homes in the outer boroughs or, more often, in the suburbs.[6]

It was not inevitable, in Newark or Boston or New York or anywhere else, that the middle and upper classes would gravitate to the urban edges. But, as we have seen, an unusual set of circumstances in the United States helped to insure that suburban areas in the second half of the twentieth century would be segregated by income, race, and lifestyle. The core has become identified in the popular mind with poor people, crime, minorities, deterioration, older dwellings, and abandoned buildings. Middle- and upper-income suburbs convey the opposite impression. The result has been as detrimental to older cities as it has been beneficial to the newer suburbs.

After World War II, the racial and economic polarization of large

American metropolitan areas became so pronounced that downtown areas lost their commercial hold on the middle class. Cities became identified with fear and danger rather than with glamour and pleasure. In Memphis the homicide rate in 1984, although deplorably high, was only about half what it was in 1915, when the Bluff City was regionally famous as a "murder capital." Yet Memphis's central business district was bustling and vibrant in 1915, in 1985 it is quiet and forlorn. In Newark the streets are also dismal and deserted in the evening, and a businessman there recently complained: "Since the riots (1967), many people—both black and white—have been afraid to come to Newark at night. We've lost many customers."

The Breakdown of Annexation

A second cause of metropolitan fragmentation has been the inability of cities to extend their boundaries through annexation and consolidation. In the nineteenth century, as we have seen in Chapter 9, suburbs typically lost their separate identities because municipal governments adopted the philosophy that "bigger is better" and expanded their populations and area by moving their boundaries outward.

In some metropolitan areas, notably those in the South and West, the addition of new land to the central city has continued. In 1970 Indianapolis became the tenth-largest city in the United States by virtue of its absorption of most of Marion County. In Memphis, Jacksonville, Oklahoma City, Houston, Phoenix, and Dallas a similar annexation process has continued. Thus, these cities registered startling population gains between 1960 and 1980. In actuality, what was called urban growth was the building up and annexation of new residential communities on the edges.[7]

This has not been the experience of most of the large, older cities of the United States. They are no longer bordered by nondescript settlements that can be amalgamated without difficulty. A new suburban consciousness has developed, and residents of outlying areas are now worried about real-estate values, educational quality, and personal safety. On all three counts, they regard cities as inferior. Articulate, affluent, and against big municipal governments, they have chosen to reject political absorption into the larger metropolis. In St. Louis, New York, Pittsburgh, Cleveland, San Francisco, and Philadelphia, for example, city boundaries have not been altered in at least half a century, and the core areas are being strangled by a tight ring of suburbs. In 1972 there were 22,185 local governmental units in the nation's SMSA's, with an average of 86 per Standard Metropolitan Statistical Area. The New York metropolitan region was bordered by an incredible number of 1,400 governments, many with overlapping responsibilities. Chicago followed with 1,100 separate units of government.[8]

The negative consequences of governmental fragmentation are especially evident in Newark. Along with Washington, D.C., Newark is unusual in having lost more territory than it has gained, and its miniscule 24-square mile size is the dominant cause of many of its contemporary problems. Like every other large city, Newark sought to annex its suburbs. And its leaders confidently expected that New Jersey's largest city would follow the example set by many other metropolitan areas. As the mayor of Newark remarked in 1900: "East Orange, Vailsburg, Harrison, Kearny, and Belleville would be desirable acquisitions. By an exercise of discretion we can enlarge the city from decade to decade without unnecessarily taxing the property within our limits, which has already paid the cost of public improvements."[9]

But Newark was stifled. While nearby suburbs prospered, the city increasingly became the home of poor minorities. Perceptive observers realized by the 1930s that the future was bleak. In that decade, Princeton University economist James G. Smith noted that Newark had potential "comparable to the phenomenal growth of Los Angeles." But he predicted, "Newark must create a hegemony over her lesser neighbors or find her great destiny aborted." No such hegemony was forthcoming. The suburbs wanted no part of the industrial city's problems, and in 1933 a Newark City Commissioner told the local Optimist Club:

> Newark is not like the city of old. The old, quiet residential community is a thing of the past, and in its place has come a city teeming with activity. With the change has come something unfortunate—the large number of outstanding citizens who used to live within the community's boundaries has dwindled. Many of them have moved to the suburbs and their home interests are there.[10]

The fact that the peripheral neighborhoods had then and usually have now the legal status of separate communities has given them the capacity to zone out the poor, to refuse public housing, and to resist the integrative forces of the modern metropolis. Thus, the problems of core neighborhoods are usually more serious than those of adjacent suburbs. And because the suburbs are independent and have their own traditions and history, residents of Brookline, or Bronxville, or Lake Forest, or Arcadia, or Ladue tend to offer their primary loyalties to their suburb and to deny responsibility for those who reside a few miles away. In Philadelphia, an angry 1968 letter to the editor expressed just such a view: "It is ridiculous to suppose that those of us in the suburbs have any responsibility to help in the current Philadelphia school crisis. We did not create the problems of the inner city and we are not obligated to help in their solution."

The most conspicuous city-suburban contrast in the United States runs along Detroit's Alter Road. Locals call the street the "Berlin Wall," or the "barrier," or the "Mason-Dixon Line." It divides the suburban Grosse Pointe communities, which are among the most genteel towns anywhere, from the East Side of Detroit, which is poor and mostly black. The Detroit side is studded with abandoned cars, graffiti-covered schools, and burned-out buildings. Two blocks away, within view, are neatly-clipped hedges and immaculate houses—a world of servants and charity balls, two-car garages and expensive clothes. On the one side, says John Kelly, a Democratic state senator whose district awkwardly straddles both neighborhoods, is "West Beirut;" on the other side, "Disneyland."[11]

The answer to America's urban ills obviously does not lie solely in absorbing the Grosse Pointes into the Detroits. As the New York City Department of Sanitation sadly illustrates, mere size is no guarantee of excellence or of efficiency. More governmental functions than we perhaps realize can be handled only on a decentralized, almost neighborhood, basis. But our cities also face problems in transportation, pollution, and unemployment that are genuinely metropolitan or even regional in scope and that cannot be solved by having each community go it alone or by creating additional monstrous and self-serving public agencies like the Port Authority of New York and New Jersey. Some sort of metropolitan or federated government whose planners recognize both the need to keep government human in scale as well as the need to develop citizen awareness of responsibilities beyond the local neighborhood or village is necessary if we are to continue to have great cities. As columnist Tom Wicker noted in the *New York Times* of August 11, 1969:

> The choice, in general, is not between the impersonal coldness of remote bureaucracy and a New England town government for every twenty city blocks. The choice is between a dangerously outmoded concept of the city, leading to abandonment and decay, and a rational development that would restore a congruence between the reach of government and the location of the governed.

The Changing Nature of Modern Entertainment

New attitudes toward leisure and especially the establishment of the home as a self-sufficient entertainment center have also contributed to the weakening of the "sense of community' in metropolitan America. In the nineteenth century, leisure was a precious and rare commodity, and retirement was a little-known concept. But men and women have always had some time of their own, and the use of that time provides one indication of their attitudes toward community life.

Cities, by their very nature, ought to encourage the elevation of the human spirit. Anyone who has ever visited the Piazza San Marco in Venice, shared the happy conviviality of Tivoli Gardens in Copenhagen, witnessed the temptations of the Reeperbahn in Hamburg, strolled at midnight along the Ramblas in Barcelona, or bicycled on Sunday in New York's Central Park knows something of the potentialities and varieties of urban experience. They remind one of Samuel Johnson's telling phrase: "When a man is tired of London, he is tired of life."[12]

American cities boast of concert halls, opera houses, ballet companies, museums, and shopping streets as distinguished as any in the world. But in the United States, as Robert C. Wood has observed, what is most significant is not the influence of urban culture, but the general suburban resistance to it. What is striking in the lives of most residents is the frequency with which they choose not to avail themselves of the variety of experiences the metropolis affords, the manner in which they voluntarily restrict their interests and associations to the immediate vicinity, and the way in which they decline contacts with the larger society.[13]

Suburbanites are of course not completely private in their associations. They participate in an incredible variety of charitable and voluntary activities and campaign and vote in numbers greater than those of either urban or rural dwellers. The more affluent join country clubs, which are almost never in the country, or hunt clubs, which almost never run a fox to ground. In general, however, they focus their energies and their leisure on the home. Indeed, homeownership tends to involve a shift in the maintenance and repair of buildings away from professionals and toward more direct work on the part of the householder. Thus, the rise of a do-it-yourself industry and a shift in leisure-time allocation is a consequence of an increase in homeownership. Our idea of the good life is to build a three-bathroom colonial house such as was never seen in any of the colonies or a ranch-style home such as was never seen on a nineteenth-century ranch. Then, we install a patio or a swimming pool for friendly outdoor living. Many back yards are overequipped, even sybaritic, with hot tubs, gas-fired barbecue grills, and changing cabanas.[14]

The real shift, however, is the way in which our lives are now centered inside the house, rather than on the neighborhood or the community. With increased use of automobiles, the life of the sidewalk and the front yard has largely disappeared, and the social intercourse that used to be the main characteristic of urban life has vanished. Residential neighborhoods have become a mass of small, private islands; with the back yard functioning as a wholesome, family-oriented, and reclusive place. There are few places as desolate and lonely as a suburban street on a hot afternoon.

A century and more ago, despite the vigor with which people such as Andrew Jackson Downing and Catharine Beecher extolled the virtues of suburban life, a house was a place of toil, a scene of production, the locus of food preparation and of laundering and of personal hygiene. During free hours, it was a place to get out of. The ventilation, heat, and lighting were atrocious; it was hot in summer and cold in winter. Window screening, which one observer termed "the most humane contribution the 19th Century made to the preservation of sanity and good temper," was not introduced until the late 1880s; before that time swarms of gnats, mosquitoes, June-bugs, and beetles moved at will through domestic quarters. The result, in both the Old World and the New, was an enthusiasm for the commonality of neighborhood life. To be within the four walls of a house was to be away from the action. Among its few pleasures were reading and making love. The miracles of modern technology have changed some of that.[15]

The evolution of the front porch is a microcosm of the decline of community. In the half-century before World War II verandas were simply de regueur. They were places for observing the world, for meeting friends, for talking, for knitting, for shelling peas, for courting, and for half a hundred other human activities. The front porch was the physical expression of neighborliness and community. With a much-used front porch, one could live on Andy Hardy's street, where doors need not be locked, where everyone was like family, and where the iceman would forever make deliveries. With a front porch, one could live in Brigadoon, Shangri-La, and Camelot, all in one.[16]

When the automobile appeared, however, the slow-motion world of the front porch began yielding to a new pace. With a car at the curb, youngsters no longer had to sit at home and wait for things to happen; cars could quickly whisk them off to the action. Home industries have largely vanished, and child-rearing, religious instruction, education, and care for the sick have been passed to public institutions. But entertainment has moved indoors. First, with the crank-up phonographs and crystal sets, and more recently with the wide availability of stereophonic music, color television, and video cassette recorders, the private dwelling offers a range of comforts and possibilities, and with the expansion of telephone service, easy and quick communication with outsiders.

Air-conditioning in particular has coincided with a general withdrawal into self-pursuit and privatism. It seduced families into retreating behind closed doors and shut windows, making sidewalk society obsolete and altering the country's character and folkways. Invented in 1906 by textile engineer Willis H. Carrier, who predicted that every day could be a good day, air-conditioning became commonplace in public places and private dwellings in the 1960s. General Electric, Westinghouse, and Carrier designed a sophisticated marketing strategy to appeal to women. Air-conditioned homes, they suggested, were happier, healthier, cleaner homes. By 1980, when the United States (with a mere 5 percent of the world's population) was consuming as much man-made coolness as the whole rest of the earth put together, air-conditioning had become an inalienable right of the middle class. Indeed, the growth of the South and Southwest has been closely linked to the technology of dehumidifiers and air-conditioners. Houston, for example, is an accidental city that was founded on a swamp by speculators who could not have dreamed that it would one day be the nation's fifth largest city and the global headquarters of the energy business. Indeed, without air-conditioning it would not have done so.[17]

No longer forced outside by the heat and humidity, no longer attracted by the corner drugstore, and no longer within walking distance of relatives, suburbanites often choose to remain in the family room. When they do venture out, it is often through a garage into an air-conditioned automobile. Streets are no longer places to promenade and to meet, but passageways for high-powered machines. The cult of domestic privatism, the desire to escape from the warm crowds of city streets, and the turning inward on the family have fully evolved since their articulation by Downing, Beecher, and Vaux a century and more earlier. In fact, more and more people now regard it as a waste of time to go out to a game or a movie, what with the Washington Redskins or *Dynasty* available in the family room. In 1976 the median number of hours per week of television-watching was a staggering twenty-eight hours per week. No large population anywhere has ever spent so much time being entertained. Because the action of the individual is passive and private rather than active and communal, the late Margaret Mead referred to such rooms as giant playpens into which the parents had crawled.[18]

The shift in residential behavior was reflected by several aspects of American popular culture. *Mr. Blandings Builds His Dream House* was perhaps the most widely read of the many postwar novels that explored the world of rising executives and long-distance transfers, of kaffee klatsches and barbeque pits, of split-levels and station wagons. Eric Hodgins's best-selling 1946 book, which was quickly made into a movie starring Cary Grant and Myrna Loy, was about the pain of city-bred innocents as they encountered the problems and frustrations of battles with pipe-fitters and painters en route to a bloated $56,000 house.

Mr. Blandings Builds His Dream House dealt with the efforts of an upper-class New Yorker to build a country house in exurban Connecticut and was hardly typical of the postwar experience. But the book and the movie obviously struck responsive chords among a population that was buying new houses in record numbers. Other books (many of them also made into movies) took up a somewhat similar theme. Sloan Wilson's *The Man in the Gray Flannel Suit,* Jean Kerr's *Please Don't Eat the Daisies,* John Cheever's *Bullet Park,* John Marquand's *Point of No Return,* Peter De Vries's *The Mackeral Plaza,* and Max Shulman's *Rally Round the Flag, Boys* all poked, either hilariously or scaldingly, at the dream of a green and pleasant oasis not far from the office. If the stereotypical portrait of the suburb as affluent, Republican, and WASPish was somewhat overdrawn, they did at least recognize that the great American land rush after 1945 was one of the largest mass movements in our history.[19]

In the 1950s, when television replaced the motion picture as the primary medium of entertainment in the United States, it was appropriate that one of the longest-running and most popular of all the programs dealt with the trials and pleasures of suburban life. Ozzie and Harriet Nelson became staples across the land as households followed the progress of David and Rickie from infancy through young manhood. The tradition continued in subsequent decades as popular series like *My Three Sons, Father Knows Best, The Brady Bunch, Leave It to Beaver,* and *Life with Riley* suggested that the appropriate setting for family life was the detached home and that the ideal symbol of making it in America was to trade a small suburban house for a large one. Even the hugely successful *I Love Lucy* program shifted locales from urban apartment to suburban house. The front yard, the porch, the street, and the corner grocery had insignificant roles in the new private environment.[20]

16

Retrospect and Prospect

The American people have absolutely refused to accept a simple fact. We have an energy crisis. We have shortages of oil. The shortages are going to get worse in the future.

—PRESIDENT JAMES EARL CARTER, May, 1979

Urban living has certain advantages from the standpoint of our energy plan because urban areas are more energy efficient than our suburban or rural areas.

—SECRETARY OF ENERGY JAMES SCHLESINGER, April 27, 1978

In 1968 Spiro Agnew became the first suburban politician to rise to national office. Although he was presented to the Republican National Convention in Miami Beach as an expert on urban problems—an important consideration when deadly riots were regularly tearing apart America's cities—his actual experience was as chief executive officer of suburban Baltimore County, a 610-square-mile swatch of green and rolling countryside that surrounds the city of Baltimore like a well-pitched horseshoe on a stake. Between 1950 and 1970, the affluent county doubled its population while its proportion of blacks fell from about 7 percent to 3 percent. Thus Agnew's real claim to expertise, as Gary Wills noted in *Nixon Agonistes,* was that he "early grasped and overcame what white suburbanites take to be their main city problem—how to escape the city."[1]

Escape Americans did. Eighteen of the nation's twenty-five largest cities in 1950 suffered a net *loss* of population over the next three decades, a circumstance which many observers have taken as the most compelling evidence that our cities are dying. By contrast, during the same years the independent suburbs of the United States *gained* more than 60 million persons. Between 1950 and 1970, the suburban population doubled from 36 to 74 million, and 83 percent of the nation's total growth took place in the suburbs. In 1970, for the first time in the history of the world, a nation-state counted more suburbanites than city dwellers or farmers. Perhaps the most remarkable statistic of all is contained in TABLE 16-1, which reveals that of the fifteen largest metropolitan areas in the United States in 1980, only in Houston did a majority of residents live in the central city.

TABLE 16-1
Suburban Proportion of the Fifteen Largest Metropolitan Areas in the United States, 1980

Metropolitan Area	*Metropolitan Population*	*Suburban Population*	*Suburban Percentage*
1. Boston	3,448,122	2,885,128	83.7%
2. Pittsburgh	2,263,894	1,839,956	81.3
3. St. Louis	2,355,276	1,902,191	80.8
4. Washington	3,060,240	2,422,589	79.2
5. Atlanta	2,029,618	1,604,596	79.1
6. Detroit	4,618,161	3,414,822	73.9
7. Cleveland/Akron	2,834,062	2,023,063	71.4
8. Philadelphia	5,547,902	3,859,682	69.6
9. San Francisco Bay	5,179,784	3,524,972	68.1
10. Los Angeles/Anaheim	11,497,568	7,620,560	66.3
11. Baltimore	2,174,023	1,387,248	63.8
12. Chicago	7,869,542	4,864,470	61.8
13. Dallas/Fort Worth	2,974,878	1,685,659	56.7
14. New York/New Jersey	16,121,297	8,721,019	54.1
15. Houston	2,905,350	1,311,264	45.1

SOURCE: Calculations Based upon 1980 United States Census Data.

Housing construction and population growth were most rapid on the moving fringe—the closer automobile suburbs exploded in the 1950s (Nassau County on Long Island, for example), while those farther on the exurban fringe (Suffolk County on Long Island, Buffalo Grove and Schaumburg near Chicago, Raleigh and Germantown near Memphis, Creve Couer and Chesterfield near St. Louis) experienced the largest influx in the 1970s. The term "sprawl" became a new Americanism as subdivisions and shopping centers sprouted across the landscape. In the late 1970s, when high interest rates and double-digit inflation were pushing home prices beyond the reach of young families, the Department of Agriculture announced that three million acres of prime farmland were being lost each year to suburban development. By 1985 reasonable people could debate whether the United States was a racist nation, an imperalist nation, or a religious nation, but scarcely anyone could quarrel with its designation as a suburban nation.[2]

The move to the suburbs was almost self-generating. As larger numbers of affluent citizens moved out, jobs followed. In turn, this attracted more families, more roads, and more industries. The cities were often caught in a reverse cycle. As businesses and taxpayers left, the demand for middle- to upper-income dwelling units in older neighborhoods declined. At the same time, population increases among low-income minorities, coupled with the demolition of inner-city housing for new expressways, produced an increase in the demand for low-income housing. The new residents required more health care and social-welfare services from the city government than did the old, but they were less able to pay for them. To increase expenditures, municipal authorities levied higher property taxes, thus encouraging middle-class homeowners to leave, causing the cycle to repeat. In contrast, suburbs were often able to keep tax burdens low by having private trash collection, volunteer fire departments, and unpaid ambulance services. In particular, they benefitted from having a small percentage of population living at the poverty level and so requiring government assistance. High quality municipal services, and especially well-funded public schools that offered racial homogeneity and harmony, attracted still more high-status residents, which in turn made select suburbs even wealthier and more attractive. As early as the 1950s, suburban real-estate advertisements were harping on the themes of race, crime, drugs, congestion, and filth. Thus, the well-to-do could avoid the local costs of urban old age by simply stepping over the border, leaving the poor to support the poor. "Escape to Scarborough Manor. Escape from cities too big, too polluted, too crowded, too strident, to call home."[3]

The positive result of this process is called "filtering." The construction of new housing in the suburbs puts competitive pressure on the older housing stock, depressing its price. In European culture age lends prestige, and older structures often command the highest prices. Americans, however, feel less reverence and pride in the past and instead pay a premium for newness. Thus, the typical model of urban growth in the United States has been the sequential reuse of housing by progressively lower-income households. Were it not for the subdivisions of the periphery, the shelter available to the poor would be even more limited and expensive.[4]

The negative results of the urban cycle are the stripped automobiles, burned-out buildings, boarded-up houses, rotting sewers, and glass-littered streets that are common in so many of America's inner cities. In parts of Brooklyn, the Bronx, Detroit, Chicago, St. Louis, Los Angeles, Atlanta, and Cleveland, whole blocks of stores and houses lie vacant. As one federal official noted: "There are some parts of these cities so empty they look as though someone had dropped nerve gas."[5]

The experience of urban population loss has not been confined to the

United States or to the twentieth century. At the height of its imperial power, the city of Rome counted almost a million inhabitants. Over the course of the next millenia, however, Rome declined to relative insignificance in the world, and by the eighteenth century its population had fallen all the way to 40,000. During the Middle Ages, other cities suffered a similar fate, as epidemic pestilence and war periodically thinned the ranks of the citizenry.[6]

In the United States, population loss was once associated only with depopulated rural districts or with mining towns after the extractive resources had played out. City neighborhoods often changed hands from one ethnic or racial group to another, and as is pointed out in the Appendix, population declines occasionally were caused by the expansion of the central business district. But until the 1950s, there was no American precedent for the complete and absolute abandonment that has become so pervasive in the 1980s. Some streets are now so devoid of life that even the rats and roaming dogs have gone.[7]

Abandonment makes it possible to speak of ten phases of land conversion in the United States. According to this idealized model, an area is first the home of Indians, still later of carefully-tended farms, still later of suburbs, then, typically after annexation to a city, of solid middle-class residences. With successive transitions to lower-income occupancy, the parcel utlimately enters into advanced neglect and is then totally abandoned. The various phases might be summarized as follows:

Phase One	Habitation by nomadic Indians
Phase Two	Agricultural Settlement
Phase Three	Suburban Development
Phase Four	Annexation to Large City; Emergence as Neighborhood
Phase Five	Maturation and Stabilization of Density
Phase Six	Aging of Population; Decline in Density
Phase Seven	Population Succession by Lower-Income Groups
Phase Eight	Abandonment of Some Residences; Crisis Mentality
Phase Nine	Emptying of Neighborhood; Reversion to Recreation
Phase Ten	Urban Redevelopment or Gentrification

The Brooklyn neighborhood of Red Hook has passed through many of the phases of the complete cycle. Once an uninhabited marsh along the eastern portion of the immense New York harbor, it was part of the region occupied by the peaceful Canarsie Indians. As the port of New York became dominant in the New World, and as commerce spilled over to the Brooklyn side, Red Hook was covered by wharves, warehouses, and houses, and by the Civil War it was a thriving community. Its homes were successively occupied by lower income persons, however, and by 1925 only the poorest of immigrants remained in the area. By 1960 it was being abandoned by everyone except the residents of the Red Hook Public Housing Project, and by 1985 the docks, industries, and homes of the neighborhood were still. Now occupied mostly by baseball diamonds and soccer fields, it seems to be returning to something like its original function.

This cycle of decline and renewal is not unique to the United States or to the twentieth century. Between 1837 and 1901, the reign of a single sovereign, the London suburb of Camberwell experienced all the vicissitudes of residential and economic change. The European continent also provides examples of neighborhoods that have experienced numerous ups and downs. In North America, however, the pace of change has typically been quicker and the shifts in ethnicity and socioeconomic status more marked than has been the case in Europe and the Orient, where urban population growth has usually been slower and stability more prized.[8]

This volume began as an attempt to account for the strange, ever-changing physical arrangement of the United States. A map of the English countryside in 1985 shows remarkably little change from a map of the same area a hundred years earlier. The farms, castles, villages, and country estates are much the same, and though the twentieth-century map might include an occasional industrial park or airport, these are relatively slight intrusions on an immemorial landscape. As noted, the English regard age as an asset, not a liability. In the United States, by contrast, the bulldozers are always at work, and a mighty engine of change seems destined to convert every farm into a shopping center, a subdivision, or a highway.

Foreign visitors are fascinated both by American suburbs and by American slums and see in them a fundamental paradox in our national life. How is it that a rich, powerful, and technologically sophisticated country can tolerate such inefficiency, such poverty, and such contrasts? Why have we neglected our cities and concentrated so much of our energy, our creativity, and our vitality in the suburbs? Clearly, no single answer can be held accountable for such an important phenomenon, but I will argue that there were two necessary conditions for American residential deconcentration—the suburban ideal and population growth—and two fundamental causes—racial prejudice and cheap housing.

The first necessary condition for the unusual residential dispersal of the American people is a national distrust of urban life and of communal living. This search for the proper balance between country and city has been part of an intellectual tradition that dates back more than two centuries. The aspirations and satisfactions experienced by the family in one of Andrew Jackson Downing's pre-Civil War Gothic cottages were not so different from those of the contemporary suburbanite. The dream of a detached house in a safe, quiet, and peaceful place has been an important part of the Anglo-American past and a potent force in the development of suburbs. In the 1850s, for example, Ralph Waldo Emerson celebrated the benefits of suburban living in rhetoric with a very modern flavor, "The aesthetic value of railroads is to unite the advantages of town and country life, neither of which we can spare." In comparison with German, Dutch, Japanese, Italian, and Spanish cultures, for example, Americans have never placed a high value on urbanity and group interaction. "The dream house is a uniquely American form because for the first time in history, a civilization has created a utopian ideal based on the house rather than the city or the nation."[9]

The suburban ideal, while a necessary precondition for deconcentration, is not unique to natives of the United States, however. A professed reverence for country life is common to many peoples and cultures. Immigrants from many lands exhibited a similar propensity for detached housing, and Polish, German, and Irish-Americans occasionally had higher homeownership rates than native-born Americans. Oliver Zunz, in particular, has demonstrated that native white American workers owned their homes substantially less often than immigrants employed in the same occupation and of the same age group. Such achievements required enormous personal sacrifices, even though as a financial investment the single-family home was a poor value throughout most of American history. The private dwelling, therefore, is a more universal aspiration than suburban historians have previously been willing to admit. Was it not Voltaire who in the eighteenth century said that the serene ideal of human existence is to cultivate one's garden? What was important is not that American expectations and attitudes about houses differed from Europeans, for they appear not to have differed. Rather, the significant fact is that in the United States the average family was more able to realize its dream of a private home.[10]

The second necessary condition for American suburbanization was massive and sustained urban population growth. In 1800 only Philadelphia and New York had as many as 50,000 residents, and both were concentrated within areas of less than two square miles. But American cities virtually exploded in the nineteenth century, when only Great Britain experienced experienced an urbanization rate comparable to the United States. Between 1800 and 1910, the percentage of Europeans who could be considered townsmen tripled, but in America it increased seven-fold. In 1890, when the Bureau of the Census announced that the Western frontier was no more, the Reverend Samuel Lane Loomis told the graduating class at Andover Theological Seminary that Americans were living in "an age of great cities." In that year, the nation's population was already one-third urban, and the population in the Northeast was over one-half urban. With two million inhabitants, New York was already the second largest city in the world, and Chicago and Philadelphia each contained about a million inhabitants. Places like Minneapolis, Denver, Seattle, San Francisco, and Birmingham, which hardly existed in 1840, had become major regional metropolises. The urban trend continued in the twentieth century, and by 1980 the Census Bureau credited thirty-eight metropolitan areas with populations of more than a million persons. Instead of growing vertically and residentially more dense, American cities spread horizontally.

Rapid population growth does not, however, explain the deconcentration of the United States. Dozens of Latin American and African cities have experienced urban explosions as spectacular as any in this country without duplicating the dispersed pattern of a Chicago or a Los Angeles. Rather, the fundamental causes of suburbanization have been racial and economic.

No discussion of the settlement patterns of the American people can ignore the overriding significance of race. In comparison with the relative homogeneity of Denmark, Germany, England, or Japan, the cities of the United States, and particularly the larger metropolises, have long been extraordinarily diverse. In suburban terms, this has provided an extra incentive for persons to move away from their older domiciles—fear. After the mass migration of blacks from the South gained momentum during World War I, and especially after the Supreme Court decision in 1954 that school segregation was unconstitutional, millions of families moved out of the city "for the kids" and especially for the educational (as measured by standardized test scores) and social (as measured by family income) superiority of smaller and more homogeneous suburban school systems. The sprawling, single-story public schools of outlying towns, surrounded by playing fields and parking lots and offering superb facilities, new laboratories, and well-paid teachers, became familiar symbols of suburban life and educational manifestations of tract developments. Unlike the locked doors and grated windows of city institutions, they reflected an openness to nature. More importantly, the suburban school promised some relief from the pervasive fear of racial integration and its two presumed fellow-travelers—interracial violence and interracial sex.

In their understandable preoccupation with improving the status and opportunities of minorities, civil-rights activists and federal judges were unconcerned with the impact of racial change on cities and suburbs. Ordinary people, however, were concerned, and because they loved and feared for their children, they simply speeded up a process that probably would have occurred a few decades later in any case—the middle-class-white abandonment of the inner city. Court-ordered school busing usually involved city neighborhoods only and left the suburbs free to practice their own form of racial exclusivity.

Economic causes have been even more important than skin color in the suburbanization of the United States. Practically every adult American complains about the high cost of housing. The impression of unaffordability is reinforced by television and newspaper commentators who regularly report that the American dream is moving out of reach. In May 1984, for example, the average price of a new, single-family home in the United States crossed the six-figure threshold for the first time.

The simple fact, however, is that the real cost of shelter in the United States has been relatively low and affordable in comparison with housing costs elsewhere in the world. There are essentially six components to this interpretation.

The first element has been per capita wealth. From its earliest days as a nation, the United States has been peculiarly blessed by nature. Explorers marveled at wealth previously undiscovered. American land and sea and forest and mineral resources were immense. Although income comparisons are not available for the nineteenth century and are crude even for our own time, the best information we have suggests that the American people, at least in material terms, were the wealthiest in the world between 1790 and 1970, and especially after 1870, when living standards in the United States began an unprecedented rise that lasted for about a century. The prosperity of the nation was such that disposable personal income nearly doubled between 1897 and 1911. In 1949, for example, the American republic, with about 7 percent of the earth's population, had 42 percent of its wealth. Sweden, West Germany, and Switzerland, as well as oil-producing Kuwait and Saudi Arabia, have recently overtaken the United States, but their prosperity is a post-World War II phenomenon, and the notion of abundance has not been burned into their national psyches. The real message of America to the world, as David Potter has noted, has always been economic abundance, not freedom. With its vast middle class, the United States was the first society in the history of the world in which the distribution of wealth did not resemble the shape of a pyramid or a tree. As a "people of plenty," Americans could afford the wastefulness of low-density housing on the metropolitan fringe. In contrast, the Japanese have experienced centuries of scarcity, and their national psyche abhors waste. Even the droppings of humans are recycled as fertilizer on the tiny farms of that island nation.[11]

The second component of low cost has been inexpensive land. Although purchasers rarely regarded real estate as cheap, and although some developers quite literally made millions of dollars on land speculation (Henry Morgenthau in the Bronx and Otis Chandler in Los Angeles are well-known examples), a broader view would reveal that the most fundamental difference between the United States and its industrial rivals was that American real estate was affordable and available whereas elsewhere it was expensive and scarce. Building lots in North America have typically been priced from one-fourth to one-half comparably sized and located parcels in Europe and Japan. This is largely a function of population density (TABLE 16-2). America was a land of spaciousness and openness in contrast to the cramped circumstances of the Old World. Although the United States population has grown rapidly in the twentieth century, even in 1980 its gross population density was far less than any of its primary competitors. Abundant land has meant cheap land. Forests and farms have never been in short supply in the United States, and large tracts of land within two or three miles of the construction

TABLE 16-2
Population Density Per Square Mile of Major Industrial Nations at Sixty Year Intervals, 1860–1980

Nation	1860	1920	1980
Netherlands	263	589	1003
Belgium	385	670	842
Japan	NA	402	801
West Germany	221	343	643
United Kingdom	260	469	593
Italy	242	335	491
France	179	192	256
United States	11	36	63
Sweden	24	38	48
Canada	1	2	6
Australia	1	2	5

SOURCE: These calculations were derived simply by dividing total square miles into the population.

zone and adjacent to roads and rail lines could typically be purchased at bargain prices, encouraging large-scale, speculative investment. In 1984, for example, large tracts of land could still be purchased for less than one thousand dollars per acre within a one-hundred mile radius of-Dallas-Fort Worth. At the same time, land prices in Japan were impossibly high, reaching $100 per square foot (or $4 million per acre) in most suburban areas.

The third component of affordable price has been inexpensive transport, which has brought home sites within easy commuting range of workplaces. Americans have always manifested a remarkable enthusiasm for mechanical gadgetry, and although the omnibus, the steam railroad, the subway, and the automobile were all developed first in Europe, it was in the United States that they were most enthusiastically adopted and where they most immediately affected the lives of ordinary citizens. Especially before World War I, the subways, commuter railroads, elevated trains, and electric trolleys of American cities were faster, more frequent, more efficient, and more cost-effective than transportation options elsewhere in the world. The mass production of automobiles reinforced the pattern because for the first seventy years of the twentieth century the real price of both cars and fuel fell. With prices low because the cost of pumping crude oil from prolific wells in the Middle East or the Gulf of Mexico was usually less than 25 cents a barrel, American gasoline consumption doubled in the 1950s and again in the 1960s. Even in 1985, the cost of operating an automobile remains cheaper in the United States than in other advanced nations. As Homer Hoyt, the nation's most distinguished demographer before World War II, noted thirty-five years ago: "The location, size and shape of our cities has thus been a function of the transportation system prevailing during is main period of growth."[12]

The fourth component of low cost has been the balloon-frame house. The development of an inexpensive and peculiarly American method of building houses with 2-by-4-inch wooden studs simplified construction and brought the price of a private dwelling within the reach of most citizens.* Each region of the United States has a favored exterior material for new single-family houses—wood clapboard in the Northeast and North Central states, brick in the South, and stucco in the West—just as the British prefer brick and the French choose stucco. But in the United States more than 90 percent of all single-family houses have been of the balloon-frame type for the past century, regardless of exterior sheathing.

*Although called "two by fours," their actual dimensions are 1½-by-3½-inches.

Such structures are uncommon in other countries, in part because their citizens regard the balloon-frame as flimsy, and in part because they lack the timber resources of the heavily forested United States.[13]

A fifth component of low price has been the role of government, particularly at the federal level. The prevailing myth is that the postwar suburbs blossomed because of the preference of consumers who made free choices in an open environment. Actually, as Barry Checkoway has noted, most postwar families were not free to choose among several residential alternatives. Because of public policies favoring the suburbs, only one possibility was economically feasible. The result, if not the intent, of Washington programs has been to encourage decentralization. FHA and VA mortgage insurance, the highway system, the financing of sewers, the placement of public housing at the center of ghetto neighborhoods, and the locational decisions of federal agencies and the Department of Defense, to name only the most obvious examples, encouraged scattered development in the open countryside. While it was a national purpose to build subsidized highways and utilities outside of cities, it was not national policy to help cities repair and rebuild aging transit systems, bridges, streets, and water and sewer lines. Thus, suburbanization was not an historical inevitability created by geography, technology, and culture, but rather the product of government policies. In effect, the social costs of low-density living have been paid by the general taxpayer rather than only by suburban residents.[14]

The most important inducements to single-family residences have been contained in the murky provisions of the Internal Revenue Code and especially in the unusual American practice of allowing taxpayers to deduct mortgage interest and property taxes from their total taxable income. The present income tax in the United States began as the Underwood-Simmons Tariff Act and was signed into law by President Woodrow Wilson in 1913. From the beginning, "interest on indebtedness" and "other taxes" were exempt from taxation. The first tax measure, however, allowed a $4,000 exemption for married persons. Thus, only one-half of one percent of the American people were obliged to file returns.

As an inducement to homeownership, residential tax benefits first became important in the 1940s, when tax rates were raised substantially to offset the cost of World War II. Their importance in the suburban trend occurs because rental payments are not deductible whereas many housing costs are. Thus, a person renting an apartment (or house) must pay the rent from his net or after-tax income, with no deductions of any sort. If the individual is subject to a marginal tax rate of only 25 percent, a rent of $600 per month requires $800 of pretax income per month. At a marginal rate of 50 percent, a rent of $900 per month requires $1,800 of pretax income.

Homeowners are treated more generously. Everyone with a mortgage is allowed to deduct the interest. But most of the benefits flow to a relative few. According to a 1981 Congressional Budget Office study, about 60 percent of all homeowners either had no mortgage or used the standard deduction, thus gaining nothing from the mortgage privilege. Moreover, because persons with the largest houses typically have the highest interest payments and property taxes, it is they who receive the largest subsidies. The system works in such a way that a $20,000-a-year-bank teller living in a private apartment earns no housing subsidy. But the $250,000-per-year bank president living in a $400,000 home in the suburbs has a veritable laundry list of deductions. All $38,000 in interest payments would be subtracted from income, as well as all $7,000 in property taxes. His $45,000 subtraction would save him approximately $22,500 in taxes, or almost $2,000 per month. Indeed, for high-income professionals it is almost prohibitively expensive to rent and thereby fail to receive the advantages of favorable tax treatment. Thus, it happens that the average housing subsidy in an elite suburb will exceed by several times the average subsidy to a welfare family in the inner city.[15]

The size of this subsidy to homeownership is staggering and exceeds by four or five times all the direct expenditures Congress grants to housing. In 1981 deductions for property taxes and for interest payments on mortgages added up to a federal subsidy of $35 billion, and by 1984 the total had risen to $53 billion per year. Economists even argue that if the homeowner were governed by the rules applicable to an investor in other assets, then he would have to pay taxes on the imputed rent he receives as a landlord from a tenant, who happens to be himself but who could be anyone. These tax benefits make it increasingly likely, as taxable income rises, that homeownership will be preferred to renting. And with each passing year, the personal income tax becomes a more important factor in the financing and location of housing. Simply put, the Internal Revenue Code finances the continued growth of suburbia.[16]

In addition, Washington tax policies encourage Americans to over-invest in shelter and to under-invest in productive enterprise, especially in comparison with Japan. The deductibility of interest distorts American investment incentives in major ways. Investment in business is disadvantaged at a time of high inflation because businesses must pay income tax on illusory paper profits. American homeowners, by contrast, gain from inflation. They service their mortgage in shrinking dollars but are not obliged to pay taxes on the inflated value of their real estate.

One understandable result has been a shift from low-yield investment in plant and machinery to high-yield investment in housing. The most obvious example of this came in 1975, when Washington granted a $2,000 tax credit to purchasers of *new* housing. A particularly telling statistic reveals that the typical Japanese family saves 21 percent of its disposable income and spends 5 percent on housing. Americans squirrel away less than 6 percent and spend 15 percent on housing.

Federal, state, and local governments in the United States have also been influential through what they have not done. In Europe land is regarded as a scarce resource to be controlled in the public interest rather than exploited for private gain. Thus, the national and municipal governments have traditionally exercised stringent controls over land development, and they have operated on the theory that the preservation of farms and open space is an appropriate national goal and that suburban sprawl is undesirable. In Germany, by 1900 most municipal governments not only had extensive restrictions on the use of private land, but they also had large land holdings of their own that they used to control development on the city's periphery. The residue of this policy can be seen in 1985, when truck farmers tend their crops within two thousand yards of the skyscrapers of Dusseldorf, the richest city on the continent, not because alternative land uses would not yield a higher return, but because the government rejects the very possibility of development. In Great Britain, the Town and Country Planning Act (and subsequent legislation) laid the foundations for effective land planning by restricting suburban growth. The English have been so successful in this regard that the rate of farmland conversion to residential use has actually been lower since World War II than it was in the 1930s. In Sweden the 1947 Building and Planning Act effectively eliminated private land use decisions in metropolitan areas and abolished the free right to build in urbanized sections.[17]

In the United States, by contrast, the essential agents of growth have been individual investors working with privately owned parcels. The building effort has been unregulated in the European sense of that term. Because American social attitudes reflect a deep suspicion of government activities, municipalities and states have traditionally imposed as few restrictions as possible on land developers, and in the nineteenth century they granted streetcar franchises for minimal fees, or sometimes none at all. Both policies resulted in cheaper land and encouraged residential development after the extension of streetcar lines, not before. Long columns of tracks stretching into the countryside inspired the building up of areas beyond the corporate limits of the city. And in the United States, unlike Europe, transit moguls were free to speculate in suburban land. This gave them an extra incentive to extend streetcar and railroad service as fast as possible.

The sixth and final economic factor encouraging suburbanization has been the capitalistic system itself. One need not be a Marxist to observe that outward residential growth in North America coincided with the rise of industrial capitalism and the separation of the population into extremes of wealth and poverty. The "free enterprise" system provided incentives to land speculators, subdivision developers, building contractors, realtors, and lending institutions. When the economic system went into cyclical decline, as in the 1890s, 1930s, and 1970s, the construction of new housing and the movement of population to peripheral areas slowed.[18]

These six economic factors, along with racial prejudice and a pervasive fondness for grass and solitude, made private and detached houses affordable and desirable to the middle class, and they produced a spread-out environment of work, residence, and consumption that has thus far been more pronounced in the United States than elsewhere. The residential behavior of the American people, therefore, can be viewed primarily as the result of market forces and government policies. Some suburban families may have acted out of ignorance or irrationality, but most moved to a single-family house because it maximized their utility from a stable set of preferences. In other words, low-density housing was a good deal.

The Future

In 1968 the distinguished Columbia sociologist Herbert J. Gans wrote in the *New York Times Magazine* that "Nothing can be predicted quite so easily as the continued proliferation of suburbia." More recently, futurists have been predicting that electronic information technology will make traditional cities obsolete as communication costs drop (whether through computer modems or cheaper telephone rates) while transportation costs remain relatively high. A 1982 report by the Institute for the Future (under contract with the National Science Foundation) suggested that one-way and two-way home-information systems, called teletext and videotext, will penetrate deeply into daily life, making the home a place of employment as both men and women conduct much of their work at a computer terminal. Jack Niles, a senior research associate at the University of Southern California, said that, while there were only 10,000 to 20,000 telecommuters in the country in 1983, most of them part-time, the number would jump to almost ten million by the end of the 1980s. And in 1980, futurist Alvin Toffler's *The Third Wave* portrayed a 21st-century world in which the computer revolution would cancel out many of the fundamental changes wrought by the Industrial Revolution: the centralization and standardization of work in the factory, the office, and the assembly line. Instead, he predicted that concentrations of population will be unnecessary as more and more social interaction takes place through the medium of electronic circuitry.[19]

Such predictions, I think, will be proved wrong in the next quarter-century. The trading of stocks by home computer has been a dismal failure, and while there are some thriving pockets of telecommuters, these tend to be populated by highly individualistic writers and programmers, or by consultants who are just striking out on their own and need the low overhead of a home office. Companies that have tried systematically installing terminals in workers' homes largely have abandoned their experiments in the face of employee discontent. Most have discovered that their best employees miss the normal social interchange with their "second family" in the office. The companies themselves have found that by removing executives from the mainstream they miss hearing rumors and news and stumbling into old friends and potential business contacts.[20]

Instead of an even more deconcentrated nation, I would argue that the long process of suburbanization, which has been operative in the United States since about 1815, will slow over the next two decades and that a new kind of spatial equilibrium will result early in the next century. Quite simply, there are powerful economic and demographic forces that will tend to undercut the decentralizing process.

The most important limiting factor involves the rising real cost of energy and the reduced availability of fluid fuels. Until October 17, 1973, when the Organization of Petroleum Exporting Countries (OPEC), meeting in Kuwait, decided to raise the price of their crude oil by 70 percent, suburban homeowners took for granted that energy, like the wind and the rain, would always be there when they needed it. So did their leaders. In 1955 Admiral Lewis Strauss, chairman of the Atomic Energy Commission, predicted that energy would be so cheap by the 1970s that electricity would no longer have to be metered. In 1973 when the Arabs imposed an embargo on oil supplies to the United States and the Netherlands, Israel's two strongest allies, however, Americans first realized that their low-density life style was dangerously dependent upon cheap fuel. In fact, as noted previously, the real cost of energy, adjusted for inflation, fell by about 5 percent per year between 1950 and 1973.[21]

Predictably, low prices meant high consumption. In 1979 after yet another and larger OPEC price hike, Americans still consumed an average of 1.4 gallons per person per day, compared with 0.3 for Western Eu-

TABLE 16-3
American Gasoline Consumption Per Capita,
Ten Most Populous States, 1978
(in gallons)

Texas	671	Ohio	513
Florida	574	New Jersey	479
Michigan	549	Pennsylvania	442
California	534	Massachusetts	427
Ilinois	520	New York	349
United States Average	532		

SOURCE: *New York Times*, May 28, 1979.

rope and 0.2 for Japan. In 1983 after the federal government imposed an additional five-cent-per-gallon levy, United States gasoline taxes remained below twenty cents per gallon, compared to figures typically eight times that high in other advanced countries. And most of the nineteen cents was going to the Highway Trust Fund to maintain existing roads.[22]

Usage varies considerably by state. In Texas, where the pick-up truck is a substitute for the cowboy's horse, where distances are great, and where railroads cross urban streets at grade (Houston has one thousand such crossings while New York City has none), gasoline consumption was 671 gallons per person in 1980, the highest among major states. In New York, where the population is more concentrated and where public transportation is heavily patronized, gasoline consumption was only 349 gallons per capita, the lowest of the states. These figures illustrate the obvious—the higher the density, the lower the per capita energy use. Conversely, the closer residences are to workplaces and services, the greater will be the savings on fossil fuels for transportation. One estimate for the District of Columbia suggests that continued urban sprawl

TABLE 16-4
Registered Automobiles Per One Thousand
Residents, Ten Most Populous States, 1980

1. Illinois	659	6. Michigan	566
2. Florida	651	7. California	541
3. New Jersey	586	8. Texas	516
4. Ohio	579	9. Pennsylvania	489
5. Massachusetts	579	10. New York	408
United States Average	544		

SOURCE: United States Bureau of the Census, 1980.

in the region by 1997 would lead to a 60 percent increase in gasoline consumption over 1975.[23]

The nature of the dwelling also has a substantial impact upon energy consumption. The basic principle is that the less exterior wall and roof space per unit, the less energy will be needed for heating and cooling. Thus, a row house with party walls will consume 30 percent less energy and an apartment 50 percent less than a fully detached house with all four walls exposed to the elements. Together with the high cost of land, such calculations have led to a late twentieth-century American return to the row house, now given the more pretentious title of town house or condominium.[24]

In 1983 the word ''glut replaced ''crisis'' in most discussions of oil, and the memory of predawn gasoline lines faded. But the world's reserves of oil are definitely limited, and, according to Humberto Calderon Berti, energy minister of Venezuela and president of OPEC in 1980, may be exhausted as early as 2010. However much remains under the earth, however, the really important questions can be reduced to one: Is the real cost of energy in relation to disposable income going to go up or down? The answer, unfortunately, is up, because the development costs of finding and extracting oil are increasing dramatically as the world draws off oil from the easily established oil fields. The search for new supplies is taking major companies to the most remote, inhospitable regions of the globe at astronomical costs. Moreover, because most fossil fuels lie beneath the politically volatile Middle East, it is probably a delusion to believe that the industrial world can ever return to the days of energy abundance. Coal is plentiful in the United States and can be used directly for generating electricity. But under present economic conditions, synthetic fuels derived from coal will cost from two to four times as much as natural fluid fuels, and then only after years of development. Thus, the maintenance of our current suburban lifestyle will not only require a dependence upon attenuated oil supplies, but an increasing reliance upon such hazardous energy sources as nuclear power plants and liquid natural gas.

A second economic constraint on the continued proliferation of suburbia involves the cost of land. In the United States, the percentage of the purchase price devoted to the purchase cost of real estate itself rose from 11 percent in 1948 to 29 percent in 1982. As we noted earlier, the availability of inexpensive building plots was an essential precondition for the rapid deconcentration of the American metropolis.[25]

A third economic constraint on future suburbanization involves the cost of money. For generations, passbook depositors essentially subsidized homeownership because they had no alternative to low-interest passbook accounts. Their money went to savings banks, which then lent the funds to home buyers. So long as banks were paying savers less than 5 percent they could afford to issue mortages at below-market rates. After 1977, however, federal legislation guaranteed that small depositors could have access to high-yielding returns. Soon thereafter, droves of savers began to withdraw their funds from savings associations, opting instead for Treasury bills, money-market mutual funds, and other investments paying significantly higher returns. The passbook savings that had provided almost 90 percent of home loans as recently as the mid-1960s accounted for only 25 percent by 1980. Without the large pools of low-cost savings, the whole American system of providing low-cost mortgage money came to a screeching halt. It is not likely that Americans will ever again have access to home loans at below market rates.

A fourth factor inhibiting further deconcentration involves building technology. Since the development of the ballroom-frame house in the nineteenth century, technological change in the residential construction sector has lagged behind other sectors of the economy, and restrictive building codes continue to retard the introduction of most prefabricated techniques in the United States, which now is less advanced than Europe in developing inexpensive construction methods. There have been some evolutionary changes—like prefabricated roof trusses, kitchen cabinets, and whole bathroom drop-in cores—but their impact upon overall costs has been slight. Not surprisingly, the median price paid for a new home in the United States tripled between 1970 and 1982, rising from $30,000 to $88,800.[26]

Fifth, it is appropriate to note that the federal government, which did so much to spur suburbanization after World War II, began to stress conservation, rehabilitation, and mass transit with the passage of the Community Development Act of 1974. Although President Reagan has attempted to reverse some of these initiatives, proposals to build new highways and to encourage dispersal do not find a sympathetic ear even in the White House. The interestate system is 99 percent complete, and there are no new circumferential expressways on the drawing boards.

The final constraint on the future of suburbia is the changing structure of the American family. The dream house was designed around the needs of a bread-winning male and a full-time housewife who would provide her prince with a haven from the cold outside world, a goal warmly endorsed by such nineteenth-century theorists of suburbia as Andrew Jackson Downing, Catharine Beecher, and Calvert Vaux. Alas, the American population no longer fits the stereotype of the nuclear family. The rise in the divorce rate, the decline of average household size (2.75 in 1980), and the movement of women into the permanent labor force have reduced the appeal of the big suburban house—miles from work, neighbors, and support services.

In addition to these economic factors, race and fear will become less important as a stimulus to white flight. Black suburbanization became a major phenomenon in the 1970s as more minorities entered the American middle class, and by 1980 the proportion of suburban blacks to all blacks in the United States reached 23.3 percent. Many suburban communities have tried to reduce the speed of the shift by upzoning—or raising the minimum lot size for a single-family house from one unit per half acre to one per two acres. In California, an increasing number of small suburbs have taken the unusual step of returning to the medieval method of building walls and of denying entrance to all but their residents, employees, and visitors. Access to Rolling Hills, Indian Wells, Bradbury, and Hidden Hills is gained only after a variety of security checks—passing uniformed guards and closed circuit television monitors, displaying an automobile sticker, and showing an identification card. But recent legal decisions, such as the Mount Laurel judgment in New Jersey and the New Castle case in New York, now require suburban communities to accept a "fair share" of the disadvantaged populations in their areas and to make "an affirmative effort to provide housing for lower-income groups. Judicial fiat has thus allowed mobile homes to creep into Manalapan, New Jersey, and apartments to go up in exclusive Chappaqua, New York.[27]

As the black percentage in suburbia rises, it can be expected to stabilize or fall in large cities. The migration of blacks from the rural South has come to a virtual end. Indeed, more than 80 percent of the black population is already urban, indicating that shifts on the scale of that experienced earlier in the century will not be repeated. Moreover, a rise in the average age of city residents, and in particular a reduction in the number of male teenagers most disposed to felonious assault, will lead to a further drop in the crime rates of central cities. The impetus to suburban movement caused by these forces will thus be proportionately reduced (TABLES 16-5 and A-14).

The cycle of decline has recently caught up with the inner suburbs. Some, like Oak Park, Michigan, which sprang up in the 1950s across Detroit's northern boundary, are prospering because of their extraordinary religious and racial diversity. Others, however, are already encountering fiscal, educational, racial, and housing crises as severe as those which troubled major cities in the 1960s and 1970s. In these aging areas, a stable tax base coupled with increased service costs necessitated by a more elderly and less affluent population have put heavy pressure on revenues. Although there has never been a monolithic suburban experi-

TABLE 16-5
The Black Suburban Population in the Fifteen Largest Metropolitan Areas of the United States, 1980

Metropolitan Area	*Black Suburban Population*	*Black Percentage*
1. Washington	404,814	16.7%
2. Atlanta	215,909	13.5
3. St. Louis	201,348	10.6
4. Los Angeles	398,069	9.6
5. Baltimore	125,721	9.1
6. New York/New Jersey	544,545	8.7
7. Philadelphia	245,527	8.1
8. Cleveland	94,285	7.1
9. Houston	88,256	6.7
10. San Francisco Bay	145,566	6.5
11. Chicago	230,827	5.6
12. Detroit	131,478	4.2
13. Pittsburgh	73,790	4.0
14. Dallas/Fort Worth	65,955	3.9
15. Boston	34,205	1.6

Source: 1980 Census

ence in the United States, from Pasadena and Santa Ana in California to New Rochelle and Mount Vernon in New York, major twentieth-century problems can now be seen to have an important suburban dimension. Even in the booming Washington metropolitan area, every single jurisdiction within the Capital Beltway lost population in the 1970s, while every area beyond the circumferential roadway continued to grow. The shibboleth of newness contained the seeds of its own destruction.

As the suburban world begins to experience unmistakable signs of decay, both central cities and rural areas are making a comeback. Although statistics of population, jobs, and income do not support the thesis that a back-to-the-city movement has reversed the century-and-a-half-old suburban trend, the gentrification of older neighborhoods may be the harbinger of major demographic changes in the next two decades. The earliest indications of an urban revival came in the Georgetown section of Washington in the 1920s, and soon thereafter the renovation of brownstones in Brooklyn Heights became fashionable. Since 1965 important sections of Boston, San Francisco, Philadelphia, Savannah, and Baltimore have witnessed a reversal of the typical urban growth model as upper-status newcomers have replaced lower-income households. In 1976 a survey of the Harvard College Class of 1968 revealed that more than 60 percent were engaged in restoring old houses. In New York the former stables of Washington Mews have become chic residences; in Brooklyn renovation has spread from Park Slope to Fort Greene and Carroll Gardens; and in Charleston, former slave quarters have become fancy apartments. The transition has been so marked in Washington, especially on Capitol Hill, that black citizens have begun to form block associations to resist white invasions. Even Los Angeles, which has long been satirized as a city without a center, has begun developing a true, high-rise downtown business district. Population density in central Los Angeles increased almost 40 percent between 1950 and 1980, and there is a palpable sense of energy downtown, where more than five thousand new apartments were built in 1982 alone.[28]

Another signal that the United States may be entering a post-suburban age comes from the countryside. After 170 years of relative decline, rural areas grew by more than 11 percent in the 1970s, to nearly 60 million. This new growth is often called exurbia or ruburbia, terms which refer to scattered residential developments or houses set in agricultural districts without most urban amenities. A typical example of this phenomenon is Nash County, North Carolina. Within its 580 square miles are 120 prosperous industries and no cities. During rush hour, hundreds of workers may be seen hurrying along networks of good blacktop highways and feeder roads free of traffic jams. They come from farm houses, rural homes, small subdivisions, and trailer parks spread out along the rural roads.

I would argue, therefore, that suburbanization can best be seen as part of an urban growth developmental model. The spatial arrangement of cities depends less on ideology than on economics, less on national idiosyncrasies than on industrial development, technological achievement, and racial integration. Thus, American cities are not so much different from those of other countries as ahead of them, and we might expect cities elsewhere to follow the "North American" pattern just as soon as they have enough automobiles, highways, and disposable wealth to make it work.

Recent changes in Europe support the thesis that suburbanization is a common human aspiration and that its achievement is dependent upon technology and affluence. Since William Levitt erected his first houses outside Paris in 1965, the European landscape has become littered with all the trappings of suburban America: tract homes of sheet-rock construction, in place of the more traditional brick and plaster; massive roadside shopping centers that are damaging village-square businesses; and *Autobahns* and highways lined with gas stations and fast-food outlets. Split-level and colonial houses have been popular in England since the mid-1970s, Houston-style traffic congestion is choking European cities from Helsinki to Barcelona, and rich Turin families have been leaving the city for the safety of the neighboring hills. In Sweden single-family home construction has recently begun to dominate the market, and in Paris, middle- and upper-income professionals are now suburbanizing in the direction of Versailles. Even the fashionable Copacabana district on Rio de Janeiro's oceanfront has been yielding prestige to the newer neighborhoods of Ipanema, Leblon, and Barra Da Tijuca farther out from the central city. This conforms to the prediction of Leo F. Schnore in 1965 that Latin Americans would tend in the direction of their United States counterparts.[29]

No other nation, however, is likely ever to be as suburban as the United States is now, if only because their economic resources and prospects are even more limited than those of the American republic. Thus, the United States is not only the world's first suburban nation, but it will also be its last. By 2025 the energy-inefficient and automobile-dependent suburban system of the American republic must give way to patterns of human activity and living structures that are energy efficient. The extensive deconcentration of the American people was the result of a set of circumstances that will not be duplicated elsewhere.

Whatever the shape of the future, either in the Old World or in the New, no amount of urban gentrification or rural revival can obscure the fact that suburbanization has been the outstanding residential characteristic of American life. The process may slow in the next half-century as rising energy costs encourage higher population densities and less sprawl, and as "urban" problems of crime and obsolescence become typical of the inevitably aging suburbs. But the national cultural preference for privacy, for the detached home on its own plot, will not easily be eroded. Charles F. Kettering of General Motors, after Thomas Edison perhaps America's most important inventor and engineer, thought studying history was a waste of time. Arguing, "You never get anywhere looking in your rearview mirror," he preferred to focus on the future because, he said, "we will have to spend the rest of our lives there." Kettering was wrong. Decisions made in the past impose powerful restraints on the future. The location of buildings, of streets, and of highway systems imposes a measure of permanence on the form of community. Quite simply, the investment costs of building are so great that no generation can afford to replace old fabric with new; adaptation of the old has always been dominant. As H. J. Dyos and M. Wolff have noted "Inertia is part of the dynamic of urban change: the structures outlast the people who put them there, and impose constraints on those who have to adapt them later to their own use."[30]

The framework of growth, however hastily devised, tends to become the permanent structure. For better or for worse, the American suburb is a remarkable and probably lasting achievement. The words of an anonymous English jingle of the 1870s are apposite:

> *The richest crop for any field*
> *Is a crop of bricks for it to yield.*
> *The richest crop that it can grow,*
> *Is a crop of houses in a row.*

Appendix

Chapter 2 focused on the redistribution of the highest socio-economic groups outward in American cities in the half-century after 1820. In addition to this shift, however, four other patterns were also developing during the same period, and each pointed to the same result. First, peripheral areas were growing more rapidly than core cities. Philadelphia affords a good example of the trend. The highest population jumps between 1810 and 1830 were registered along the city's northern edges by the independent suburbs of Northern Liberties and Spring Garden.[1] In the next twenty-year period the areas of most spectacular growth moved beyond the inner suburbs to Moyamensing, Penn District, Richmond, and Kensington.[2]

By midcentury, numerous district and borough governments had been created to offer rudimentary urban services to the early suburbanites, and the Philadelphia Board of Trade was noting with pride that "the open grounds and commons of the suburbs are fast vanishing before the march of enterprise and construction."[3] When in the late 1850s good horsecar and steam-railroad service began to be offered to Germantown, that once-isolated village became one of the areas of high-percentage growth, and by 1870 all of the older suburbs were in relative decline. Throughout the remainder of the century, the point at which growth rates in the Philadelphia area tended to level and then decrease moved centrifugally at the speed of about one-half mile per decade.[4]

Two general observations can be made from these early Philadelphia growth patterns. The first is that they are not noticeably different from those of other large cities in the same period. As TABLE A-1 indicates, New York, Boston, Cleveland, and St. Louis experienced consistently higher suburban than central-city growth rates by 1840, and in some cases considerably earlier. The second point is that the pre-Civil War Philadelphia population experience differed in degree but not in substance from the more widely publicized suburban explosion of the post-World War II era. As FIGURE A-1 clearly shows, the pattern of centrifugal movement of high growth rates has simply moved outward from Philadelphia, Delaware, Montgomery, and Camden Counties to more recently developed Burlington, Chester, Gloucester, and Bucks Counties.

The second major suburbanization process that can be dated before 1860 is that the density curve in larger American cities had begun to level out. In 1830, for example, urban populations tended to be very concentrated at the center and to drop off sharply within a few miles. TABLE A-2 provides a comparison of the intensity of residential land use for minor civil divisions in Philadelphia County between 1800 and 1890. It indicates that the leveling of densities began there about 1850, when densities per square mile ranged from about 85,000 in Northern Liberties to about 100 in farming areas in the northernmost part of the county. Similarly, the density level at which the average citizen lived began to decline about 1850. In that year, less than 25 percent of the population lived at densities of less than 50 per acre (32,000 per square mile); by 1890 about 45 percent of the citizenry lived at this level of dispersal. Thus, as Philadelphia spread itself out in the middle of the nineteenth century, the density curve representing the variation in density from one ward to another became less steep.

In other words, the sharp distinctions between city and country which had characterized human settlement patterns until 1815 were becoming more blurred as suburban houses began to be scattered about the countryside. Writing in *The Crayon* in 1855, William J. Stillman lamented the absence of walls around American cities, noting, "There is something in a wall which divided the city from the country, and while it shuts them into the former, by a kind of stimulant to contrariness drives him out to the latter. Here city grows into country; we never know when we leave or enter into the other."[5]

Thirdly, the innermost sections of large American cities were experiencing an absolute loss in population by 1860. Philadelphia again provides evidence that residential decline in the urban core is not peculiar to the twentieth century.[6] There, the old city east of Seventh Street reached its all-time population peak in 1830, while the original Chestnut and Walnut Wards reached maximum residential densities in 1820 (TABLE A-3), and the old High Street Ward actually recorded its peak population in 1800, the first census in which ward totals were reported. The process accelerated after the Civil War, and between 1860 and 1890 the innermost wards were joined by nine additional wards in population decline.[7]

In New York City, the entire area south of Canal Street began declining about 1850; the population of the Second Ward dropped particularly dramatically from 9,300 in 1825 to 3,300 in 1860.[8] There, the expansion of commerce proceeded to such an extent that the new warehouse and retail buildings contained no residential quarters at all. Property values escalated rapidly as the lower part of Manhattan was consigned to business. As early as 1843, Caleb Woodhull, a prosperous Whig businessman, noted that:

> A few years ago no building in this city was uninhabited—now whole quarters are devoted to stores alone in which no one dwells. This will continue to be the case until, in the course of time, it is probable that the whole island will be covered with store houses, and the residences of those doing business there will be on the opposite shores and in contiguous counties.[9]

This phenomenon was not unique to antebellum Philadelphia and New York. In Boston six out of twelve wards showed a decrease in population between 1857 and 1867. In St. Louis and Baltimore, the central areas were also losing people when Abraham Lincoln was an unknown Springfield lawyer.[10] Newer and smaller cities such as Memphis exhibited the same tendency after 1890.[11] Thus the post-World War II population decline of such places as St. Louis, Philadelphia, Detroit, and Chicago, which is often cited as the most dramatic evidence that our cities are dying, can be seen to have an important nineteenth-century dimension.[12]

The fourth suburbanization process clearly evident before the Civil War was that middle-class citizens were commuting increasing distances to work. In Philadelphia, as FIGURES A-2 and A-3 illustrate, the three decades between 1829 and 1862 witnessed a significant movement of merchants toward the Schuylkill River in the western part of the city or toward Germantown west of Vine Street.[13] During those years the proportion of merchants living away from their stores and the average length of commutation approximately doubled. Similarly, the typical bank president, who had to reside away from his office under any circumstances, also traveled about twice as far to work in 1862 as in 1829. Banks and large stores remained concentrated east of Seventh Street, but the fine residential districts moved progressively farther away.[14] The inveterate Philadelphia diarist Sidney George Fisher was quick to identify this rapidly increasing ''taste for country life'' among his social peers, noting in 1847, "New and tasteful homes are being built every year," especially in Germantown, which was "perfectly healthy and the scenery very handsome."[15]

In New York City, where the transit system was more extensive and population growth more rapid than Philadelphia, the separation of work and residence proceeded even more swiftly, at least among the middle and upper classes.[16] An examination of the journey-to-work experiences of Manhattan attorneys at ten-year intervals between 1825 and 1973 (TABLE A-4) reveals that in terms of percentage increase in length of commutation, the most important decade was 1835 to 1845. Although the average journey-to-work tripled between 1835 and 1865, it required another half-century for the figure to triple again. The three decades between 1865 and 1895 were a time of little increase, but the electrification of the streetcar lines in the 1890s and the rapid extension of the elevated and subway systems led to a doubling of the distance between place of work and residence between 1895 and 1915.[17]

In Boston prosperous wholesale merchants in the 1840s were the first important group to migrate from the city, but they were soon joined by bankers and attorneys, both of whom had short hours and relatively high incomes. By 1851 almost half of the Hub's bank personnel and lawyers lived in the suburbs. In particular, outlying residence seemed to have appealed to those just below the highest rungs of the socioeconomic ladder, those who could not afford the prestigious homes of Boston's hills but whose incomes nevertheless offered the opportunity for substantial choice (TABLE A-7).[18]

The increase in the typical journey-to-work was more dramatic in New York than elsewhere. Using attorneys as the reference group because they were economically tied to the central business district and were sufficiently affluent to be able to live anywhere, the evidence in FIGURE A-4 indicates that in Memphis, Binghamton, and Paterson the average lawyer did not live as much as two miles from his office until 1925, a figure that had been exceeded by their New York City counterparts by 1860. And even in the huge Hudson River metropolis, most people continued to live and to work in close proximity until the construction of the elevated lines in the 1870s. But the process of suburbanization had begun among the trend-setters, and the foundation had been laid for more dramatic changes in the spatial arrangements of American cities in the decades to come (TABLE A-9).[19]

The process of suburbanization began between 1815 and 1835 and was well-advanced by the advent of the Civil War. Cities increased in size and area and also in internal structure as speculators constructed class-segregated residential suburbs for white-collar workers and managers at the circumference of the city. Whereas poorer housing had been hidden, quite literally, behind the agreeable facades of the more prosperous dwellings, and whereas blacks had lived in back-alley slums, close to the houses of whites, the new suburban areas, with discriminatory barriers, attractive topography, and inflated real-estate values, attracted wealthier residents (TABLE A-12).

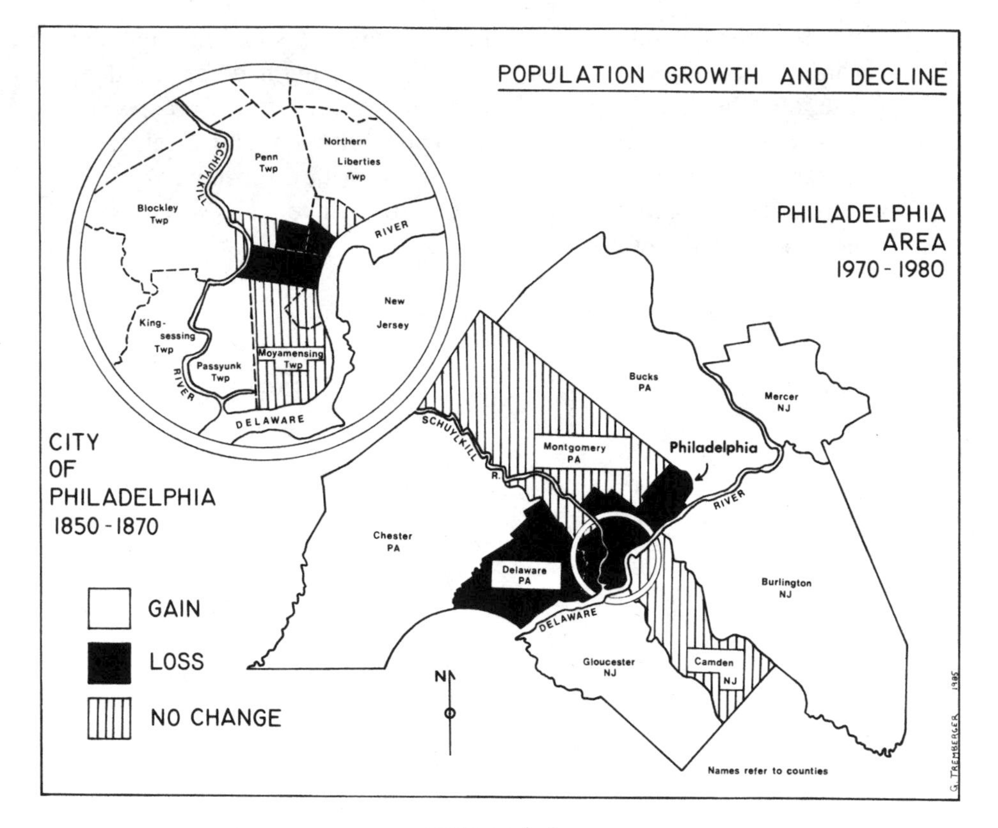
POPULATION GROWTH AND DECLINE
PHILADELPHIA AREA 1970 - 1980
CITY OF PHILADELPHIA 1850 - 1870
GAIN
LOSS
NO CHANGE
Penn Twp
Northern Liberties Twp
Blockley Twp
King-sessing Twp
Passyunk Twp
Moyamensing Twp
New Jersey
SCHUYLKILL
RIVER
DELAWARE
Bucks PA
Mercer NJ
Montgomery PA
Philadelphia
Chester PA
Delaware PA
Burlington NJ
Gloucester NJ
Camden NJ
SCHUYLKILL R.
N
Names refer to counties
G. TREMBERGER 1985

FIGURE A-1.

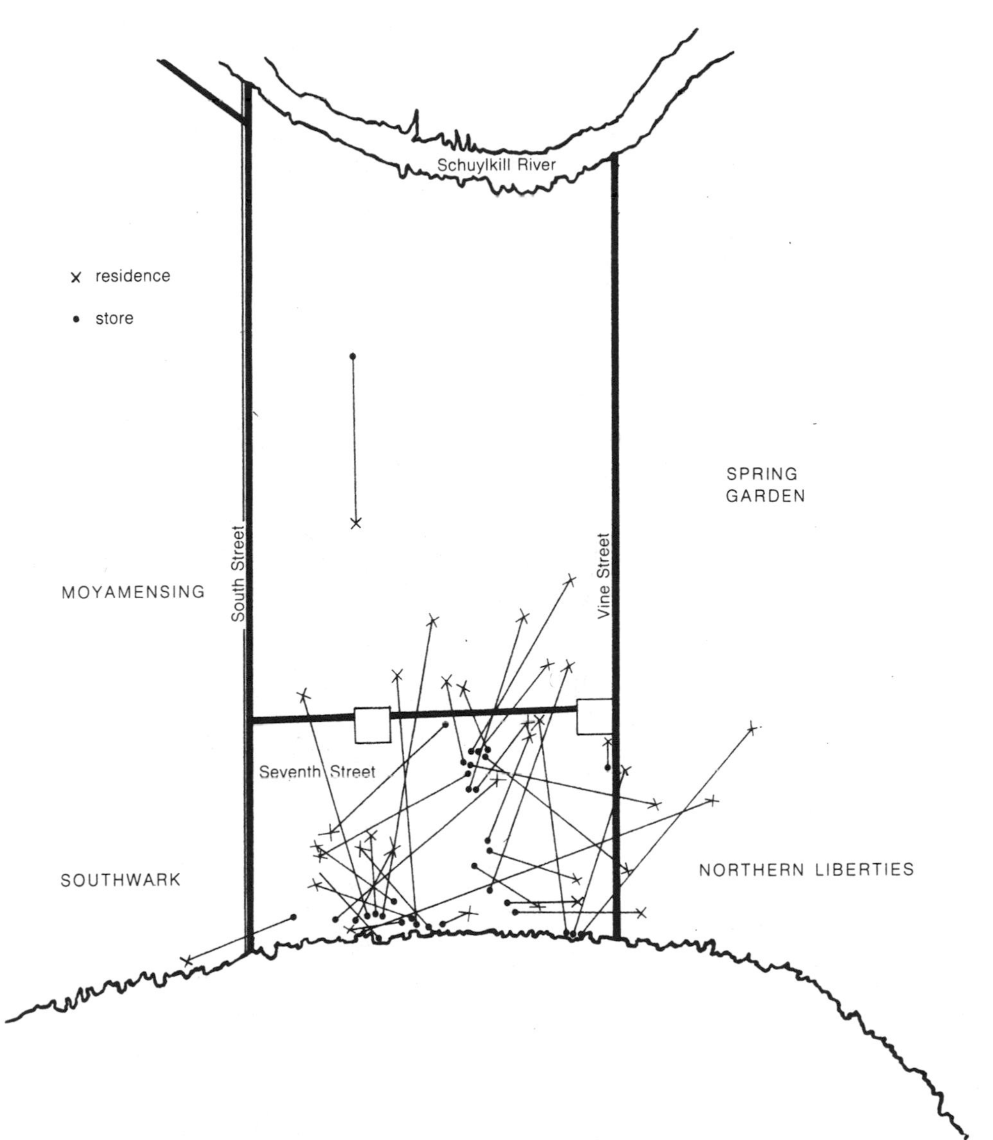

FIGURE A-2. The journey-to-work of Philadelphia merchants, 1829. Forty-two persons of a sample size of eighty-six commuted and were plotted.
SOURCE: *DeSilver's Philadephia Directory and Stranger's Guide, 1829* (Philadelphia: Robert DeSilver, 1829).

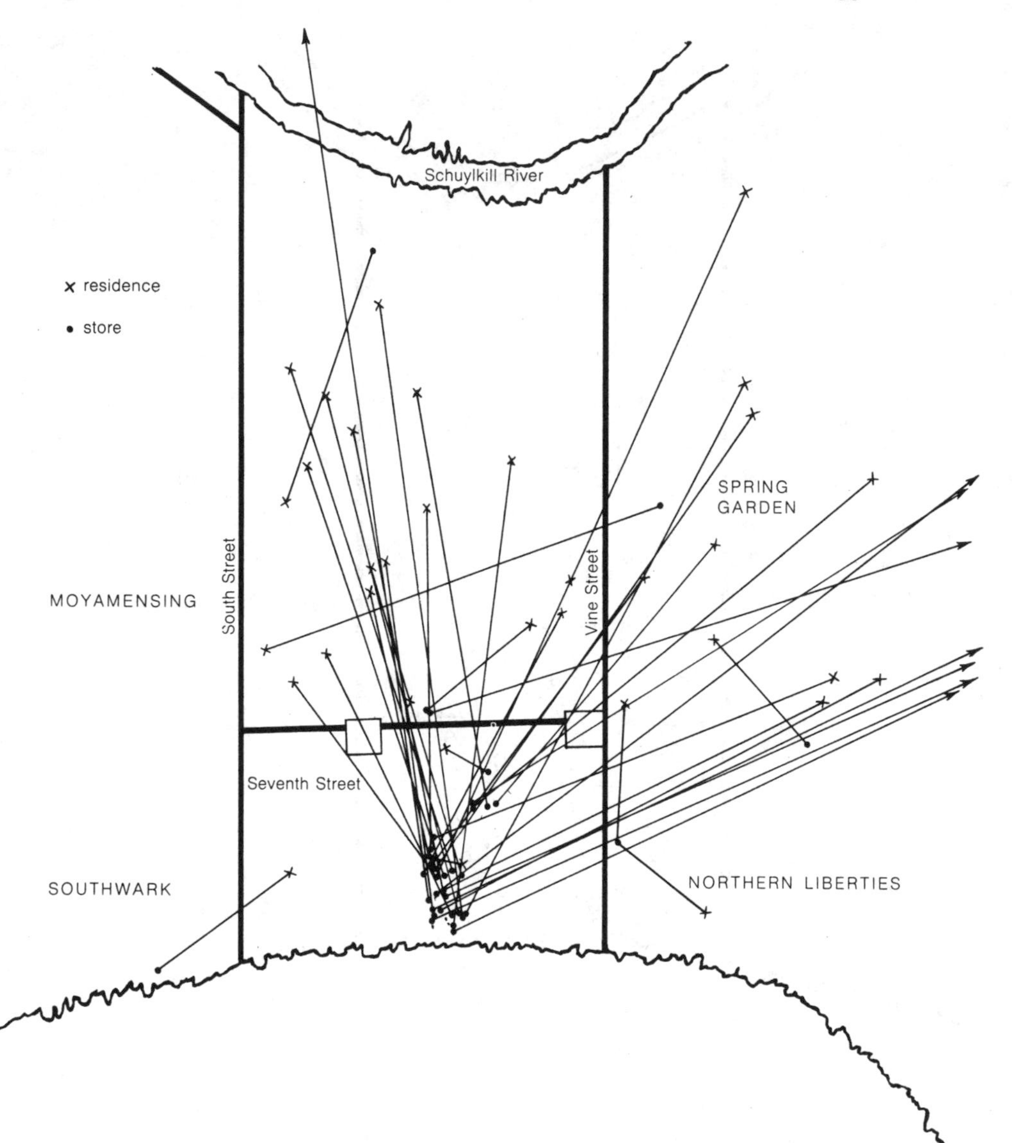

FIGURE A-3. The journey-to-work of fifty-six randomly selected Philadelphia lumber, china, glass, cotton, dry goods, and iron merchants. Seven men lived and worked at the same address, and eight could not be plotted.
SOURCE: *McElroy's Philadelphia City Directory for 1862* (Philadelphia: E. C. and J. Biddle and Company, 1862).

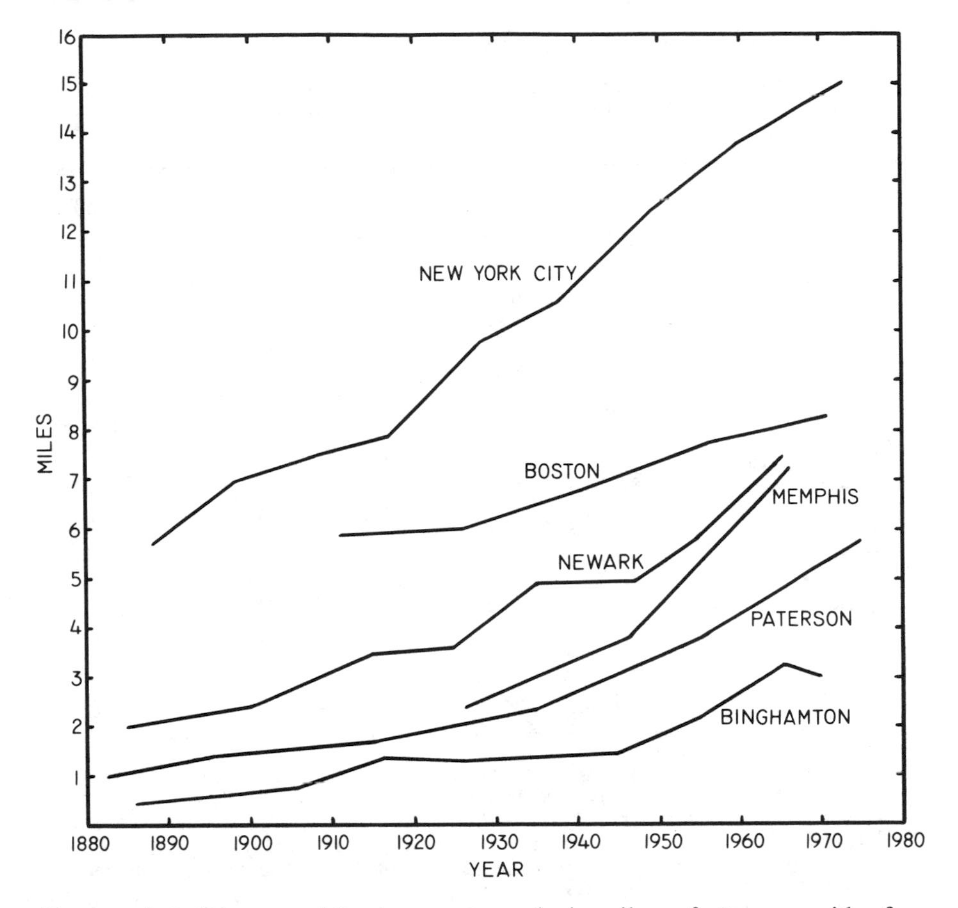

FIGURE A-4. Distance of the journey-to-work, in miles, of attorneys with offices in selected cities, 1880-1890.

Table A-1

Comparative Decennial Rates of Population Increase For Five Cities and Their Suburbs, 1810-1860

Metropolis	1810-20	1820-30	1830-40	1840-50	1850-60
New York					
City	28.4%	63.8%	54.4%	64.9%	57.8%
Suburbs	63.0	114.6	135.4	167.3	88.5
Boston					
City	28.1	41.8	38.5	61.0	29.9
Suburbs	25.7	35.2	44.8	84.7	53.8
Philadelphia					
City	18.8	26.1	16.4	29.6	11.0
Suburbs	25.3	47.8	51.7	74.8	48.8
Cleveland					
City			393.7	50.4	10.8
Suburbs			128.0	89.2	72.4
St. Louis					
City	187.4	27.3	181.4	51.8	20.0
Suburbs	33.3	52.6	135.9	309.9	100.7

The New York periphery here includes only Brooklyn, although the addition of Hudson County, New Jersey would not have altered the pattern. Boston's suburbs here include Chelsea, Charlestown, Cambridge, Brighton, Brookline, Roxbury, and Dorchester. The Philadelphia periphery is defined as everything in Philadelphia County outside the city before 1854. The Cleveland periphery is defined as Cuyahoga County minus the city of 1830. The St. Louis suburbs are defined as those sections of St. Louis County that were outside the city in 1840. In the New York and Philadelphia areas, the suburban growth advantage actually dates to 1800 rather than 1810.

Source: Decennial United States Census Returns, 1810-1860.

Table A-2

Comparative Density Per Square Mile of Selected Minor Civil Divisions, Philadelphia City and County, 1800-1890

Civil Division	Square Miles	1800	1830	1860	1890
East Central City	.683	50,000	65,000	59,000	42,000
West Central City	1.594	12,000	25,000	65,000	50,000
Northern Liberties	.556		50,000	75,000	65,000
Southwark	1.050	9,000	19,000	50,000	72,000
Kensington	1.899		1,800	25,000	65,000
East Spring Garden	.639		14,000	65,000	63,000
West Spring Garden	1.000		1,500	32,000	62,000
Penn District	1.984		500	16,000	60,000
Kingsessing	8.923	60	110	250	2,880
Moreland	4.779	75	85	110	800
Byberry	9.045	69	108	130	850
County Total	129.583	625	1,460	4,367	8,600

Source: The square mile totals are from William Bucke Campbell, "Old Towns and Districts of Philadelphia," Philadelphia History, V (1942), 94. The calculations of density, which are rounded off in the case of heavily populated areas, are based on the United States Census for the years indicated.

Table A-3

Pre Civil War Population Losses in Philadelphia's Inner City Wards

Ward	1800	1820	1830	1850
Before Adjustment				
Lower Delaware	3377	3237	6863	6425
High Street	2792	2529	4427	3549
Chestnut	2693	2930	4115	2443
Walnut	2169	2817	3428	2544
Dock	2235	2415	5378	5734
Upper Delaware	3067	3396	5763	7224
New Market	4865	5892	12983	14405
Total	21,198	23,216	42,957	42,324
After Adjustment				
Lower Delaware	3377	3237	3920	3672
High Street	2792	2529	2528	2028
Chestnut	2693	2930	2352	1396
Walnut	2169	2817	1960	1454
Dock	2235	2415	3074	3276
Upper Delaware	3067	3396	3295	4128
New Market	4865	5892	7418	8232
Total	21,198	23,216	24,548	24,185

The ward totals are underlined for the census in which an all-time maximum population density was reached. The totals given in the before adjustment column are those reported by the United States Census. The figures in the after adjustment column equal four-sevenths of the official total as listed from 1830 through 1850. This alteration, which was first applied by William Hastings for a smaller number of wards over a shorter period, takes account of the extension of ward boundaries from Fourth Street to Seventh Street in 1825.

Source: United States Census Returns and John Daly and Allen Weinberg, Genealogy of Philadelphia County Subdivisions (Philadelphia Department of Records, 1966), 92-100.

Table A-4

Average Length of the Journey-To-Work, In Miles, of Attorneys With Offices in Manhattan, By Decades, 1825-1973

Year	Sample Size	% Outside Manhattan	Avg. JTW Man. Res.	Avg. JTW Others	Avg. JTW All Attys
1825	88	1%	.67 miles		
1835	84	2	.81		
1845	84	14	1.39		
1855	97	23	1.81		
1865	99	35	2.43		
1875	100	44	2.91		
1888	123	41	4.02	8.47 miles	5.74 miles
1898	118	45	5.10	9.21	6.95
1908	129	38	5.41	10.76	7.49
1917	108	47	6.50	9.47	7.87
1928	120	61	5.17	12.55	9.72
1938	120	64	4.43	14.01	10.56
1949	120	62	4.09	17.40	12.41
1960	120	68	3.36	18.62	13.74
1973	120	66	3.49	20.97	15.03

The average journey-to-work of attorneys who both lived and worked in Manhattan began to decline after 1908 because some law offices began to move away from the financial district toward the mid-town area, a trend which has continued into the 1980s. William Meyers, Robin Lewis, and Dorothy Freeman assisted with sample selections and calculations.

Source: The samples were selected from Longworth's City Directories for 1825 to 1875, from Trow's New York City Directory for 1888 through 1917, from the New York City Directory of Lawyers for 1928, and from the Columbia Law Register for 1938 through 1973.

Table A-5

Population Density Per Square Mile in 1980 of the Twelve Largest Declining Cities and the Twelve Largest Growing Cities

Declining Cities	Density PSM	Growing Cities	Density PSM
1. New York City	23,283	1. Los Angeles	6,395
2. Chicago	13,173	2. Houston	2,867
3. Philadelphia	12,986	3. Dallas	2,416
4. Detroit	8,620	4. San Diego	2,711
5. Baltimore	8,646	5. San Antonio	2,980
6. San Francisco	14,955	6. Phoenix	2,428
7. Washington	10,385	7. Indianapolis	1,862
8. Milwaukee	6,641	8. Memphis	2,226
9. Cleveland	7,560	9. San Jose	4,216
10. Boston	12,239	10. Columbus	3,087
11. St. Louis	7,403	11. Jacksonville	654
12. Pittsburgh	7,639	12. Seattle	5,391

Source: The population of the city was divided by the area in square miles of the municipal corporation in 1980.

TABLE A-6

NINETEENTH CENTURY VILLAGE INCORPORATIONS IN WESTCHESTER COUNTY, NEW YORK

Village	In the Town Of	Incorporation Date
Ardsley	Greenburgh	1896
Bronxville	Eastchester	1898
Croton	Cortlandt	1898
Dobbs Ferry	Greenburgh	1873
Hastings	Greenburgh	1879
Irvington	Greenburgh	1872
Larchmont	Mamaroneck	1891
Mamaroneck	Mamaroneck	1895
Mount Kisco	Bedford	1875
New Rochelle	New Rochelle	1858
North Pelham	Pelham	1896
North Tarrytown	Mount Pleasant	1875
Pelham	Pelham	1896
Pelham Manor	Pelham	1891
Peekskill	Cortlandt	1827
Pleasantville	Mount Pleasant	1897
Port Chester	Rye	1868
Sing Sing	Ossining	1813
Tarrytown	Greenburgh	1870
White Plains	White Plains	1866

Table A-7

Average Length of the Journey-To-Work, in Miles, of Attorneys With Offices in Boston, at Fifteen Year Intervals, 1911-1971

Year	Sample Size	% Outside Boston	Avg. JTW Boston Res.	Avg. JTW Others	Avg. JTW All Attys
1911	76	56.6%	2.7 miles	8.6 miles	5.9 miles
1926	77	58.4	2.9	8.5	6.0
1941	78	62.8	3.2	9.0	6.8
1956	77	72.7	3.6	9.2	7.7
1971	78	78.2	3.1	9.8	8.3

The sample was chosen by taking every tenth attorney in the directory of a given year until a total of 76-78 was reached.

Source: Boston City Directories for 1911, 1926, 1941, 1956, and 1971.

Table A-8

Average Length of the Journey-To-Work, in Miles, of Attorneys with Offices in Newark, New Jersey, by Decades, 1915-1965

Year	Sample Size	% Living Outside Newark	Avg JTW All Attys
1915	200	39.5%	3.47 miles
1925	200	41.0	3.64
1935	200	57.5	4.90
1947	100	62.5	4.98
1955	100	63.5	5.85
1965	100	78.0	7.40

The sample was taken by selecting every tenth attorney in the city directory until the desired total was reached. Milton Hess and Robin Lewis assisted in the calculations.

Source: Newark City Directories for the above years.

Table A-9

Average Length of the Journey-To-Work, in Miles, of Attorneys with Offices in Binghamton, New York, 1859-1970

Year	Sample Size	% Outside Binghamton	Avg. JTW Bing Res	Avg. JTW Others	Avg. JTW All Attys
1859	23	0%	.36 miles		.36 miles
1870	25	0	.32		.32
1886	40	0	.46		.46
1896	47	0	.60		.60
1906	40	0	.74		.74
1916	45	2	1.28	2.57 miles	1.31
1926	44	11	1.05	3.28	1.29
1935	45	13	1.08	3.18	1.32
1945	50	12	1.11	3.73	1.42
1955	50	18	1.45	5.15	2.12
1965	49	45	1.59	5.11	3.27
1970	49	42	1.47	5.09	2.99

Source: Binghamton City Directories for the years listed. The calculations were made for me by Robert Davis.

Table A-10

Workplace Destinations of Randomly Selected Residents of South Orange, New Jersey, By Percentages, 1914-1954

Destination	1914	1934	1954
South Orange	44.4%	19.4%	19.7%
New York City	25.9	33.3	15.8
Newark	7.4	23.6	37.1
Other Suburbs	0	12.5	15.7
Widows	7.4	2.7	5.6
Unknown	14.8	8.3	6.1
Sample Size	35	72	74

Source: The method consisted on selecting every twentieth name from the South Orange City Directories for the years indicated.

Table A-11

Home Values and Home Owners Loan Corporation Residential Security Ratings for Selected St. Louis Area Communities, 1930

Community	% Home Owners	Value of Owned Homes in thousands of dollars				Dominant HOLC Rating, 1940
		Below 3	3-7	8-15	Over 15	
St. Louis	31.6%	11.6%	49.6%	30.1%	7.7%	C
Univ. City	50.1	4.8	19.9	23.2	37.5	A
Web. Groves	78.5	7.3	29.5	39.2	23.2	A
Maplewood	56.3	7.0	60.9	28.2	12.0	C
Kirkwood	68.3	13.0	41.9	32.1	11.2	
Richmond Hts	57.2	6.5	28.3	50.6	13.8	B
Clayton	49.8	2.5	7.5	17.4	72.3	A
Ferguson	72.7	9.7	52.2	29.6	7.4	B
Brentwood	66.3	14.6	70.5	13.2	1.5	C
Ladue	84.6	1.2	4.8	14.9	79.1	A

Source: 1930 United States Census Tracts for St. Louis; HOLC City Survey Files in the National Archives, and National Resources Committee, Regional Planning, Part II - St. Louis Region (Washington: Government Printing Office, 1936), 52.

Table A-12

Comparative Degree of Metropolitanization of Selected American Cities, 1850 and 1980

City	1850	1980
New York		
City Population	515,547	7,071,030
Metro Population	868,918	16,121,297
City Percentage	59.3%	43.9%
Philadelphia		
City Population	121,376	1,688.210
Metro Population	371,860	4,700,996
City Percentage	32.6	35.9
Pittsburgh		
City Population	46,601	423,928
Metro Population	150,000	2,260,919
City Percentage	31.1	18.8
Cleveland		
City Population	7,977	573,822
Metro Population	29,000	1,895,997
City Percentage	27.5	30.3
St. Louis		
City Population	25,000	453,085
Metro Population	77,860	2,344,912
City Percentage	32.1	19.3

The New York City population in 1850 is defined as Manhattan; the metropolitan population is defined as the other four boroughs of the modern city plus adjacent counties in New Jersey. The 1850 Philadelphia metropolitan population is defined as Kensington, Northern Liberties, Southwark, Spring Garden, Moyamensing, South Penn, Richmond, West Philadelphia, Germantown, and Frankford. The 1850 Pittsburgh city figures reflect the situation before the substantial annexation of that year. The 1850 city of Cleveland is defined as the First Ward. The annexations of the 1840s areincluded in the suburban total.

Source: United States Census, 1850 and 1980; Paul Studenski, ed., The Government of Metropolitan Areas in the United States (New York, 1930), 17-8; and Roderick D. McKenzie, The Metropolitan Community (New York, 1933), 194-97.

Table A-13

New Housing Starts in the United States, 1935-68
(In Thousands)

Year	Total Starts	FHA Starts	VA Starts	Public Housing
1935	216	14	0	5
1936	304	49	0	15
1937	332	60	0	4
1938	339	119	0	7
1939	458	158	0	57
1940	530	180	0	73
1941	619	220	0	87
1942	301	166	0	55
1943	184	146	0	7
1944	139	93	0	3
1945	325	41	9	1
1946	1015	69	92	8
1947	1265	229	160	3
1948	1344	294	71	18
1949	1430	364	91	36
1950	1408	487	191	44
1951	1420	264	149	71
1952	1446	280	141	59
1953	1402	252	157	36
1954	1532	276	307	19
1955	1627	277	393	20
1956	1325	192	271	24
1957	1175	168	128	49
1958	1314	295	102	68
1959	1495	332	109	37
1960	1230	261	75	44
1961	1284	244	83	52
1962	1439	259	78	30
1963	1582	221	71	32
1964	1502	205	59	32
1965	1451	196	49	37
1966	1142	158	37	31
1967	1268	180	52	30
1968	1484	220	56	38

Source: U. S. Department of Housing and Urban Development, HUD Trends: Annual Summary (Washington, 1970); and U. S. Bureau of the Census, Housing Construction Statistics, 1889-1964 (Washington, 1966), Table A-2.

Table A-14

American Cities and Suburbs in Which Afro-Americans Constituted a Majority of the Population, 1980

Community	1970 Percentage	1980 Percentage
East St. Louis, Illinois	69.1%	95.6%
East Cleveland, Ohio	58.6	86.5
East Orange, New Jersey	53.1	83.5
Compton, California	71.0	74.8
Prichard, Alabama	50.5	73.7
Gary, Indiana	52.8	70.8
Washington, D. C.	71.1	70.3
Atlanta, Georgia	51.3	66.6
Detroit, Michigan	43.6	63.1
Newark, New Jersey	54.2	58.2
Inglewood, California	11.2	57.3
Birmingham, Alabama	42.0	55.6
New Orleans, Louisiana	45.0	55.3
Baltimore, Maryland	46.4	54.8
Camden, New Jersey	39.1	53.1
Richmond, Virginia	42.0	51.3
Wilmington, Delaware	43.6	51.1

Source: United States Bureau of the Census, 1980. Only cities of 35,000 or more residents in 1980 were included.

Notes

INTRODUCTION

1. Museum of Modern Art, *Modern Architecture: International Exhibition* (New York, 1932), 179; Amos Rapoport, *House Form and Culture* (Englewood Cliffs, N.J., 1968); Marcel Griaule and Germaine Dieterlen, "The Dogon of the French Sudan," in Daryll Forde, ed., *African Worlds* (London, 1965); and *New York Times,* January 15, 1984.

2. One of the most perceptive interpretations of the meaning and origin of suburbanization is Robert Fishman, "The Origins of the Suburban Idea in England," *Chicago History,* XIII (Summer 1984), 26–35. See also, Arthur Edwards, *The Design of Suburbia: A Critical Study of Environmental History* (London, 1981).

3. On the problems of definition, see Herbert J. Gans, "Urbanism and Suburbanism as Ways of Life: A Re-evaluation of Definitions," in Arnold M. Rose, ed., *Human Behavior and Social Processes* (Boston, 1962), 625–48; Robin J. Pryor, "Defining the Rural-Urban Fringe," *Social Forces,* XLVII (December 1968), 202–15; and Chauncey D. Harris, "Suburbs," *American Journal of Sociology,* IL (1943), 1–13.

4. Because much of Newark is swamp or port, only about seventeen square miles of the city are in fact usable to the residents.

5. A disappointing recent attempt at a broad suburban synthesis is John R. Stilgoe, "The Suburbs," *American Heritage,* XXXV (February–March 1984), 21–36. Much more useful are Stilgoe's comprehensive *Common Landscape of America, 1580–1845* (New Haven, 1982); and Peter O. Muller's interdisciplinary *Contemporary Suburban America* (Englewood Cliffs, N.J., 1981), and *The Outer City* (Washington, 1976).

6. Ronald R. Boyce and Dlilp K. Pal, "Changing Urban Densities," *Annals of the Association of American Geographers,* LII (1962), 321–22; Noel P. Gist, "Developing Patterns of Urban Decentralization," XXX (March 1952), 257–67; and Colin Clark, "Urban Population Densities," *Journal of the Royal Statistical Society,* Series A, Part IV (1951), 490–96.

7. David Popenoe, *The Suburban Environment: Sweden and the United States* (Chicago, 1977) compares an American suburb (Levittown in Pennsylvania) with a Swedish suburb (Vallingby) to show how design affects life. See also, David Goldfield, "Suburban Development in Stockholm and the United States: A Comparison of Form and Function," in Ingrid Hammarstrom and Thomas Hall, eds., *Growth and Transformation*

of the Modern City (Stockholm, 1979), 139–56; David Goldfield, "National Urban Policy in Sweden," *APA Journal,* Winter 1982, pp. 24–38. For the argument that Sweden is now following the "North American" pattern, see Thomas Falk, *Urban Sweden: Changes in the Distribution of Population in the 1960s in Focus* (Stockholm, 1977).

8. *Time* magazine exaggerated only slightly on September 12, 1977, when it reported that "the United States is the only nation in which a private house has been brought within the reach of the broad middle class." See also, Jim Kemeny, "Forms of Tenure and Social Structure," *British Journal of Sociology,* XXIX (March 1978), 44–46; Kemeny, "Urban Homeownership in Sweden," *Urban Studies,* XV (October 1978), 313–20; Colin Duly, *The Houses of Mankind* (London, 1979), 5–27; *Survey of AFL/CIO Members' Housing: 1975* (Washington, 1975), 16; and Theodore Caplow, *Middletown Families: 50 Years of Change and Continuity* (Minneapolis, 1982), 14.

9. Among the many studies of this topic, see Leo F. Schnore, "Measuring City-Suburban Status Differences," *Urban Affairs Quarterly,* III (September 1967), 95–108; Schnore, "The Socio-Economic Status of Cities and Suburbs," *American Sociological Review,* XXVII (1963), 76–85; Peter O. Muller, "Toward a Geography of the Suburbs," *Proceedings of the Association of American Geographers,* VI (1974), 36–40; and Richard F. Muth, "Urban Residential Land and Housing Markets," in Harvey S. Perloff and Lowden Wingo, Jr., eds., *Issues in Urban Economics* (Baltimore, 1968), pp. 285–333.

10. A fascinating recent analysis of the Cape Town slums is in *New York Times,* January 27, 1984.

11. Delhi is an exception that conforms to the North American pattern. Ved Mehta, "City of Dreadful Night (Calcutta)," *The New Yorker,* March 21, 1970, pp. 47–112; *Sofia News,* August 8, 1979. The suburban character of Lebanese society is the subject of Fuad I. Khuri, *From Village to Suburb: Order and Change in Greater Beirut* (Chicago, 1975).

12. In the conclusion of this volume, I will acknowledge that Parisian suburbs in the direction of Versailles appear to be following the "North American pattern."

13. The attempt of *favelados* to improve their surroundings with crude water systems, sewers, and small businesses is recounted in Peter Lloyd, *Slums of Hope: Shanty Towns of the Third World* (New York, 1979). See also, Peter W. Amato, "A Comparison: Population Densities, Land Values, and Socio-economic Class in Four Latin American Cities," *Land Economics,* XLVI (November 1970), 447–55; and Walter Harris, *The Growth of Latin American Cities* (Athens, Ohio, 1971).

14. By focusing on the city as a physical container, I acknowledge my debt to urban geographers as well as to Roy Lubove, who called for such an approach in his essay, "The Urbanization Process: An Approach to Historical Research," *Journal of the American Institute of Planners,* XXXIII (January 1967), 32–6. The best overview of the geographical approach to urban form is Michael P. Conzen, "Analytical Approaches to the Urban Landscape," in Karl W. Butzer, ed., *Dimensions in Urban Geography: Essays on Some Familiar and Neglected Themes* (Chicago: University of Chicago Department of Geography Research Paper No. 186, 1978), 128–65.

15. Oscar Handlin, *Truth in History* (Cambridge, 1979), 291.

CHAPTER 1: *Suburbs as Slums*

1. Ivar Lissner, *The Living Past: 7,000 Years of Civilization* (New York, 1957), 44; and C. Leonard Woolley, *Ur of the Chaldees* (Ithaca, N.Y., 1982), 31–32, 112.

2. *The Oxford English Dictionary* provides the best and most complete history of the word "suburb." In a contemporary-sounding practice, patricians sometimes lived just outside the Toulouse gates in order to avoid paying local taxes. John Mundy, *Liberty and Political Power in Toulouse* (New York, 1954), passim.

3. By 1775 the Philadelphia suburbs of Northern Liberties and Southwark contained more than 7,000 people, compared with 16,560 in the city proper. By 1790 the suburbs contained 14,000 and the city 28,000. Carl Bridenbaugh, *Cities in the Wilderness: The First Hundred Years of Urban Growth in America* (New York, 1938), 146, 267–68, 306, 411.

4. New Orleans suburbs appear on the 1815 plan of the city surveyor. Early Parisian suburbs are discussed in Orest Ranum, *Paris in the Age of Absolutism: An Essay* (New York, 1968), 2–4.

5. Kenneth T. Jackson, "Urban Deconcentration in the Nineteenth Century: A Statistical Inquiry," in Leo F. Schnore, ed., *The New Urban History: Quantitative Explorations by American Historians* (Princeton, N.J., 1975), 110–42.

6. In 1700 Edo (Tokyo) had over one million inhabitants and was the world's largest city. By 1815, however, it had entered a period of relative decline and was not as large as London. On London deconcentration, see Harold James Dyos, *Victorian Suburb: A Study of the Growth of Camberwell* (Leicester, 1961).

7. This residential pattern was reflected in the steep drop in land values as one moved away from the core. In Cincinnati in 1815, an acre of land sold for $500 to $1,000 within two miles of the center. Three miles farther out the price dropped to less than $50 per acre. Thomas Senior Berry, *Western Land Prices Before 1861: A Study of the Cincinnati Market* (Cambridge, 1943), 11.

8. Fortified St. Augustine, Pensacola, Mobile, and Santo Domingo, like most other Spanish settlements in the New World, were military and religious outposts rather than cities in the early centuries of their existence.

9. Because cities were so compact, the only possibility of traveling much more than a mile to work was to live and labor on opposite edges of the built-up area. Allan R. Pred, "Manufacturing in the American Mercantile City, 1800–1840," *Annals of the Association of American Geographers,* LVI (July 1966), 307–38.

10. Gerardus Antonius Wissink, *American Cities in Perspective, With Special Reference to the Development of Their Fringe Areas* (Assen, Sweden, 1962), 76–77; Dyos, *Victorian Suburb,* 34–35; and Asa Briggs, *Victorian Cities* (London, 1963), 280–81.

11. I have borrowed this turn of phrase from K. H. Schaeffer and Elliot Sclar, *Access for All: Transportation and Urban Growth* (Baltimore, 1975), 10.

12. Evidence on the relative status of various Philadelphia neighborhoods and suburbs in 1775 is provided by Dee Blomstedt, "Wealth Distribution in Colonial Philadelphia, 1774–1775" (Master's thesis, Columbia University, 1974); and by Sam Bass Warner, Jr., *The Private City: Philadelphia in Three Periods of Its Growth* (Philadelphia, 1968), 12–15.

13. Quoted in David Schuyler, "Public Landscapes and American Urban Culture, 1800–1870: Rural Cemeteries, City Parks, and Suburbs," (Ph.D. dissertation, Columbia University, 1979), chapter 6.

14. Named for the London suburb guarding the "south work" of London Bridge, Southwark was settled by Swedes in 1638, forty-four years before Philadelphia itself was born. Southwark was created a municipality in 1762 and incorporated in 1794 before becoming part of Philadelphia in 1854. Norman J. Johnston, "The Caste and Class of the Urban Form of Historic Philadelphia," *Journal of the American Institute of Planners,* XXXII (November 1966), 334–49. See also, M. Antonia Lynch, "The Old Dis-

trict of Southwark in the County of Philadelphia," *Philadelphia History,* I (1909), 83–126; and *Southwark, Moyamensing, Wecacoe, Passyunk, Dock Ward for Two Hundred and Seventy Years* (Philadelphia, 1892).

15. I did not carry my complete occupational analysis of Southwark beyond 1790 because Philadelphia grew mainly to the north and west, and Southwark was no longer on the developing fringe. The 1790 comparison with Philadelphia was made by John K. Alexander, "The City of Brotherly Fear: The Poor in Late Eighteenth-Century Philadelphia," in Kenneth T. Jackson and Stanley K. Schultz, eds., *Cities in American History* (New York, 1972), 93.

16. The suburban prostitution tradition was carried to the cow towns as well as to the largest cities. Thus the dance halls and brothels of Wichita and Abilene in Kansas and of Calgary in Alberta were on the outskirts. Robert R. Dykstra, *The Cattle Towns* (New York, 1969), 233; and Max Foran, "Land Development Patterns in Calgary, 1884–1945," in Alan F. J. Artibise and Gilbert A. Stelter, eds., *The Usable Urban Past: Planning and Politics in the Modern Canadian City* (Toronto, 1979), 295.

17. Quoted in Nelson Manfred Blake, *Water for the Cities: A History of the Urban Water Supply Problem in the United States* (Syracuse, 1956), 8.

18. George Rogers Taylor, ed., "Philadelphia in Slices: The Diary of George G. Foster," *The Pennsylvania Magazine of History and Biography,* XCIII (January 1969), 34, 39, 41.

19. In 1703 the Dutch accounted for 80 percent of New York's North Ward, the poorest in the city, while the English, who had taken the city from the Dutch in 1664, dominated the more fashionable streets. Thomas Archdeacon, "New York City During the Leislerian Period, 1689–1710: A Social and Demographic Interpretation" (Ph.D. dissertation, Columbia University, 1971), 228.

20. The best general analyses of urban spatial structure in the nineteenth century are: David Ward, "The Internal Spatial Structure of Immigrant Residential Districts in the Late Nineteenth Century," *Geographical Analysis,* I (October 1969), 337–53; David Ward, "A Comparative Historical Geography of Streetcar Suburbs in Boston, Massachusetts and Leeds, England, 1850–1920," *Annals of the Association of American Geographers,* LIV (1964), 477–89; Sam Bass Warner, Jr., *Streetcar Suburbs: The Process of Growth in Boston, 1870–1900* (Cambridge, 1962), 16–20; Thomas K. Peucker, "Johann Georg Kohl, A Theoretical Geographer of the Nineteenth Century," *Professional Geographer,* XX (1968), 247–50; Jay W. Forrester, *Urban Dynamics* (Cambridge, 1969); Edward Gross, "The Role of Density as a Factor in Metropolitan Growth in the United States," *Population Studies,* VII (November 1954), 113–20; David Harrison, Jr., and John F. Kain, "An Historical Model of Urban Form," *Harvard University Program on Regional and Urban Economics Discussion Paper No. 63* (Cambridge, 1970); Leo F. Schnore, "The Timing of Metropolitan Decentralization: A Contribution to the Debate," *Journal of the American Institute of Planners,* XXV (November 1959), 200–206; and Peter Goheen, *Victorian Toronto* (Chicago, 1972), 8–10, 150–52.

21. Richard C. Wade, *Slavery in the Cities: The South, 1820–1860* (New York, 1964), 274–77.

22. Wade, *Slavery in the Cities,* 78–79, 275–76.

23. The best analysis of the social and functional meaning of the villa is David R. Coffin, *The Villa in the Life of Renaissance Rome* (Princeton, 1979). See also, Lewis Mumford, "Suburbia: The End of a Dream," *Horizon,* II (July 1961), 62.

24. Carl Bridenbaugh, *Cities in Revolt: Urban Life in America, 1743–1776* (New York, 1964), 24–25, 144.

25. Similar country homes around Philadelphia included those of Robert Strettel at

Germantown, James Logan at Stenton, Andrew Hamilton at Bush Hill, and George McCall at "Chevy Chase.,, Bridenbaugh, *Cities in the Wilderness,* 306, 411.

26. What is now the Bronx was Westchester County until 1874, when the southern portion became the "annexed district" of New York City. In 1898 an even greater annexation resulted in the remainder of the present borough breaking away from Westchester.

27. A discussion of commuting issues is in Hans Huth, *Nature and the American: Three Centuries of Changing Attitudes* (Berkeley, 1957), 121–25.

CHAPTER 2: *The Transportation Revolution*

1. Quoted in David Schuyler, "Public Landscapes and American Urban Culture, 1800–1870: Rural Cemeteries, City Parks, and Suburbs" (Ph.D. dissertation, Columbia University, 1979), chapter 7.

2. The most crowded neighborhoods, including the Lower East Side and East Harlem in New York, did not reach maximum densities until the 1890s, but they were not typical of urban America, and they were both losing population by 1910. A fine economic analysis of the shift is Raymond L. Fales and Leon Moses, "Land-Use Theory and the Spatial Structure of the Nineteenth-Century City," *The Regional Science Association Papers,* XXXVII (1972), 49–80.

3. Elliott D. Sclar, et. al, *Shaky Palaces: Home Ownership and Social Mobility in Boston, 1890–1970* (New York, 1984), chapter 3.

4. *The Diary of Phillip Hone, 1828–1851,* ed. Allan Nevins (New York, 1936), 202.

5. *New York Herald-Tribune,* May 17, 1931; *New York Times,* April 29, 1951; and John B. Pine, *The Story of Gramercy Park* (New York, 1921),

6. In 1860 Penn District was a net exporter of workers, whereas Spring Garden was a net importer. Sam Bass Warner, Jr., *The Private City: Philadelphia in Three Periods of Its Growth* (Philadelphia, 1968), 60, 135.

7. Stuart Blumin's sophisticated analysis of occupational and residential mobility in Philadelphia between 1820 and 1860 errs, I believe, in suggesting that movement toward the periphery represented *downward* mobility while movement toward the center certified increasing status. Stuart Blumin, "Mobility and Change in Ante-Bellum Philadelphia," in Stephen Thernstrom and Richard Sennett, eds., *Nineteenth Century Cities: Essays in the New Urban History* (New Haven, 1969), 47–69.

8. Daniel Bown, *A History of Philadelphia, with a notice of villages in the vicinity* (Philadelphia, 1839), 183.

9. Roger Lotchin, "San Francisco, 1846–1856: The Patterns and Chaos of Growth," in Kenneth T. Jackson and Stanley K. Schultz, eds., *Cities in American History* (New York, 1972), 151–60; Homer Hoyt, *One Hundred Years of Land Values in Chicago: The Relationship of the Growth of Chicago to the Rise in Its Land Values, 1830–1933* (Chicago, 1933), chapter 6; and Mark B. Riley, "Edgefield: A Study of an Early Nashville Suburb," *Tennessee Historical Quarterly,* XXXVII (Summer 1978), 133–54.)

10. Max Foran, "Land Development Patterns in Calgary, 1884–1945," in Alan F. J. Artibise and Gilbert A. Stelter, eds., *The Usable Urban Past: Planning and Politics in the Modern Canadian City* (Toronto, 1979), 297. Zane L. Miller has carefully described the spatial patterns of residential growth in Cincinnati and has interpreted the city's political experience in terms of the interaction of those areas. The outlying "hilltop" section was the high-status sector. Miller, *Boss Cox's Cincinnati: Urban Politics in the Progressive Era* (New York, 1968), 18–43.

11. Park Village East, designed by John Nash in the Regents Development of London in 1825, was not as easily distinguished from the core city as was Brooklyn. Robert A. M. Stern makes the claim that it is the first suburb in "The Suburban Alternative for the 'Middle City,'" *Architectural Record,* August 1978, pp. 93–100; and Stern, *The Anglo-American Suburb* (London, 1981), 4–19.

12. The best study of the neighborhood is Clay Lancaster, *Old Brooklyn Heights* (Rutland, Vt., 1961). The best recent survey of the entire borough is David Ment, *A Brief History of Brooklyn* (Brooklyn, 1979). In 1770, when Brooklyn Heights was known as "Clover Hill," three residents were already commuting via private ferry to lower Manhattan.

13. The *town* of Brooklyn was one of six towns that were later to make up the *city* and then the *borough* of Brooklyn. It was the *town* of Brooklyn that experienced early suburbanization; outlying towns such as Flatlands remained rural throughout the nineteenth century. Walt Whitman, *I Sit and Look Out* (New York, 1932), 145.

14. The consolidation of New York ferry companies into a few large conglomerates was a harbinger of a similar process with ground transportation firms later in the century. Henry Evelyn Pierrepont, *Historical Sketch of the Fulton Ferry* (Brooklyn, 1879), 161–63; and George Hilton, *The Staten Island Ferry* (Berkeley, 1964).

15. New York *Evening Post,* November 17, 1854. Quoted in Edward Spann, *The New Metropolis: New York, 1840–1857* (New York, 1981), 185; David Ment, *The Shaping of a City* (Brooklyn, 1980), 28–36.

16. Lois P. Gilman, "The Development of a Neighborhood: Bedford, 1850–1880: A Case Study," (Master's thesis, Columbia University, 1971), 8–31.

17. Spann, *The New Metropolis,* 185.

18. Brooklyn early developed a reputation as a city of churches. In 1855, when it had only about one-fourth as many people as New York City, it had half as many churches (111 as compared to 218). *Census of the State of New York for 1855* (Albany, 1857).

19. Harold Coffin Syrett, *The City of Brooklyn, 1865–1898: A Political History* (New York, 1944), 241–42; James Wilson, *History of the City of New York* (New York, 1893), IV, 23–25.

20. Although Hezekiah Pierrepont often used the anglicized version of his name, "Pierpont," his children used the original spelling, as does Pierrepont Street, which runs through his former estate.

21. Pierrepont Papers, Long Island Historical Society, Brooklyn. The best sketch of Pierrepont's life is in Henry R. Stiles, *History of the City of Brooklyn,* II (Brooklyn, 1869), 145–51.

22. Although "Four Chimneys" no longer stands, Brooklyn Heights still contains more than seven hundred houses that were built between 1820 and 1865. The neighborhood blocks, still quiet and tree-lined, became New York City's first official Historic District in 1966.

23. The Pierrepont holdings extended from the waterline almost to Fulton Street, a distance of almost a half-mile at its extreme.

24. Although a public ferry had been established between New York and Brooklyn in 1642, regularly scheduled steam service at frequent intervals did not begin until 1814. Ralph Foster Weld, *Brooklyn Village* (New York, 1938), 28.

25. My analysis is based upon an examination of every name on Pierrepont, Remsen, and Montague Streets listed in the 1841 Brooklyn City Directory. About 50 percent of the heads of households on these streets worked in Gotham, and about 30 percent were merchants. In London, by contrast, in 1831 about 20 percent of the inhabitants of aristocratic Paddington and Kensington were capitalists, merchants, or professionals, indi-

cating that those neighborhoods were relatively of much lower status than Brooklyn Heights. D. A. Reeder, "A Theater of Suburbs: Some Patterns of Development in West London, 1801–1911," in H. J. Dyos, ed., *The Study of Urban History* (London, 1968), 255.

26. The promenade, which offers a magnificent view of New York harbor, was finally built under the aegis of Robert Moses shortly after World War II.

27. On an irregular basis, a steam ferry began service to Hoboken in 1814.

28. The most important stimulus to the suburbanization of New Jersey was the system of tunnels. In 1910 the Hudson Tubes were completed and Pennsylvania Station was opened, beginning a service that would make it possible to travel from Newark to midtown Manhattan in less time than it took from most of Brooklyn and the Bronx. John E. Bebout and Ronald J. Grele *Where Cities Meet: The Urbanization of New Jersey* (Princeton, 1964), 43.

29. Lotchin, "San Francisco," 151–60.

30. I am discounting the *voiture omnibus,* or "vehicle for all," which some scholars claim began in Paris in 1819, and the *carosse à cinq sous,* the brainchild of Blaise Pascal, which made its debut in Paris in 1662. The best comparative analysis is John P. McKay, *Tramways and Trolleys: The Rise of Urban Mass Transport in Europe* (Princeton, 1976). The best surveys of the literature are Glen E. Holt, "Urban Mass Transit History: Where We Have Been and Where We are Going," in Jerome Finster, ed., *The National Archives and Urban Research* (Athens, Ohio, 1974), 81–105, and Foster M. Palmer, "The Literature of the Street Railway," *Harvard Library Bulletin,* XII (1948), 117–38.

31. McKay, *Tramways and Trolleys,* 10–12.

32. The best overviews of United States public transportation in the nineteenth century are: Glen E. Holt, "The Changing Perception of Urban Pathology: An Essay on the Development of Mass Transit in the United States," in Kenneth T. Jackson and Stanley K. Schultz, eds. *Cities in American History* (New York, 1972), 324–43; Joel A. Tarr, "From City to Suburb: The 'Moral' Influence of Transportation Technology," in Alexander B. Callow, Jr., ed., *American Urban History: An Interpretive Reader with Commentaries* (New York, second edition, 1973), 202–12; and George Rogers Taylor, "The Beginnings of Mass Transportation in Urban America: Parts I and II," *Smithsonian Journal of History,* I (Summer and Autumn 1966), 35–52, 31–54.

33. Quoted in Holt, "Changing Perception," 325.

34. The condition of urban road services in the nineteenth century is considered in the chapter on "The New Age of Automobility."

35. This calculation is based upon a ridership of about six million per year. In Europe, the omnibus was less successful; in Brussells, the various lines attempted between 1835 and 1867 failed for lack of patronage.

36. On October 9, 1826, a three-mile-long railroad began hauling granite in Massachusetts. Another small claimant operated out of Charleston. The first passenger train in public service in the world opened as the Stockton and Darlington Railway in England in September, 1825.

37. Harlem had been a Dutch farming settlement since the seventeenth century, but for two hundred years it was a relatively isolated outpost far from the city.

38. Joseph Warren Greene, Jr., "New York City's First Railroad: The New York and Harlem, 1832 to 1867," *New York Historical Society Quarterly Bulletin,* IX (January 1926), 107–23.

39. These towns, as well as Yonkers and Mamaroneck, predated the railroad by almost as long as Harlem. But they also were agricultural villages before real-estate de-

velopments for the express purpose of escaping high city prices and rents made their appearance in the 1850s.

40. W. E. Baxter, *America and the Americans* (London, 1855), .

41. Mildred H. Smith, *Early History of the Long Island Railroad* (New York, 1958), 2–8.

42. Spann, *The New Metropolis,* 189.

43. The chapter on the "Main Line" will develop this theme.

44. The Mayor of Boston boasted of a commuting population of 100,000, but a persuasive estimate by Henry C. Binford has recently put the figure at about 20,000 in 1860. Charles J. Kennedy has placed the commuting population of the city at the time of the Civil War at 10 percent of working persons. Henry C. Binford, "The Suburban Enterprise: Jacksonian Towns and Boston Commuters," (Ph.D. dissertation, Harvard University, 1975); and Charles J. Kennedy, "Commuter Services in the Boston Area, 1835–1860," *Business History Review,* XXXVI (Summer 1962), 153–70.

45. *Boston Post,* June 24, 1850. Quoted in Binford, "The Suburban Enterprise," 125–28.

46. Assuming three hundred round trips per year into the city, the cost per ride would have approximated five cents, or even less than the omnibus. In many cases, however, the commuter would have connected with the omnibus at the railroad terminal, and thus used both modes.

47. Kennedy, "Commuter Services," 169; and Robert A. Gross, "Transcendentalism and Urbanism: Concord, Boston, and the Wider World," (Paper presented at biennial conference of the Nordic Association for American Studies, Copenhagen, Denmark, Jan. 27, 1982).

48. Acts of sabotage were particularly virulent in Suffolk County on Long Island, but incidents were reported on more than a dozen lines before 1860.

49. For examples of such legislation, see George Rogers Taylor, "The Beginnings of Mass Transportation, in Urban America: Part II," *The Smithsonian Journal of History,* I (Autumn 1966), 39–43.

50. New York *Tribune,* June 11, 1849. Quoted in Spann, *The New Metropolis,* 192. See also, Tarr, "From City to Suburb," 202–12.

51. The official opening of the first horsecar line in 1836 was hailed by New York City Mayor Walter Browne as an event which "will go down in the history of our country as the greatest achievement of man." An excellent introduction to transit history is K. H. Schaeffer and Elliot Sclar, *Access for All: Transportation and Urban Growth* (Baltimore, 1975).

52. On some omnibus lines it was possible to buy an annual ticket that reduced the per ride cost to four cents, an option occasionally available with the horsecar also. The overwhelming proportion of riders paid in cash whenever they roade. Frederic W. Speirs, *The Street Railway System of Philadelphia: Its History and Present Condition* (Baltimore, 1897), 9–11.

53. Dallas Smythe, "Economic History of the East Bay," (Ph.D. dissertation, University of California, Berkeley, 1937); and Binford, "The Suburban Enterprise," 140–49.

54. George Rogers Taylor, ed., "The Diary of Sidney George Fisher, 1859–1860," *Pennsylvania Magazine of History and Biography,* LXXXVII (April 1963), 191–92.

55. Quoted in Holt, "Changing Perception," 329. See also, Harry J. Carman, *The Street Surface Railway Franchises in New York City* (New York, 1919); John Anderson Miller, *Fares Please! From Horse Carts to Streamliners* (New York, 1941), and Harold

E. Cox, "Daily Except Sunday: Blue Laws and the Operation of Philadelphia Horsecars," *Business History Review,* XXXIX (September 1965), 228–42.

56. In the same year, Boston had 13.5 million riders, or about 35,000 per day. Because Boston had about a third as many inhabitants as New York, market penetration by the horse railways was about the same.

57. Spann, *The New Metropolis,* 417.

58. Taylor, "Diary of Sidney George Fisher," 191–92.

59. Miller, *Fares Please,* 33–37.

60. McKay, *Tramways and Trolleys,* 15–26.

61. Omnibuses may have been more successful in England than in the United States because of excellent street paving, which allowed them to operate at almost the same speed as a vehicle on rails. David Ward, "A Comparative Historical Geography of Streetcar Suburbs in Boston, Massachusetts and Leeds, England, 1850–1920," *Annals of the Association of American Geographers,* LIV (1964), 477–89.

62. Robert Fishman, "The Origins of the Suburban Idea in England," *Chicago History,* XIII (Summer 1984), 26–35; Reeder, "A Theater of Suburbs," 253–55; and David M. Gordon, "Capitalist Development and the History of American Cities," in William K. Tabb and Larry Sawyers, eds., *Marxism and the Metropolis: New Perspectives in Urban Political Economy* (New York, 1978), 25–63.

63. Adna Ferrin Weber, *The Growth of Cities in the Nineteenth Century: A Study in Statistics* (New York, 1899), 468–69.

CHAPTER 3: *Home, Sweet Home*

1. A careful study of the pre-Civil War suburban economy is Henry Claxton Binford, "The Suburban Enterprise: Jacksonian Towns and Boston Commuters, 1815–1860" (Ph.D. dissertation, Harvard University, 1973).

2. William Dean Howells, *Suburban Sketches* (New York, 1871), 11–12.

3. Philippe Aries, *Centuries of Childhood* (New York, 1965), 8–12; Aries, "The Family and the City," in Alice S. Rossi, ed., *The Family* (New York, 1978), 227–35; and Elizabeth Janeway, *Man's World, Woman's Place* (New York, 1971), 9–26.

4. One thrust of the recent work of French sociologists and cultural historians has been to deny that the family is the meeting ground for social and biological necessity. Instead, they have viewed it as an instrument of oppression and disaster. Jacques Donzelot, *The Policing of Families,* trans. Robert Harley (New York, 1979), 3–87. The notion of privacy and overcrowding, as Colin Duly has noted, is relative and in different cultures cannot simply be measured by counting the number of individuals sharing a house. Duly, *The Houses of Mankind* (London, 1979), 5–27.

5. Heinrich Engel, *The Japanese House: A Tradition for Contemporary Architecture* (Rutland, Vermont, 1964), 221–29; and Lewis Mumford, *The City in History; Its Origins, Its Transformations, and Its Prospects* (New York, 1961), passim.

6. For especially perceptive inquiries into these changes, see Kirk Jeffrey, "The Family as Utopian Retreat From the City: The Nineteenth Century Contribution," *Soundings,* LV (Spring 1972), 21–42; Philippe Aries, "The Family and the City," *Daedalus,* CVI (Spring 1977), 227–35; and David P. Handlin, *The American Home: Architecture and Society, 1815–1915* (Boston, 1979), passim.

7. William G. Eliot, Jr., *Lectures to Young Women* (Boston, 1880, first published in 1853), 55–56. Quoted in Jeffrey, "Family as Utopian Retreat," 21.

8. Gwendolyn Wright, *Building the Dream: A Social History of Housing in America* (New York, 1981), chapters 5 and 6.

9. Quoted in Wright, *Building the Dream,* chapter 5. On the other side of the issue, see Betty Friedan, *The Feminine Mystique* (New York, 1963), 307, which called the home "a comfortable concentration camp."

10. *The American Builder,* September 1869, p. 180. Although faith in rising property values is traditional in the United States, housing prices have not kept pace with inflation over the past century. Matthew Edel, Elliott D. Sclar, and Daniel Luria, *Shaky Palaces: Homeownership and Social Mobility in Boston, 1870–1970* (New York: Columbia University Press, 1984).

11. Stephan A. Thernstorm and Peter R. Knights, "Men in Motion: Some Data and Speculation About Urban Population Mobility in Nineteenth Century America," *Journal of Interdisciplinary History,* I (Autumn 1970), 7–35. Theodore Caplow, et al., *Middletown Families: 50 Years of Change and Continuity* (Minneapolis, 1982), 104.

12. Quoted from Edel, *Shaky Palaces,* chapter 8.

13. Dolores Hayden, *The Grand Domestic Revolution: A History of Feminist Designs for American Homes, Neighborhoods, and Cities* (Cambridge, Mass., 1981), 34–38. Friedrich Engels, *The Origin of the Family, Private Property, and the State* (Moscow: Progress Publishers, 1977), 73–75. Jonathan Beecher and Richard Bienvenu, eds., *The Utopian Vision of Charles Fourier* (Boston, 1971).

14. Clare Cooper, "The House as Symbol of the Self," in Lan, Jen, et al., eds., *Designing for Human Behavior: Architecture and Behavioral Sciences* (Stroudsburg, Pa., 1974), 130–46.

15. In Rome there was a ban against accumulating too much land in one place. It was considered socially irresponsible to use one's wealth to build up a huge estate because that would deprive others of the chance to live in the area. Lecture of Professor Paul B. Harvey of Pennsylvania State University at Harvard University, April 28, 1980.

16. Mark Girouard, *Life in the English Country House: A Social and Architectural History* (New Haven, 1978), 2; and Orest Ranum, *Paris in the Age of Absolutism* (New York, 1969), 197.

17. Donna Merwick, "Dutch Townsmen and Land Use: A Spatial Perspective on Seventeenth Century Albany, New York," *William and Mary Quarterly,* XXXVII (1980), 53–78; and Robert Ostergren, "A Community Transplanted: The Formative Years of a Swedish Immigrant Community in the Upper Middle West," *Journal of Historical Geography,* V (1979), 190–212.

18. William Alexander Percy, *Lanterns on the Levee: Recollections of a Planter's Son* (New York, 1941), 272.

19. A perceptive essay on ideas toward land is Suzannah Lessard, "The Suburban Landscape: Oyster Bay, Long Island," *The New Yorker,* LII (Oct. 11, 1976), 44–79.

20. *The Crayon,* IV (1857), 304.

21. Mary Mix Foley, *The American House* (New York, 1980); William John Murtagh, "The Philadelphia Row House," *Journal of the Society of Architectural Historians,* XVI (December 1957), 9; Sam Bass Warner, Jr., *Streetcar Suburbs: The Process of Growth in Boston, 1870–1900* (Cambridge, Mass., 1962), 136–41; and David B. Hanna, "Creation of an Early Victorian Suburb in Montreal," *Urban History Review,* IX (October 1980), 38–64.

22. Recent excavation in Brooklyn Heights has provided evidence of the disreputable nature of back yards. *New York Times,* June 11, 1978.

23. Even in neighborhoods where houses were completely detached, there was no distinctly suburban form. For example, an 1857 series of design schemes for small, middle-

class suburban plots executed by the noted firm of Downing and Vaux included some with setbacks of less than ten feet and others with carriage drives looping around large, tree-lined lawns. Calvert Vaux, *Villas and Cottages* (New York, 1857), passim.

24. Bayard Taylor, *Hannah Thurston* (New York, 1891), 129. Quoted in Edward Spann, *The New Metroplis: New York, 1840–1857* (New York, 1981), 418.

25. In Manhattan, Brooklyn, Baltimore, Philadelphia, and more surprisingly, Winnipeg and Calgary, whole neighborhoods were developed in widths of less than twenty-five feet, and sometimes the transition was from detached to row-house construction. For example, in the 250-year-old farming community of Flatlands in Brooklyn, the 1910 shift was from detached frame houses to brick row houses. In Park Slope and on the Upper East and West sides of Manhattan, the predominant structural type at the turn of the twentieth century was the large, attached brownstone.

26. Charles Rosenberg, *The Cholera Years* (Chicago, 1962).

27. William H. Pierson, Jr., *American Buildings and Their Architects: Technology and the Picturesque; The Corporate and Early Gothic Styles* (Garden City, 1978), 296–98; Anthony D. King, *The Bungalow: The Production of a Global Culture* (London, 1984).

28. Mumford, *The City in History*, 495.

29. Mumford, *The City in History*, 497.

30. J. Weidenmann, *Beautifying Country Homes: A Handbook of Landscape Gardening* (New York, 1870), recently issued as facsimile edition by David Schuyler, ed., *Victorian Landscape Gardening* (Watkins Glen, N.Y., 1978). A very useful study is James Underwood Crockett, *Lawns and Ground Covers*, (New York, 1971), 7–13. Gardeners interested in the period might enjoy a book entitled *Victorian Gardens*, by Frank J. Scott. It was published in 1870 and recently reprinted by the Library of Victorian Culture of the American Life Foundation in Watkins Glen, New York.

31. Handlin, *American Home*, 180–82.

32. Frederick Law Olmsted and Alexander Jackson Davis, who had a somewhat similar influence, will be discussed in the next chapter.

33. On the entire group, see Milton Rugoff, *The Beechers: An American Family in the 19th Century* (New York, 1981). Other leading writers on the role of the female in the home were Lydia Maria Child and Caroline Howard Gilman. The best biography of Catharine Beecher is Kathryn Kish Sklar, *Catharine Beecher: A Study in American Domesticity* (New Haven, 1973). See also Russell Lynes, *The Domesticated Americans* (New York, 1957), 59–63.

34. Most of the *Treatise* had to do with the arrangement of rooms and functions within the house. For example, Beecher recommended that nurseries, kitchens, and living areas all be on the same floor. Catharine Beecher, *A Treatise on Domestic Economy* (New York, 1847), 256–62. See also, Hayden, *Grand Domestic Revolution*, 55–58; and Cathy Alexander, "Architecture and Philosophy in Catharine Beecher's *Treatise on Domestic Economy*" (Seminar paper, Columbia University, 1976).

35. Catharine Beecher and Harriet Beecher Stowe, *The American Woman's Home* (New York, 1869), 13–19.

36. The persistence of such notions in books about domestic life is so obvious as to require no special preachment. This chapter began with a quote from a best-selling high-school textbook in 1931 by Mary Lockwood Matthews, *Elementary Home Economics* (Boston, second edition, 1931), 1. Beecher, *The American Woman's Home*, 24.

37. The most complete source on Downing is George Tatum, "Andrew Jackson Downing: Arbiter of American Taste, 1815–1852" (Ph.D. dissertation, Princeton University, 1950). Also excellent is Handlin, *American Home*, 29–40.

38. Lynes, *Domesticated Americans,* 13–14; Handlin, *American Home,* 37–47.

39. Mary Mix Foley, *The American House* (New York, 1980), 155.

40. The extraordinary significance of the balloon-frame construction method will be discussed in Chapter 7.

41. Downing's pattern books, with their wood engravings, were modest in price compared to the expensive offerings of other authors who used colored lithographic plates. A collection of Downing's writings as editor of *The Horticulturist* was published a year after his death under the editorship of George W. Curtis as *Rural Essays* (New York, 1853). Foley, *The American House,* 153–57; and *The Horticulturist,* III (1848), 10.

42. Andrew Jackson Downing, *A Treatise on the Theory and Practice of Landscape Gardening* (New York, 1859), ix.

43. Downing, "Our Country Cottages," *The Horticulturist,* IV (June 1850), 539. See also, Schuyler, "Public Landscapes," chapter 6; and John William Ward, "The Politics of Design," in L. B. Holland, ed., *Who Designs America* (Garden City, 1965), 54–79.

44. Downing, *Rural Essays,* 241–43; and Downing, *Cottage Residences, Rural Architecture, and Landscape Gardening* (New York, 1847); 3–28.

45. Andrew Jackson Downing, *The Architecture of Country Houses* (New York, 1850), 38–41 and 276–78.

46. Handlin, *American Home,* 46.

47. Calvert Vaux, *Villas and Cottages* (New York, 1853), xi–xii.

48. There is no adequate biography, to my knowledge, of Calvert Vaux, which is a major reason that he is not given due credit for his primary role in the planning of Central Park.

49. Vaux, *Villas and Cottages,* 27.

50. Vaux, *Villas and Cottages,* 51.

51. The standard work, somewhat overdrawn, on this topic is Morton and Lucia White, *The Intellectual Versus the City: From Thomas Jefferson to Frank Lloyd Wright* (Cambridge, 1962). A much more complete and incisive analysis is Andrew Lees, *Cities Perceived: Urban Society in European and American Thought, 1820–1940* (New York, 1985). See also, Henry David Thoreau, *Walden and Other Writings* (New York, 1950), 615–18.

52. Nicholas B. Wainwright, ed., *A Philadelphia Perspective: The Diary of Sidney George Fisher Covering the Years 1834–1871* (Philadelphia, 1967), 202.

53. *Louisville Courier-Journal,* March 31, 1871; and *Louisville Courier-Journal,* September 25, 1873.

54. This information comes from the yet unfinished dissertation of Evelyn Gonzalez, who is now completing a study of population and land development in the Bronx for her Ph.D. at Columbia University.

55. Despite the popularity of Beecher, Downing, and Vaux, the most widely read theorist, or indeed writer, on architecture in America was John Ruskin. *The Seven Lamps of Architecture* and *The Stones of Venice,* which American publishers printed almost as soon as they appeared in London, went through twenty-five printings before 1895. Yet Ruskin's books had an extremely limited influence on American architecture. Woods, "*The American Architect* and *Building News*" (Ph.D. diss., Columbia University, 1980), 27. See also, Mumford, *The City in History,* 496–97.

56. *The Crayon,* i (January 3, 1855), 11; and Mumford, *The City in History,* 485–86.

57. Wright, *Building the Dream,* 73–113; and Hayden, *Grand Domestic Revolution,* passim.

CHAPTER 4: *Romantic Suburbs*

1. Middle Eastern cities were an exception. They substituted the camel for the wheel for a thousand years before 1850. The only individual to have any sort of carriage or cart in Egypt in the first part of the nineteenth century was Muhammed Ali. Most studies of the street deal primarily with construction and maintenance; an important recent departure is Francois Bedarida and Anthony Sutcliffe, "The Street in the Life of the Modern City: Reflections on Nineteenth Century London and Paris," in Bruce Stave, ed., *Modern Industrial Cities: History, Policy, and Survival* (Beverly Hills, 1981), 21–38.

2. R. E. Wycherly, *How the Greeks Built Cities* (New York, second edition, 1962), 17–37.

3. D. Stanislawski, "The Origin of the Grid-Pattern Town," *Geographical Review,* XXXVI (1946), 105–20; and David Schuyler, "Public Landscapes and American Urban Culture, 1800–1870: Rural Cemeteries, City Parks, and Suburbs" (Ph.D. dissertation, Columbia University, 1979), 221–22.

4. Roy Lubove's introduction to the recent reprint of H. W. S. Cleveland's *Landscape Architecture* is particularly useful on the grid. Cleveland, *Landscape Architecture as Applied to the Wants of the West* (Pittsburgh, reprint of 1873 edition, 1965), ix–xi. See also, Schuyler, "Public Landscapes," 198–244; and Richard C. Wade, *The Urban Frontier: The Rise of Western Cities, 1790–1830* (Cambridge, 1959), 27–28.

5. Cleveland, *Landscape Architecture,* 18; Schuyler, "Public Landscapes," 261–62.

6. Ronald Dale Karr, "The Evolution of an Elite Suburb: Community Structure and Control in Brookline, Massachusetts, 1770–1900" (Ph.D. dissertation, Boston University, 1981), 264–66.

7. The best example of such a courtyard apartment building is the Dakota, which was built in 1879 on the western rim of Central Park at 72nd Street. Other enduring examples of this form include the Apthorp, the Belnord, Astor Court, and Knickerbocker Village in New York.

8. Cleveland, *Landscape Architecture,* 31. For a discussion of the Boston grid, see Sam Bass Warner, Jr., *Streetcar Suburbs: The Process of Growth in Boston, 1870–1900* (Cambridge, Mass., 1962), 157–59. Innovative communties like Llewellyn Park, Riverside, Lake Forest, and Roland Park, to be discussed later, adopted curvilinear streets in the nineteenth century, but the grid remained dominant in most suburbs until after 1900.

9. "Landscape Gardening, Llewellyn Park, Orange Mountain, New Jersey," *The Crayon,* IV (August 1857), 247–48. The best recent work is Richard Wilson, "Idealism and the Origin of the First American Suburb: Llewellyn Park, New Jersey," *American Art Journal,* XI (Fall 1979), 79–93. Also excellent is Christopher Tunnard, "The Romantic Suburb in America," *Magazine of Art,* XL (May 1947), 184–87.

10. For examples of Davis's work, see Alexander Jackson Davis and other architects, *Rural Residences, etc., Consisting of Designs, Original and Selected, for Cottages, Farm Houses, Villas, and Village Churches* (New York, 1837).

11. The best work on Davis is William H. Pierson, Jr., *American Buildings and Their Architects: Technology and the Picturesque: The Corporate and the Early Gothic Styles* (Garden City, 1978), 270–348.

12. Many other communities, Boston and West Morrisania in the Bronx among them, had curving roads before Llewellyn Park, but they were not consciously planned to undulate, and in fact many dated to Indian and farm paths before urban spread.

13. Pierson, *American Buildings,* 426.

14. The gatehouse and private guards remained a prominent feature of Llewellyn Park in 1985, when few houses sold for less than $200,000. For early praise of Llewellyn Park, see Calvert Vaux, *Villas and Cottages* (New York, 1864, second edition), 339.

15. Olmsted has been the subject of an avalanche of publications in recent years, among them Laura Wood Roper, *FLO: A Biography of Frederick Law Olmsted* (Baltimore, 1973); and the less useful, Elizabeth Stevenson, *Park Maker: A Life of Frederick Law Olmsted* (New York, 1977).

16. S. B. Sutton, ed., *Civilizing American Cities: A Selection of Frederick Law Olmsted's Writings on City Landscape* (Cambridge, 1971), 293–96.

17. Schuyler, "Public Landscapes," chapter 6.

18. Olmsted, Vaux, and Co., "Report upon the Proposed Suburban Village at Riverside," in Theodora Kimball Hubbard, ed., "Riverside, Illinois: A Residential Neighborhood Designed Over Sixty Years Ago," *Landscape Architecture,* XXI (July 1931), 257–91.

19. Unlike Llewellyn Park, Riverside had a town center that included a hotel, commercial buildings, and a railroad station.

20. Albert Fein, *Frederick Law Olmsted and the American Environmental Tradition* (New York, 1969), 34.

21. Olmsted, Vaux and Company, *Preliminary Report on the Proposed Suburban Village at Riverside, Near Chicago* (New York, 1868), 7.

22. The best work on Garden City is Deborah S. Gardner, "Architecture and Society in the Nineteenth Century City: John Kellum's Career and His Work for A. T. Stewart" (Ph.D. dissertation, Columbia University, 1978). See also, Roger A. Wines, "A. T. Stewart and Garden City," *Nassau County Historical Journal,* XIX (Winter 1958), 1–15; Donald S. Richards, "A History of Garden City, New York," *Yesteryears,* VIII (June 1965), 197–201; and G. L. Hubbell, "The Making of Garden City," *New York History,* XXI (January 1940), 32. Mildred H. Smith, *History of Garden City* (Garden City, 1963), is unsatisfactory.

23. Hubbell, "Making of Garden City," 31–38.

24. Quoted in Gene Gleason, "Garden City: A Dream Come True," *New York Herald-Tribune,* May 26, 1963.

25. Deborah S. Gardner, "The Sources and Influences of Garden City, Long Island" (Seminar paper, Columbia University, 1973).

26. Stewart was personally disappointed in Garden City's progress, and at the time of his death in 1876 he was planning to build railroad repair shops and transfer yards in Garden City.

27. Horace Greeley practiced what he preached and purchased a small farm in Chappaqua, New York, which he used as a weekend and summer residence in the 1850s and 1860s.

28. There is no adequate history of Mount Vernon, which is currently wrestling with problems of an increasingly minority population, a declining tax base, and an aging housing stock.

29. Gardner, "Sources and Influences," 7–12; and Charles K. Landis, *Souvenir Program: Pageant of Progress Commemorating the Founding of Vineland,* 1861 (Vineland, 1935).

30. Gardner, "Sources and Influences," passim.

CHAPTER 5: *The Main Line*

1. Mark Girouard, *The Victorian Country House* (New Haven, revised edition, 1979), passim.

2. Barr Ferree, *American Estates and Gardens* (New York, 1904), 1.

3. The Phipps house in Old Westbury, for example, has been a nonprofit museum and botanical garden since 1959. Monica Randall, *The Mansions of Long Island's Gold Coast* (New York, 1979), 3–27.

4. The best work on land conversion in cities from elite to working-class occupancy is David Ward. "The Emergence of Central Immigrant Ghettoes in American Cities, 1840–1920," *Annals, Association of American Geographers,* LVIII (June 1968), 343–59; and David Ward, "The Internal Spatial Structure of Immigrant Residential Districts in the Late Nineteenth Century," *Geographical Analysis,* I (October 1969), 337–53.

5. The word "tenement" did not specifically refer to slum buildings until after the Civil War.

6. The Hotel Pelham in Boston (1859) had facilities for transients and was not a true apartment house. Andrew Alpern, *Apartments for the Affluent: A Historical Survey of Buildings in New York* (New York, 1975), 1. See also, St. James Richardson, "The New Homes of New York," *Scribner's Monthly* VIII (May 1874), 68.

7. Quoted in Ronald Dale Karr, "The Evolution of an Elite Suburb: Community Structure and Control in Brookline, Massachusetts, 1770–1900" (Ph.D. dissertation, Boston University, 1981), 144.

8. Nathaniel Burt, *The Perennial Philadelphians: The Anatomy of an American Aristocracy* (Boston, 1963), 1–21.

9. Everett Chamberlin, *Chicago and Its Suburbs* (Chicago, 1874), 1–11. The best recent work on Chicago suburbanization is Michael H. Ebner, "In the Suburbs of the Town: Chicago's North Shore to 1871," *Chicago History,* XI (Summer 1982), 66–77. See also, Paul Frederick Cressey, "The Population Succession in Chicago, 1898–1930," *American Journal of Sociology,* XLIV (July 1938), 59–69; and William D. Middleton, *North Shore: America's Fastest Interurban* (San Marino, Calif. 1964).

10. Quoted in Ebner, "In the Suburbs," 66–77. Kenilworth was as exclusive as Lake Forest but much smaller. Located on three-fifths of a square mile of drained swampland along the shore of Lake Michigan, Kenilworth was laid out by land developer Joseph Sears as a model suburb for the elite.

11. An excellent comparison of two architectural styles, which focuses on Lake Forest as a traditional enclave for the wealthy, is Leonard K. Eaton, *Two Chicago Architects and Their Clients: Frank Lloyd Wright and Howard Van Doren Shaw* (Cambridge, 1969). See also, Carl Abbott, "Necessary Adjuncts to its Growth: The Railroad Suburbs of Chicago," *Journal of the Illinois State Historical Society,* LXXIII (Summer 1980), 117–31.

12. A typology of Westchester suburbs is included in George A. Lundberg, Mirra Komarovsky, and Mary A. McInemy, *Leisure: A Suburban Study* (New York, 1934). See also, Richard F. Crandell, *This is Westchester: A Study of Suburban Living* (New York, 1954); Nancy Mayer, "Westchester: The Land of Let's Pretend," *New York Magazine,* (February 9, 1970); and George Howe, *Memoirs of a Westchester Realtor: Half a Century of Property Development in Westchester County and Southwestern Connecticut* (New York, 1959); and Harry Hansen, *Scarsdale: From Colonial Manor to Modern Community* (New York, 1954).

13. The Leland mansion is now the Administration Building of The College of New

Rochelle; the Chapman home is now the clubhouse of the Rye Golf Club. The Rockefeller estate, on which all of the family members had one or more large homes and which maintains its own police force and school system, has been deeded to New York State upon the death of the last surviving brother.

14. *Railroad Gazette,* XXXVII (September 2, 1904), 290. See also, Carl W. Condit, *The Port of New York: A History of the Rail and Terminal System From the Grand Central Electrification to the Present* (Chicago, 1981).

15. The only study specifically on Bronxville is short, chatty, and unsatisfactory. Victor Mays, *Pathway to a Village: A History of Bronxville* (New York, 1961). More general but also useful are Alvah P. French, *History of Westchester County,* vol. II (New York, 1925), 677–81; Harry Hansen, *North of Manhattan* (New York, 1950), 4–7; Ernest F. Griffin, *Westchester County and Its People* (New York, 1946), 12; "Bronxville, New York; 75 years of the Good Life," Bronxville *Review Press Reporter,* October 18, 1973; and Works Progress Administration, *Historical Development of Westchester County, A Chronology* (Washington, 1939), 11–15.

16. David C. Itzkowitz, *Peculiar History: A Social History of English Fox Hunting, 1753–1885* (Sussex, England, 1977), 1–18.

17. Quoted in Benjamin Rader, *American Sports* (Englewood Cliffs, N.J., 1982), chapter 3.

18. Rader, *American Sports,* 75–101. See also, Benjamin G. Rader, "The Quest for Subcommunities and the Rise of American Sport," *American Quarterly,* XXIX (Fall 1977), 355–69.

19. The New York Yacht Club, sponsor of the America's Cup, was founded in 1844, but its clubhouse has always been in Manhattan. In 1900, when August Belmont, Andrew Carnegie, J. P. Morgan, and most of the Vanderbilts were on the registers, the Larchmont Yacht Club included forty-seven great steam yachts of more than one hundred feet in length.

20. Lundberg, *Leisure,* 24–57; and Rader, *American Sports,* passim. By 1982, over half of all Americans participated in outdoor recreation.

21. In Lake Forest, for example, the pioneer settlers of the 1840s included many Irish immigrants who had toiled on the Illinois and Michigan Canal and who remained as laborers and tradespeople. In highly exclusive Grosse Point Park Village in 1920, twenty-two of 164 sampled heads of households were gardeners, chauffeurs, and chefs living in houses separated from the main mansion. Zunz, 355.

22. The geographic area studied was constructed from the memoir of Mary Lee, *A History of Chestnut Hill Chapel* (Newton, Mass., 1937). See also, Henry K. Rowe, *Tercentenary History of Boston, 1630–1930* (Newton, Mass., 1931), 168–69; and Cleveland Amory, *The Proper Bostonians* (New York, 1947), 106.

23. The tax lists indicated not only the actual tax, but also the amount of taxable property and income of each independent adult or head of household. The compilations and tables are omitted; they were put together for me by a student, Julia Weinstein Tossell.

24. Ronald Dale Karr, "The Evolution of an Elite Suburb: Community Structure and Control in Brookline, Massachusetts, 1770–1900" (Ph.D. dissertation, Boston University, 1981), 169–87.

25. The best analysis of transit and urban growth is Joel A. Tarr, "Transportation Innovation and Changing Spatial Patterns: Pittsburgh, 1850–1910" (CMUTRI-TP-72-06) (Pittsburgh: Transportation Research Institute of Carnegie-Mellon University, 1972). Heavily detailed, but less satisfactory is Carl W. Condit, *The Railroad and the City: A Technological and Urbanistic History of Cincinnati* (Columbus, 1977).

26. Karr, "The Evolution of an Elite Suburb," 169–87.

27. David Goldfield, "Suburban Development in Stockholm and the United States: A Comparison of Form and Function" (unpublished paper, 1979), 12–14.

CHAPTER 6: The Time of the Trolley

1. An excellent study of popular and intellectual attitudes toward Thomas A. Edison is Wyn Wachhorst, *Thomas Alva Edison: An American Myth* (Cambridge, 1981.

2. Quoted in John Anderson Miller, *Fares Please! From Horse Carts to Streamliners* (New York, 1941), 51.

3. James Leslie Davis, *The Elevated System and the Growth of Northern Chicago* (Evanston: Northwestern University Studies in Geography No. 10, 1965), 25.

4. In 1979 an estimated 90 percent of the twelve million annual riders on the San Francisco cable cars were tourists. *New York Times,* September 27, 1979.

5. William Dean Howells, *Suburban Sketches* (Boston, 1871), 110–11.

6. Just a few years earlier, the horsecar had been heralded as "the improvement of the age."

7. The best work on waste disposal is by Joel A. Tarr. An introduction to his early research that remains both useful and entertaining is Joel A. Tarr, "Urban Pollution: Many Long Years Ago," *American Heritage,* XXII (October 1971), 65–69, 106.

8. Tarr, "Urban Pollution," 66–67.

9. In densely settled areas, such as Manhattan and large parts of Chicago, the public successfully resisted the hanging of overhead wires. On the history of the trolley, see John P. McKay, *Tramways and Trolleys: The Rise of Urban Mass Transport in Europe* (Princeton, 1976); Harry J. Carman, *The Street Surface Railway Franchises of New York City* (New York, 1919); Henry C. Adams, comp., *Report on the Transportation Business in the United States, Part I, Transportation by Land, Eleventh Census* (Washington, 1895); Edward S. Mason, *The Street Railway in Massachusetts: The Rise and Decline of an Industry* (Cambridge, 1932); James D. Johnson, *A Century of Chicago Streetcars, 1858–1958* (Wheaton, Ill., 1964); Frederic W. Speirs, "The Street Railway System of Philadelphia.: Its History and Present Condition," *Johns Hopkins Studies in History and Political Science,* XV (1897), nos. 3, 4, and 5; Yasuo Sakakibara, "The Influence of the Introduction of Street Railways Upon Urban and Industrial Growth in Eastern Massachusetts, 1855–1875," (Master's thesis, Amherst College, 1956); and Joel A. Tarr, *Transportation Innovation and Changing Spatial Patterns: Pittsburgh, 1850–1910* (Pittsburgh: Carnegie Mellon University Transportation Research Institute, April, 1972).

10. Among Sprague's later creations were the multiple-unit system, which made it possible to control any number of cars with a single master switch, and the dead man's button, which shut off all power when pressure on the lever was reduced. Harold C. Passer, "Frank Julian Sprague," in William Miller, ed., *Men in Business* (New York, 1952); and Miller, *Fares Please,* 62.

11. The price of a one-way fare dropped primarily because the price of electricity was less than the price of using horses to generate the motive power.

12. Peter Widener went down with the *Titanic* in 1912, leaving his fortune to Harvard University, which built Widener Library in his memory.

13. In Massachusetts, which had the most extensive streetcar mileage in proportion to area and population of any state in the union, the Board of Railway Commissioners approved every application for the consolidation of street railways that proposed no re-

duction in service. In Canada consolidation took somewhat longer. As late as 1921 Toronto had nine separate transportation systems collecting nine separate fares to serve the city. Mason, *The Street Railway in Massachusetts,* 51–52.

14. This phrase has also been attributed to Anthony N. Brady, the major figure in building the Brooklyn Rapid Transit and the president of the New York Edison Company.

15. George W. Hilton, "Transport Technology and the Urban Pattern," *Journal of Contemporary History,* IV (1969), 126. Some American horsecar lines did continue to operate until World War I. Even in New York City, the Bleeker Street-Broadway horsecar line did not make its last trip until July 26, 1917. Stan Fischler, *Uptown, Downtown: A Trip Through Time on New York's Subways* (New York, 1976), 7.

16. David Ward, "A Comparative Hostorical Geography of Streetcar Suburbs in Boston, Massachusetts, and Leeds, England, 1850–1920," *Annals of the Association of American Geographers,* LIV (1964), 477–89; McKay, *Tramways and Trolleys;* and Shunichi J. Watanabe, "Metropolitanism as a Way of Life: The Case of Tokyo, 1868–1930," in Anthony Sutcliffe, ed., *Metropolis, 1890–1940* (London, 1984), 403–29.

17. With the increased use of automobiles for pleasure during the 1920s, summer riding on the trolleys dropped off, and the companies began to rely on winter revenues to offset summer losses. Mason, *The Street Railway in Massachusetts,* 116.

18. California Press Association, *California Today* (Berkeley, 1903), 3.

19. The best overview of the topic is K. H. Schaeffer and Elliot Sclar, *Access for All: Transportation and Urban Growth* (Baltimore, 1975). A careful and incisive case study is Olivier Zunz, *The Changing Face of Inequality: Urbanization, Industrial Development, and Immigrants in Detroit, 1880–1920* (Chicago, 1982), 94–101.

20. The planning of Garden City is discussed in Chapter 5. On the origins of the department store, see Gunther Barth, *City People: The Rise of Modern City Culture in Nineteenth-Century America,* (New York, 1980), 110–47.

21. The best work on this topic is Homer Hoyt, *One Hundred Years of Land Values in Chicago* (Chicago, 1933), passim.

CHAPTER 7: Affordable Houses for the Working Man

1. Joel A. Tarr, *Transportation Innovation and Changing Spatial Patterns: Pittsburgh, 1850–1910* (CMUTRI-TOP-72-06, Pittsburgh: Transportation Research Institute, Carnegie Mellon University, April 1972), passim. As Chapter 2 indicated, the general pattern in the largest American cities was for deconcentration to commence well before 1860, but by modern standards congestion remained frightfully high and in some neighborhoods it increased until about the turn of the century.

2. "The Problem of Living in New York," *Harper's New Monthly Magazine,* LXV (November 1882), 924.

3. Zane L. Miller, *Boss Cox's Cincinnati: Urban Politics in the Progressive Era* (New York, 1968), 220–21; Dana W. Bartlett, *The Better City: A Sociological Study of a Modern City* (Los Angeles, 1907), 74.

4. See Chapter 4 for more discussion of this point.

5. Stephan Thernstrom, *Poverty and Progress: Social Mobility in a Nineteenth Century City* (Cambridge, 1964), 115–22: Olivier Zunz, *The Changing Face of Inequality: Urbanization, Industrial Development, and Immigrants in Detroit, 1880–1920* (Chicago, 1982), 142–71.

6. See for example, Thomas Kessner, *The Golden Door: Italian and Jewish Immigrant Mobility in New York City, 1880–1915* (New York, 1977); and Deborah Dash Moore, *At Home in America: Second Generation New York Jews* (New York, 1981).

7. Sam Bass Warner, Jr., *Streetcar Suburbs: The Process of Growth in Boston, 1870–1900* (Cambridge, Mass. 1962), *passim.*

8. Cynthia Howard, *Your House in the Streetcar Suburb: The History and Care of Houses in Medford, Massachusetts* (Medford, 1979).

9. Warner, *Streetcar Suburbs,* passim.

10. Max Foran, "Land Development Patterns in Calgary, 1884–1945," in Alan F. J. Artibise and Gilbert A. Stelter, eds., *The Usable Urban Past: Planning and Politics in the Modern Canadian City* (Toronto, 1979), 293–315; Richard Hurd, *Principles of City Land Values* (New York, 1903); and Edward Mason, *The Street Railway in Massachusetts: The Rise and Decline of an Industry* (Cambridge, 1932), 194. Hurd was president of the Lawyers Mortgage Company and vice president of the Mortgage Bond Company of New York.

11. Edwin H. Spengler, *Land Values in New York in Relation to Transit Facilities* (New York, 1930); Homer Hoyt, *One Hundred Years of Land Values in Chicago* (Chicago, 1933); and C. S. Sargent, "Land Speculation and Urban Morphology," in J. S. Adams, ed., *Urban Policymaking and Metropolitan Dynamics: A Comparative Geographical Analysis* (Cambridge, 1976), 21–56.

12. Dallas W. Smythe, "An Economic History of Local and Interurban Transportation in the East Bay Cities with Particular Reference to the Properties Developed by F. M. Smith" (Ph.D. dissertation, University of California, Berkeley, 1937); Thomas J. Flynn, "The Relationship Between Street Railways and Urban Land Subdivision: Oakland, 1850–1930" (Seminar paper, Columbia University, 1972); and John B. Dykstra, "A History of the Physical Development of Oakland, 1850–1930" (Ph.D. dissertation, University of California, Berkeley, 1967).

13. Perhaps because of the steep hills and the necessity for several tunnels, the speculative inducement to streetcar building was not as strong across the Bay in San Francisco. Only after 1912, when the city government began developing its own lines, were the outlying areas of San Francisco settled.

14. The Southern Pacific eventually took its case to the California Supreme Court, where the connection between Smith's transit and real-estate operations was exposed.

15. For a comparison of the origins of real-estate wealth in the Los Angeles area, see Frederic Cople Jaher, *The Urban Establishment: Upper Strata in Boston, New York, Charleston, Chicago, and Los Angeles* (Urbana, 1982).

16. In the Middle West, the numerous interurbans sometimes traveled as much as one hundred miles, but those in the Huntington system had much shorter routes. Spencer Crump, *Henry Huntington and the Pacific Electric* (Los Angeles, 1970); and Crump, *Ride the Big Red Cars* (Costa Mesa, Calif., 1962).

17. The best study of suburbanization in Los Angeles is Robert M. Fogelson, *The Fragmented Metropolis: Los Angeles, 1850–1930* (Cambridge, 1967).

18. Because Nevada's nineteenth-century senators were elected by a corrupt legislature controlled by mining interests, it was not necessary for its representatives to actually live in Nevada in anything other than a legal sense.

19. Easily the best source on this subject is Roderick S. French, "Chevy Chase Village in the Context of the National Suburban Movement, 1870–1900," *Records of the Columbia Historical Society.* XLIX (1973–1974), 300–29.

20. Warner, *Streetcar Suburbs,* passim.

21. The log cabin was not properly a frame structure because the house itself rested

on the walls. Carl W. Condit, *American Building Art: The Nineteenth Century* (New York, 1960), 10–13; and Ettore Camescasca, *History of the House*, trans. Isabel Quigley (New York, 1971).

22. An excellent treatment of the subject is Mary Mix Foley, *The American House* (New York, 1980). See also, Abbott Lowell Cummings, *The Framed Houses of Massachusetts Bay, 1625–1725* (Cambridge, 1979).

23. In the United States in 1976, more than 92 percent of new, single-family homes, and about 85 percent of existing dwellings were of balloon-frame construction. The method is less common in Canada and is almost unknown in Mexico.

24. *New York Tribune*, January 18, 1855, as quoted in Sigfried Giedion, *Space, Time and Architecture, The Growth of a New Tradition* (Cambridge, 1967), 348; Christopher Tunnard and Henry Hope Reed, *American Skyline: The Growth and Form of Our Cities and Towns* (Boston, 1953). Carl Condit also credits Taylor with the innovation.

25. Zunz, *The Changing Face of Inequality*, 153–77.

26. Giedion, *Space, Time and Architecture*, 347–49.

27. James Sanders, "We've Been Framed: The History of the Wood Frame House in the United States" (Seminar paper, Columbia University, 1976).

28. Gervase Wheeler, *Homes for the People, in Suburb and Country: The Villa, the Mansion and the Cottage, Adapted to American Climate and Wants. With Examples Showing How to Alter and Remodel Old Buildings* (New York, 1855). Also in this tradition, but aimed less at the easily built balloon frame, were Edward Shaw, *Rural Architecture* (1843); Andrew Jackson Downing, *Cottage Residences* (1842); and Alexander Jackson Davis, *Rural Residences* (1842). Earlier handbooks like Asher Benjamin, *The American Builder's Companion* (1827) and Minard Lefever's *The Modern Builder's Guide* (1833) emphasized Greek Revival details. The best survey of this literature is Mary Norman Woods, *"The American Architect and Building News*, 1876–1907," (Ph.D. dissertation, Columbia University, 1982), 22–28.

29. Friedrich Engels, *The Housing Question* (New York, 1872), 18.

30. *Louisville Courier-Journal*, April 18, 1871; and Tunnard and Reed, *American Skyline*, 69.

31. "Homes for the Industrious," *Louisville Courier-Journal*, September 3, 1861; and John F. Sutherland, "Housing the Poor in the City of Homes: Philadelphia at the Turn of the Century" in Allen F. Davis and Mark H. Haller, eds., *The Peoples of Philadelphia: A History of Ethnic Groups and Lower-Class Life, 1790–1940* (Philadelphia, 1973), 176–79. In New York City and Brooklyn, where the high cost of land inhibited the effective use of savings and loans for private building, there were only forty-eight such associations in 1888. And Olivier Zunz has reported that there were only six building-and-loan associations in Detroit in 1896, and those were not directed to or controlled by the working class. Zunz, *The Changing Face of Inequality*, 162–63. The best general work on the subject is Horace Russell, *Savings and Loan Associations* (New York, second edition, 1960).

32. The financing of homes before the New Deal is discussed in Chapter 12.

33. Jon A. Peterson, "The Impact of Sanitary Reform upon American Urban Planning, 1840–1890," *Journal of Social History*, XIII (Fall 1979), 83–103; and Harold L. Platt, *City Building in the New South: The Growth of Public Services in Houston, Texas, 1830–1915* (Philadelphia, 1983).

34. In 1869, for example, the New York State Assembly passed a law permitting New York City to pay half the cost of street openings.

35. Howard L. Preston, *Automobile Age Atlanta: The Making of a Southern Metropolis, 1900–1935* (Athens, Georgia, 1979).

36. Mark H. Rose and John G. Clark, ''Light, Heat, and Power,'' *Journal of Urban History,* V (1979), 340–64; and Mark H. Rose, ''Urban Environments and Technological Innovation: Energy Choices in Denver and Kansas City, 1900–1940,'' *Technology and Culture,* XX (July 1984), 503–39.

37. Roger Simon, *The City-building Process: Housing and Services in New Milwaukee Neighborhoods, 1880–1910* (Philadelphia, 1978); Carolyn Tyirin Kirk and Gordon W. Kirk, ''The Impact of the City on Home Ownership: A Comparison of Immigrants and Native Whites at the Turn of the Century,'' *Journal of Urban History,* VII (August 1981), 471–87; Donald R. Deskins, ''Race, Residence, and Workplace in Detroit, 1880–1965,'' *Economic Geography,* XLVIII (January 1972), 79–94.

38. Ronald Dale Karr, ''The Evolution of an Elite Suburb: Community Structure and Control in Brookline, Massachusetts, 1770–1900'' (Ph.D. dissertation, Boston University, 1981), 254–63. The best studies of the contemporary process are Marion Clawson, *Suburban Land Conversion in the United States: An Economic and Governmental Process* (Baltimore, 1971); Michael J. Doucet, ''Urban Land Development in Nineteenth Century North America: Themes in the Literature,'' *Journal of Urban History,* VIII (May 1982), 299–342; and Stanley L. McMichael, *Real Estate Subdivisions* (New York, 1949).

39. After his death, John Jacob Astor was criticized for leaving only a miniscule amount to charity. His descendents, however, especially the inimitable Brooke Astor, have lavishly supported New York City institutions. Kenneth Higgins Porter, *John Jacob Astor: Businessman* (Cambridge, 1931); and Brooke Astor, ''John Jacob Astor: From the Founder to the Foundation,'' *Seaport,* XVII (Fall 1983), 10–17.

40. Karr, ''The Evolution of an Elite Suburb,'' 256–63.

41. Zunz, *The Changing Face of Inequality,* 161–63.

42. Gwendolyn Wright, *Building the Dream: A Social History of Housing in America* (New York, 1981), 100.

43. Wright, *Building the Dream,* 102–3.

44. Dolores Hayden, *The Grand Domestic Revolution* (Cambridge, 1981), 7–9, 193–95.

45. My description follows that of Henry Seidel Canby, *The Age of Confidence* (New York, 1934), 8–10.

46. See E. W. Burgess, ''The Growth of the City,'' in Robert E. Park, Ernest W. Burgess, and Roderick D. McKenzie, *The City* (Chicago, 1925); Richard F. Muth, *Cities and Housing: The Spatial Pattern of Urban Residential Land Use* (Chicago, 1969); William Alonso, *Location and Land Use* (Cambridge, 1964).

CHAPTER 8: *Suburbs into Neighborhoods*

1. Boston and New Haven lost population temporarily in the eighteenth century, and Charleston, New London, Schenectady, and Newburyport declined during the nineteenth century, but such cases were unusual. This chapter is an expansion and revision of an earlier article, Kenneth T. Jackson, ''Metropolitan Government Versus Suburban Autonomy: Politics on the Crabgrass Frontier,'' in Kenneth T. Jackson and Stanley K. Schultz, eds., *Cities in American History* (New York, 1972), 442–62.

2. The only general history of municipal area growth is a brief volume by Jon C. Teaford, *City and Suburb: The Political Fragmentation of Metropolitan America, 1850–1970* (Baltimore, 1979). The best specialized studies are Richard Bigger and James D. Kitchen, *How the Cities Grew: A Century of Municipal Independence and Expansion in Metropolitan Los Angeles* (Los Angeles, 1952); Paul Studenski, ed., *The Government*

of Metropolitan Areas in the United States (New York, 1930); William G. Coleman, *Cities, Suburbs, and States: Governing and Financing Urban America* (New York, 1975); Robert G. Dixon, Jr., John R. Kerstetter, and Charles A. Hollister, *Adjusting Municipal Boundaries: Law and Practice* (Washington, revised edition, 1966); Thomas R. Dye, "Urban Political Integration: Conditions Associated with Annexation in American Cities," *Midwest Journal of Political Science,* VII (November 1964), 430–46; Michael P. McCarthy, "The New Metropolis: Chicago, the Annexation, and Progressive Reform," in Michael H. Ebner and Eugene M. Tobin, eds., *The Age of Urban Reform: New Perspectives on the Progressive Era* (Port Washington, N.Y., 1977), 43–54; David G. Temple, *Merger Politics: Local Government Consolidation in Tidewater Virginia* (Charlottesville, 1972); and Joseph W. Barnes, "The Annexation of Charlotte," *Rochester History,* XXXVII (January 1975), 1–28.

3. By the terms of the Dongan Charter of 1686 and colonial legislation, the City of New York has been coterminus with the island of Manhattan. This despite the fact that the upper reaches of the island, such as Harlem, Washington Heights, and Inwood, were not residentially developed for another two centuries.

4. Although annexation and consolidation technically refer to two different legal procedures, the outcome is the same and the words will be used interchangeably on the following pages.

5. As Chapter 2 demonstrates, if annexation had not occurred, the suburbs would have been growing faster than the cities by 1800 in Philadelphia, by 1810 in New York, by 1840 in Boston, and by 1850 in St. Louis and Cleveland.

6. Court-ordered public school integration between the city of St. Louis and its surrounding suburbs somewhat erased the city/county boundary after busing and taxiing began in the fall of 1983.

7. Some cities, such as Baltimore, added frequently to their boundaries before 1800, but most were stable or even lost territory.

8. Of the twenty *arrondissements* in Paris, the eleventh through the twentieth were taken in by the 1859 extension. Like Philadelphia (which did add 0.131 square miles from Montgomery County in 1916), Paris has not been extended since that time.

9. Southwark, which was settled by Swedes in 1638, is older than Philadelphia, and like Spring Garden and Kensington is named for a London suburb. It was created a municipality in 1762 and incorporated in 1794. There were 28 other governments in what is now the city of Philadelphia prior to consolidation. William Bucke Campbell, "Old Towns and Districts of Philadelphia," *Philadelphia History,* IV (1942), 94–149; M. Antonia Lynch, "The Old District of Southwark in the County of Philadelphia," *Philadelphia History,* I (1909), 83–126; Eli K. Price, *The History of the Consolidation of the City of Philadelphia* (Philadelphia, 1873).

10. There were 23 distinct communities in Hyde Park Township, which together made up less than half of the annexed area. Stanley Buder, *Pullman: An Experiment in Industrial Order and Community Planning, 1880–1930* (New York, 1967), 109; and McCarthy, "The New Metropolis," 46–51.

11. As early as 1833, New York City's mayor and aldermen opposed the incorporation of Brooklyn as a separate city on the grounds that it should ultimately be joined to Manhattan and that separate city status would make annexation more difficult. In the 1870s and 1880s a Municipal Union Society agitated in favor of a Greater New York City and sent the legislature petitions and bills. The two best studies of the Gotham consolidation are David C. Hammack, *Power and Society: Greater New York at the Turn of the Century* (New York, 1982), 185–229; and Barry J. Kaplan, "Andrew H. Green and the Creation of a Planning Rationale: The Formation of Greater New York

City, 1865–1890," *Urbanism Past and Present,* VIII (1979), 32–41. See also, Richard Stone, "The Annexation of the Bronx, 1874," *The Bronx County Historical Society Journal,* VI (January 1969), 1–24.

12. Outlying St. Louis County residents, who did not want to lose city tax revenue, voted four to one against the separation, but the total was 12,181 to 10,928 in favor of separation. It was the first home-rule charter of its kind in the United States, but it unfortunately made no provision for future annexation—a policy of omission which was followed in the charter of 1914. Howard Lee McBain, *The Law and the Practice of Municipal Home Rule* (New York, 1916), 146.

13. Slight alterations were made in the Boston boundary by the legislature in 1836, 1838, and 1859.

14. In most communities, annexation maps may be consulted in the office of the city planner, city engineer, and city tax assessor.

15. Minneapolis demanded the removal of the capital from St. Paul, while St. Paul publicly regretted its connection with a city "that stands degraded and ashamed in the eyes of the nation." Minnesota Works Progress Administration Writer's Project, *Minneapolis: The Story of a City* (Minneapolis, 1940), 66–67.

16. The opening of the Brooklyn Bridge in 1883 was another powerful stimulus. Margaret Latimer, *Two Cities: New York and Brooklyn the Year the Great Bridge Opened* (Brooklyn, 1983).

17. *Philadelphia North American,* February 4, 1854.

18. For a typical expression of this view, see Paul U. Kellogg, "The Civic Responsibilities of Democracy in an Industrial District," in Kellogg, ed., *The Pittsburgh Survey,* 6 vols. (New York, 1910).

19. As Samuel P. Hays has persuasively argued, the desire to centralize also found expression in the elimination of ward representation in favor of "at-large" elections. If the neighborhood was not already inside the city, then the plan was to annex it. Hays, "The Politics of Reform of Municipal Government in the Progressive Era," *Pacific Northwest Quarterly* (October 1964), 157–69.

20. Consolidation proponents in Cleveland argued that rents would fall, sickness would be reduced, and excess officials eliminated. *Forest City Democrat,* February 1, 1854; *Cleveland Leader,* April 3, 1844; Robert M. Fogelson, *The Fragmented Metropolis: Los Angeles, 1850–1930* (Cambridge, 1967), 115; Studenski, *Government of Metropolitan Areas,* 127.

21. In a strident little volume entitled *Neighborhood Government: The Local Foundations of Political Life* (Indianapolis, 1969), Milton Kotler argues that annexation was solely a device to exploit. The volume would have benefitted from less passion and more research.

22. Philadelphia abandoned an expensive 1820 plan to improve sanitary facilities and to rebuild rotting docks because shipping interests threatened to move their business to neighboring Southwark and Northern Liberties. J. Thomas Scharf and Thompson Westcott, *History of Philadelphia, 1606–1884* (Philadelphia, 1884), 599–600; Philadelphia Councils, Joint Special Committee on Removing the Railway on High, Third, and Dock Streets, *Report* (Philadelphia, 1841), 24–28.

23. Their intention was to divert farm traffic to themselves.

24. Jacob Judd, "A Tale of Two Cities: Brooklyn and New York, 1834–1855," *Journal of Long Island History,* III (Spring 1963), 19–23; Edward F. Williams, "Memphis' Early Triumph Over Its River Rivals," *West Tennessee Historical Society Papers,* XXII (1968), 5–27; Lois D. Bejach, "The Seven Cities Absorbed by Memphis," *West Tennessee Historical Society Papers,* VIII (1954), 95–104.

25. John Smith Kendall, *History of New Orleans* (Chicago, 1922), 172.

26. *Philadelphia Public Ledger,* November 12, 1844. On the riots and subsequent consolidation, see Vincent P. Lannie and Bernard C. Diethorn, "For the Honor and Glory of God: The Philadelphia Bible Riots of 1844," *History of Education Quarterly,* VIII (1968), 44–106; Price, *History of Consolidation of Philadelphia;* Harry Leffmann, "The Consolidation of Philadelphia," *Philadelphia History,* I (1908), 26–40; and Sam Bass Warner, Jr., *The Private City: Philadelphia in Three Periods of Its Growth* (Philadelphia, 1968), 125–57.

27. Some early opponents of annexation, such as the chief editor of the *North American,* switched over after the second riot. The newspaper then called it "vital to the future progress and welfare of the city." The volunteer fire companies typically opposed consolidation, while an organization called the "Friends of a Paid Fire Department," made up of important businessmen, supported it. *Philadelphia Public Ledger,* August 10 and August 29, 1853; *Philadelphia North American,* April 19, 1844.

28. Scharf and Westcott, *History of Philadelphia,* 691.

29. The location of the line of settlement can usually be determined either through a comparison of ward boundaries and population changes or through an examination of the Dynamic Factor Maps of Homer Hoyt or the Maps of the Enumeration Districts of the Various Censuses, both of which are located in the Cartographic Division of the National Archives in Washington. In Memphis, for example, owners of land on the periphery were the prime movers in the aggressive annexation policy of the city. In New York Mayor Havemeyer opposed the annexation of part of the Bronx (then Westchester) because it would primarily serve the interests of "speculators on both sides of the Harlem River." Seymour Mandelbaum, *Boss Tweed's New York* (New York, 1968), 109–10. Sam Bass Warner, Jr., also found that speculators were active in the annexation of West Roxbury to Boston in 1873. Warner, *Streetcar Suburbs: The Process of Growth in Boston, 1870–1900* (Cambridge, Mass., 1962), 41–42.

30. As late as 1900, Byberry and Moreland were made up of farms ranging in size from thirty to one hundred acres. Joseph C. Martindale, *A History of the Townships of Byberry and Moreland in Philadelphia* (Philadelphia, ca. 1900), 148.

31. For example, see Warner, *Streetcar Suburbs,* 163–64.

32. At the turn of the century, newcomers to suburbia were more likely to support annexation than long-time residents. For example, of the 449 men who signed a petition for the annexation of Morgan Park to Chicago in 1914, 202 were not listed as residents of Morgan Park in 1910. Hubert Morken, "The Annexation of Morgan Park to Chicago: One Village's Response to Urban Growth" (Master's thesis, University of Chicago, 1968), 67.

33. For example, see I. N. Phelps Stokes, ed., *Iconography of Manhattan Island* (New York, 1928), I, pp. 162, 197; and Chapter 1 of this volume.

34. The importance of public services in suburban development is discussed in Chapter 8. On their importance in one area, see Bradley Rice, "The Battle of Buckhead: The Plan of Improvement and Atlanta's Last Big Annexation," *The Atlanta Historical Journal,* XXV (Winter 1981), 5–22.

35. Olivier Zunz, *The Changing Face of Inequality: Urbanization, Industrial Development, and Immigrants in Detroit* (Chicago, 1983), 143.

36. Roxbury was threatened both by ocean storm tides and by high water in the Stony Brook.

37. In the areas newly annexed to Chicago in 1889, sixteen new fire companies were formed by 1890, and fire-alarm boxes were extended as far south as Pullman. Prior to annexation, most of the southern suburbs of Chicago were lighted by kerosene lamps.

Studenski, *Government of Metropolitan Areas,* 177, 129–33. Suburban water problems are highlighted by Nelson Manfred Blake, *Water for the Cities: A History of the Urban Water Supply Problem in the United States* (Syracuse, 1956), 87–89.

38. The estimate is that of John T. Scharf regarding an 1816 annexation to Baltimore. Scharf, *The Chronicles of Baltimore* (Baltimore, 1874), 61. If areas are unincorporated, the same thing occasionally happens today in the South and West. In the late 1960s, a vast unincorporated region south of Memphis known as Whitehaven (the name was apt) publicized a straw vote indicating that the residents opposed annexation to Memphis by a 19 to 1 margin. Because Whitehaven was not a legally incorporated city, however, the 55,000 person community was added to the Tennessee metropolis anyway.

39. The various types of legislative, popular, municipal, judicial, and quasi-legislative methods of annexation are discussed in Frank Sengstock, *Annexation: A Solution to the Metropolitan Area Problem* (Ann Arbor, 1960), 6–12. For relatively current annexation laws in each state, see National League of Cities, Department of Urban Studies, *Adjusting Municipal Boundaries: Law and Practice* (Washington, 1966).

40. *Daly vs. Morgan* (69 Maryland Reports, 461), quoted in Studenski, *Government of Metropolitan Areas,* 75–76.

41. An important source of resistance to annexation came from industries on the periphery that were anxious to minimize taxes. Thus, George Pullman fought for years to keep Pullman out of Chicago, and the big Birmingham consolidation of 1910 was carefully configured to leave out the largest factories so as "to relieve these huge enterprise, of the burden of municipal taxation." The quotation is from a disappointing volume by John R. Hornady, *The Book of Birmingham* (New York, 1921), 268.

42. John Smith Kendall, *History of New Orleans,* (Chicago, 1922), I, 742–59.

43. The strongest opposition in Cleveland came from the Fourth Ward, which was most distant from the area to be added. The Third Ward, which included the retail and warehouse districts of the city, was the only section to vote in favor of the merger. James Harrison Kennedy, *A History of the City of Cleveland* (Cleveland, 1896); and Bigger and Kitchen, *How the Cities Grew,* 145–46.

44. For instance, Roxbury voted for annexation to Boston in 1857 (no action taken, however); Cleveland approved the merger with Ohio City in 1854; and St. Louis was enlarged by popular vote in 1856.

45. *Oak Park Reporter,* November 16, 1899; quoted in Arthur LeGacy, "Improvers and Preservers: A History of Oak Park, Illinois" (Ph.D. dissertation, University of Chicago, 1967), 83.

46. Isaac Atwater, ed., *History of the City of Minneapolis* (New York, 1893), 87; and Lucille M. Kane, *The Waterfall That Built a City: The Falls of St. Anthony in Minneapolis* (St. Paul, 1966), 96.

47. Newark is unusual in that it has lost more territory than it has gained since 1800. Joseph Fulford Folsom, ed., *The Municipalities of Essex County, New Jersey, 1666–1924* (New York, 1925), I, 232; and John P. Snyder, "The Bounds of Newark: Tract, Township, and City," *New Jersey History,* LXXXVI (Summer 1968), 92–105.

48. Bigger and Kitchen, *How the Cities Grew,* 144–45.

49. The Alameda County decision, which was the first federation proposal to be put to a popular vote in the United States, lost in nine out of ten cities and was defeated by about 10,000 votes. Council of State Governments, *The States and the Metropolitan Problem: A Report to the Governor's Conference* (Chicago, 1956), 87. See also, Black McKelvey, *Rochester: The Quest for Quality, 1890–1925* (Cambridge, 1956), 110.

50. *The States and the Metropolitan Problem,* 72–73.

51. Quoted in Studenski, *Government of Metropolitan Areas,* 154–55. Retrenchment

generally followed a huge expansion. Thus, Philadelphia cut back for a dozen years following its 1854 consolidation because most suburban governments had brought the city enormous debts and precious little cash. *Philadelphia Public Ledger,* August 11, 1853; and *Annual Message of the Mayor of Philadelphia,* January 8, 1857.

52. In 1907 Los Angeles voters rejected merger with Hyde Park, Green Meadows, Gardena, Ivanhoe, and a half dozen other suburban communities. In 1926 Detroit voted against annexing portions of Warren and Royal Oak Townships.

53. The vote, which was authorized by a constitutional amendment in Missouri in 1924, would have permitted the annexation of the entire county and would have made St. Louis the world's largest city in area. Many state laws do not provide for annexations that cross county lines. American Municipal Association, *Changes in Municipal Boundaries Through Annexation, Consolidation, and Detachment,* American Municipal Association Report No. 127 (Chicago, 1939).

54. A good brief account of the Pittsburgh battle is Roy Lubove, *Twentieth Century Pittsburgh: Government, Business, and Environmental Change* (New York, 1969), 27, 97–101.

55. Quoted in Hubert Morken, ''The Annexation of Morgan Park to Chicago: One Village's Response to Urban Growth'' (Master's thesis, University of Chicago, 1968).

56. Zane L. Miller, *Boss Cox's Cincinnati* (New York, 1968), 57; and Warner, *Streetcar Suburbs,* 164.

57. LeGacy, ''Improvers and Preservers,'' passim.

58. Carol A. O'Connor, *A Sort of Utopia: Scarsdale, 1891–1981* (Albany, 1983), 13–16.

59. The municipal incorporation movement, or the villagification process, is alive and well in Los Angeles County, where ten incorporations per year were taking place in the 1960s. An excellent study of the Atlanta area is Bradley R. Rice, ''Mountain View, Georgia: The Rough and Not So Ready Suburb,'' *The Atlanta Historical Journal,* XXIV (Winter 1980), 26–40.

60. Betty Tableman, *Governmental Organization in Metropolitan Areas* (Ann Arbor, 1951), 61.

61. The only biography of the foremost champion of New York consolidation is John Foord, *The Life and Public Services of Andrew Haswell Green* (Garden City, 1913).

62. Teaford, *City and Suburb,* passim.

63. *Wall Street Journal,* March 8, 1983.

64. The twentieth-century growth of Dallas is the subject of a remarkably thorough and important study by William Black, ''Empire of Consensus: City Planning, Zoning, and Annexation in Dallas, 1900–1960'' (Ph.D. dissertation, Columbia University, 1982).

65. Lewis Mumford, *The City in History: Its Origins, Its Transformations, and Its Prospects* (New York, 1961), 494.

CHAPTER 9: *The New Age of Automobility*

1. I have borrowed the term ''automobility'' from Professor John C. Burnham of the Ohio State University. When the chain-driven safety bicycle first appeared, its appeal was so unanimous that it fostered a bicycle craze, and hundreds of thousands of them were in use within a few years.

2. The first self-propelled vehicle of any kind was a heavy, three-wheeled steam carriage put together by Nicolas Joseph Cugnot, a French artillery captain, in Paris in 1769. Many English inventors experimented with steam-driven vehicles in the nineteenth cen-

tury, but they were slow, clumsy, inefficient, and unreliable, and they were hampered by high bridge tolls and by stringent legislation. Other claims to the first automobile have been made for Enrico Bernardi of Italy and Edouard Delamarre-Debouteville of France.

3. In 1879 a Rochester patent attorney, George Selden, applied for a United States patent for an automobile. The best general histories of the early development of this new form of transportation in the western hemisphere are those of James J. Flink. See Flink, *American Adopts the Automobile* (Cambridge, 1970); and Flink, *The Car Culture* (Cambridge, 1975).

4. Gerald Carson, "Goggles and Side Curtains" *American Heritage,* XVIII (April 1967), 32–38, 108–12; and Charlton Ogburn, Jr., "The Motorcar vs. America," *American Heritage,* XXI (June 1970), 104–10.

5. Frederic L. Paxson, "The American Highway Movement, 1916–1935," *American Historical Review,* CI (January 1946), 236–53.

6. Auto executives did use their wartime positions in government to promote good roads and trucking. *Trolley Exploring, Brooklyn Eagle Library,* XXV (Serial No. 65), 17–21. See also, Donald F. Davis, "The Rise and Fall of Automotive Progressivism: Reform in Detroit, 1910–1929" (unpublished paper, University of Ottawa, 1978), 9.

7. Folke T. Kihlsted, "The Automobile and the Transformation of the American House, 1910–1935," *Michigan Historical Quarterly,* XIX (Fall 1980), 555.

8. An excellent interpretation of the famed industrialist is David L. Lewis, *The Public Image of Henry Ford: An American Folk Hero and His Company* (Detroit, 1976).

9. Reynold M. Wik, *Henry Ford and Grassroots America* (Ann Arbor, 1972), 233. Ford's statement was made in 1909.

10. Allan Nevins and Frank Ernest Hill, *Ford: Expansion and Challenge, 1915–1933,* II (New York, 1957); and R. Eugene Melder, "The Tin Lizzie's Golden Anniversary," *American Quarterly,* XII (Winter 1961), 469–71.

11. Robert S. and Helen M. Lynd, *Middletown: A Study of American Culture* (New York, 1929), 254.

12. Although the Volkswagen "bug" eventually broke the Model T sales record of 15,458,781 units, Volkswagen never accounted for more than 10 percent of world automotive production in any one year, while the Model T at its peak accounted for almost 50 percent. My observations about Ford are based upon John Lukacs's perceptive *Outgrowing Democracy: A History of the United States in the Twentieth Century* (Garden City, 1984), 24–27.

13. Davis, "Rise and Fall," 3. See also, G. W. Atterbury, "The Commercial Car as a Necessity," *Harper's Weekly,* LI (December 28, 1907), 19–25.

14. Homer Hoyt, "The Effects of the Automobile on Patterns of Urban Growth," *Traffic Quarterly,* XVII (April 1963), 293–301.

15. See for example, Blaine A. Brownell, "The Commercial-Civic Elite and City Planning in Atlanta, Memphis, and New Orleans in the 1920s," *The Journal of Southern History,* XLI (August 1975), 353–56; Mark H. Rose, *Interstate: Express Highway Politics, 1941–1956* (Lawrence, Ks, 1979), 5–7; *New York Times,* November 12, 1926; and Mark S. Foster, *From Streetcar to Superhighway: American City Planners and Urban Transportation, 1900–1940* (Philadelphia, 1981).

16. In 1657 a street in New Amsterdam was paved with stones, making New York the first city in North America to have paving. The first asphalt paving was laid in Newark in 1870, and the first concrete in Bellefontaine, Ohio in 1894. The best history of road surfacing is Clay McShane, "Transforming the Use of Urban Space: A Look at the Revolution in Street Pavements, 1880–1924," *Journal of Urban History,* V (May 1979),

279–307. See also, John B. Rae, *The Road and the Car in American Life* (Cambridge, 1971); and Harvey A. Kantor, *Prisoners of Progress: American Industrial Cities, 1850–1920* (New York, 1976), 164–65.

17. On the function of the street, see Chapter 5. See also, McShane, "Transforming Urban Space," 279–307.

18. LeCorbusier, *The City of Tomorrow,* trans. Frederick Etchells (Cambridge, reprint of 1929 translation of the 1924 French edition, 1971), 3.

19. Woodward Avenue is also Michigan State Highway Number 1.

20. Robert Hoover, "Policy Growth and Transportation Planning in the Detroit Metropolitan Area," *Papers and Proceedings of the Regional Science Association,* VII (1961), 223–39.

21. S. R. DeBoer, "Automobile Roads in Park Design," *Parks and Recreation,* IV (May–June 1923), 421–22; Peter M. Wolf, *Eugene Henard and the Beginning of Urbanism in Paris* (The Hague, 1968), 49–53; and Jay Downer, "How Westchester County, New York, Made Its Park System," *Papers and Discussions at the Twentieth National Conference on City Planning, 1928* (Philadelphia, 1938), 184–86.

22. Carl W. Condit considers Elm Street in Manchester, New Hampshire, as the first parkway, while Douglas Haskell regards Park Avenue in New York City as the first such road. Other candidates are Memorial Drive in Boston and the Grand Concourse in the Bronx. None was so long or so elaborate as the Bronx River Parkway, however. Carl Condit, *The Port of New York, II* (Chicago, 1982); Douglas Haskell, "Futurism With its Covers On," *Architectural Review,* CLVII (May 1975), 301; and Roger Wines, "Vanderbilt's Motor Parkway: America's First Auto Road," *Journal of Long Island History,* II (Fall 1962), 14–28.

23. *New York Times,* November 21, 1926; and *New York Times,* January 6, 1929.

24. Not until the 1930s were federal funds available for road construction *within* cities. Kenneth E. Peters, "The Good Roads Movement and the Michigan Highway Department, 1905–1917" (Ph.D. dissertation, University of Michigan, 1973).

25. John C. Burnham, "The Gasoline Tax and the Automobile Revolution," *The Mississippi Valley Historical Review,* XCVIII (December 1961), 435–56; and John C. Long, "Motor Transport and Our Radial Frontier," *The Journal of Land and Public Utility Economics,* II (January 1926), 109–18.

26. McShane, "Transforming the Use of Urban Space," 300.

27. The best comparison of American and European transit systems is the thorough study of John P. McKay, *Tramways and Trolleys: The Rise of Urban Mass Transit in Europe* (Princeton, 1976).

28. On the decline of public transportation in the United States, see especially, Edward Mason, *The Decline of Street Railways in Massachusetts: The Rise and Decline of an Industry* (Cambridge, Harvard Economic Studies No. XXXVII, 1932; Glen E. Holt, "The Changing Perception of Urban Pathology: An Essay on the Development of Mass Transit in the United States," in Kenneth T. Jackson and Stanley K. Schultz, eds., *Cities in American History* (New York, 1972), 324–43; and Donald Dewees, "The Decline of American Street Railways," *Traffic Quarterly,* XXIV (October 1970), 563–81.

29. Eleanora Schoenebaum, "Emerging Neighborhoods: The Development of Brooklyn's Fringe Areas, 1850–1930" (Ph.D. dissertation, Columbia University, 1976).

30. Robert A. M. Stern, *The Anglo-American Suburb* (London 1981), 44–45.

31. See the provocative article and recent book by Paul Barrett, "Public Policy and Private Choice: Mass Transit and the Automobile in Chicago Between the Wars," *Business History Review,* IL (Winter 1975), 473–97; and Barrett, *The Automobile and Urban Transit: The Formation of Public Policy in Chicago, 1900–1930* (Philadelphia, 1983).

32. On the collapse of urban mass-transit systems with the help of General Motors, see Bradford C. Snell, *American Ground Transport,* Presented to the Subcommittee on Antitrust and Monopoly of the Committee on the Judiciary, U. S. Senate, (February 26, 1974), 27–34. Snell was the assistant counsel of Senator Philip A. Hart's antitrust subcommittee investigating the restructuring of the automobile and the ground-transportation industries. See also, Mark Foster, "City Planners and Urban Transportation: The American Response, 1900–1940," *Journal of Urban History,* V (May 1979), 365–96; and Stanley Mallach, "The Origins of the Decline of Urban Mass Transporation in the United States, 1890–1930." *Urbanism Past and Present,* VIII (Summer 1979), 1–17.

33. General Motors disposed on National City Lines in 1949.

34. Foster, "City Planners," 388; and Reyner Banham, *Theory and Design in the First Machine Age* (New York, second edition, 1967), 103, 282.

35. George W. Hilton, "The Decline of Railroad Commutation," *Business History Review,* XXXVI (Summer 1962), 171–87; and H. I. Phillips, "The 7:58 Loses a Passenger," *Collier's,* LXXV (April 11, 1925), 11, 44.

CHAPTER 10: *Suburban Development Between the Wars*

1. The marginal cost is the direct out-of-pocket expense of a particular trip. It does not include the original cost of the vehicle, regular maintenance, insurance, or depreciation. It assumes that if an individual makes an economic decision to drive across town he considers only the gasoline expense, not a fair share of the other costs of owning and operating a car.

2. President's Commission on Recent Social Trends in the United States, *Recent Social Trends in the United States* (Washington, 1933), passim; and Robert S. and Helen M. Lynd, *Middletown: A Study in American Culture* (New York, 1929), passim.

3. G. W. Atterbury, "The Commercial Car as a Necessity," *Harper's Weekly,* LI (December 28, 1907), 19–25.

4. The best analysis of the influence of the automobile on small-town and rural America is Norman T. Moline, *Mobility and the Small Town: Transportation Change in Oregon, Illinois, 1900–1930* (Chicago: University of Chicago Department of Geography Research Report No. 132, 1971). See also, John C. Long, "Motor Transport and Our Radial Frontier," *The Journal of Land and Public Utility Economics,* II (January 1926), 109–18.

5. Blaine A. Brownell, "A Symbol of Modernity: Attitudes Toward the Automobile in Southern Cities in the 1920's," *American Quarterly,* XXIV (March 1972), passim.

6. In the early years of the twentieth century, Lewis Mumford was consistent and lonely in his denunciation of the motorcar and particularly its influence upon civilized life. Joseph Interrante, "The Road to Autopia: The Automobile and the Spatial Transformation of American Culture," *Michigan Historical Quarterly,* XIX (Fall 1980), 505–8.

7. *New York Times,* November 21, 1926. The congestion of down-town areas had, of course, been a problem in cities since the late nineteenth century, when unrully horses, numerous trolleys, and masses of people competed for limited space. Interrante, "The Road to Autopia," 507–8.

8. Interrante, "The Road to Autopia," 506–8.

9. Mark S. Foster, "The Automobile and the City," *Michigan Historical Quarterly,* XIX (Fall 1980), 460–62.

10. The first quotation is from Wolf von Eckardt, *A Place to Live* (New York, 1974), 337. The second is from Mitchell Gordon, *Sick Cities: Psychology and Pathology of American Urban Life* (Baltimore, 1965), 13. See also, James J. Flink, "Three Stages of American Automobile Consciousness, 1895–1971" (Paper presented at the American Historical Convention, New York City, December 29, 1971).

11. In 1919, for example, Wilminton, Delware, registrations were disproportionately suburban. Carol E. Hoffecker, *Wilmington: From Industrial City to Chemical Capital* (Philadelphia, 1982). And in 1970 suburbanites in the New York Metropolitan Region were more likely to own automobiles and to drive them farther than persons living at higher densities. "The Use of Automobiles," *Regional Plan News*, No. 108 (August 1981), 1–17. Ford's estate was west of Detroit out Michigan avenue, just beyond its junction with Southfield Road. Allan Nevins and Frank Ernest Hill, *Ford: Expansion and Challenge, 1915–1933*, II (New York, 1957), 479–80.

12. Paul F. Brissenden, *Earnings of Factory Workers, 1899–1927* (Washington, 1929), 392, 413.

13. Quoted in Foster, "The Automobile and the City," 470.

14. Amos H. Hawley, *The Changing Shape of Metropolitan America: Deconcentration Since 1920* (Glencoe, 1956), passim. As noted in the chapter on annexation, such statistics reflected as much about boundary changes as they did about population movement.

15. William Joseph Showalter, "The Automobile Industry," *The National Geographic Magazine*, XLIV (October 1923), 343.

16. New York City Tenement House Department, *Report, 1918–1929* (New York: Tenement House Commission, 1929), vol. X, 44–46.

17. Eleanora Schoenebaum, "Emerging Neighborhoods: The Development of Brooklyn's Fringe Areas, 1850–1930" (Ph.D. dissertation, Columbia University, 1976), 244–46.

18. The best treatments of land development are Marion Clawson, *Suburban Land Conversion in the United States: An Economic and Governmental Process* (Baltimore, 1971); Shirley Weiss, et al., *Residential Developers' Decisions: A Focused View of the Urban Growth Process* (Chapel Hill, 1966); and H. Allan Schmid, *Converting Land From Rural to Urban Uses* (Washington, 1968).

19. Martin Mayer, *The Builders: Houses, People, Neighborhoods, Government, Money* (New York, 1978), 57; J. C. Nichols, "The Planning and Control of Outlying Shopping Centers," *The Journal of Land and Public Utility Economics*, II (January 1926), 17–22; Gwendolyn Wright, *Building the Dream: A Social History of Housing in America* (New York, 1981), 202–3; William H. Wilson, *The City Beautiful Movement in Kansas City* (Columbia, 1964); and Robert A. M. Stern, *The Anglo-American Suburb* (London, 1981), 76–77.

20. Fred W. Viehe, "Black Gold Suburbs: The Influence of the Extractive Industry on the Suburbanization of Los Angeles," *Journal of Urban History*, VIII (November 1981), 3–26.

21. The best work on Los Angeles is Robert M. Fogelson, *The Fragmented Metropolis: Los Angeles, 1850–1930* (Cambridge, 1967). See also, Ashleigh E. Brilliant, "Social Effects of the Automobile in Southern California During the 1920's" (Ph.D. dissertation, University of California, Berkeley, 1964); Interrante, "The Road to Autopia," 505–8; Oscar O. Winther, "The Rise of Metropolitan Los Angeles, 1870–1910," *Huntington Library Quarterly*, X (1947), 391–405; Mark S. Foster, "The Model T, the Hard Sell, and Los Angeles Urban Growth: The Decentralization of Los Angeles During the 1920's," *Pacific Historical Review*, XLIV (1975), 459–84; Eshref Shevsky and Marilyn Williams, *The Social Areas of Los Angeles* (Berkeley, 1949), 36–43.

22. Stern, *Anglo-American Suburb*, 78–80.

23. The best study of Chandler and the Los Angeles water supply is William L. Kahrl, *Water and Power: The Conflict of Los Angeles' Water Supply in the Owens Valley* (Berkeley, 1982), esp. 184–91. See also, David Halberstam, "The California Dynasty: Otis Chandler and His Publishing Empire," *The Atlantic*, CCXLIII (April 1979), 58–59; Abraham Hoffman, *Vision or Villainy: Origins of the Owens Valley-Los Angeles Water Controversy* (College Station, Tx., 1981); Remi Nadeau, *The Water Seekers* (Garden City, 1950); Robert Gottlieb and Irene Wolt, *Thinking Big: The Story of the Los Angeles Times, Its Publishers, and Their Influence of Southern California* (New York, 1977); H. Marshall Godwin, Jr., "From Dry Gulch to Freeway," *Southern California Quarterly*, XLVII (1965), 73–102; and Gordon Miller, "Los Angeles and the Owens River Aqueduct" (Ph.D. dissertation, Claremont Graduate School, 1977).

24. Frederic Cople Jaher, *The Urban Establishment: Upper Strata in Boston, New York, Charleston, Chicago, and Los Angeles* (Urbana, 1982), 617–19.

25. Herbert Ladd Towle, "The Automobile and Its Mission," *Scribner's*, LIII (1913), 150–51.

26. Schoenebaum, "Emerging Neighborhoods," passim.

27. In 1929 only 34 percent of the people working in downtown Washington used public transportation, the lowest percentage of any major city in the nation. An excellent analysis of automobility and cities is David Owen Wise and Marguerite Duprie, "The Choice of the Automobile for Urban Passenger Transportation in Baltimore in the 1920's," *South Atlantic Urban Studies*, II (1978), 153–79. On the journey-to-work, see Edward J. Taaffe, Barry J. Garner, and Maurice H. Yeates, *The Peripheral Journey to Work: A Geographic Consideration* (Evanston: The Transportation Center, Northwestern University, 1963); John F. Kain, *Essays on Urban Spatial Structure* (Cambridge, 1975); Louis K. Loewenstein, *The Location of Residences and Work Places in Urban Areas* (New York, 1965); Kate K. Liepmann, *The Journey to Work* (New York, 1944); Richard F. Muth, *Cities and Housing: The Spatial Pattern of Urban Residential Land Use* (Chicago, 1969).

28. The population of South Orange was 6,014 in 1910; 7,274 in 1920; 13,630 in 1930; 13,742 in 1940; 15,230 in 1950; 16,175 in 1960; and 16,971 in 1970.

29. The sampling procedure was to take every twentieth name in the South Orange city directory for the three years under consideration.

30. Towle, "The Automobile and Its Mission," 149–52.

31. Joel A. Tarr, *Transportation Innovation and Changing Spatial Patterns: Pittsburgh, 1850–1910* (CMUTRI-TP-72-06, Pittsburgh: Transportation Research Institute, Carnegie Mellon University, April 1972), passim. Prior to 1840, or before the widespread adoption of steam power for manufacturing in the United States, factories tended to locate along rivers and near waterfalls. Rail lines began to influence industrial location later in the century. See, for example, Peter Goheen, *Victorian Toronto, 1850–1900: Pattern and Process of Growth* (Chicago: University of Chicago Department of Geography Research Study No. 127, 1970), 10–12; and George W. Grupp, *Economics of Motor Transportation* (New York, 1924), passim.

32. James B. Kenyon, *Industrial Location and Metropolitan Growth: The Paterson-Passaic District* (Chicago: University of Chicago Department of Geography Research Study Number 67, 1960); Victor R. Fuchs, *Changes in the Location of Manufacturing in the United States Since 1929* (New Haven, 1962), 90–95; Allan Pred, "Manufacturing in the American Mercantile City, 1800–1840," *Annals of the Association of American Geographers*, LVI (June 1966), 307–38; Allan Pred, *The Spatial Dynamics of Urban Industrial Growth, 1870–1914: Interpretive and Theoretical Essays* (Cambridge, 1966); Clawson, *Suburban Land Conversion in the United States*, 40; Leon Moses and

Harold F. Williamson, Jr., "The Location of Economic Activity in Cities," *American Economic Review,* LVII (May 1967), 211–22; and Evelyn M. Kitagawa and Donald J. Bogue, *Suburbanization of Manufacturing Within Standard Metropolitan Areas* (Oxford, Oh., 1955).

33. A minister, writer, and social worker, Graham R. Taylor was president of the Chicago School of Civics and Philanthropy from 1903 to 1920. The book was originally prepared for *Survey* magazine. Taylor, *Satellite Cities: A Study of Industrial Suburbs* (New York, 1915). See also, Mel Scott, *American City Planning Since 1890* (Berkeley, 1969), 130–31.

34. Schaeffer and Sclar, *Access for All: Transportation and Urban Growth* (London, 1975), 84–86; and "Motor Trucks Relieve Traffic Congestion," *Municipal Engineering,* XLIII (October 1912), 258–60; W. Hutchinson, "Influence of the Motor Truck in Relieving Traffic Congestion," *The American City,* VIII (May 1913), 561–62; and Phil Jacobsen, "Hauling Our Needs by Motor Truck," *Southern California Business,* July, 1922, 17.

35. The separation of production and office functions also began early in the twentieth century, as noted in Dennis P. Sobin's otherwise disappointing *The Future of the American Suburbs: Survival or Extinction* (Port Washington, N.Y., 1971), 53. See also, Harry Allen Smith, *Let the Crabgrass Grow: H. Allen Smith's Suburban Almanac* (New York, 1960), 21; and Daniel B. Creamer, *Is Industry Decentralizing* (Philadelphia, 1935); and Showalter, "The Automobile Industry," 404.

36. Consolidated Edison, *The Population of New York City, 1910–1948* (New York, 1948), 17; and Walter Laidlaw, *Population of the City of New York, 1890–1930* (New York, 1932), 236.

37. The decline of the front porch is considered in Chapter 16.

38. Robert C. Twombly, "Saving the Family: Middle Class Attraction to Wright's Prairie House, 1901–1909," *American Quarterly,* XXVII (1975), 66–69.

39. Gwendolyn Wright, *Building the Dream, A Social History of Housing in America* (New York, 1981), 162–64; and Edward Bok, *The Americanization of Edward Bok: The Autobiography of a Dutch Boy Fifty Years After* (New York, 1923), 238–43.

40. Wright, *Building the Dream,* 163–64.

41. Robert S. and Helen M. Lynd, *Middletown in Transition* (New York, 1935).

42. The movie was not a major hit in the United States either.

43. William Fielding Ogburn, "Machines and Tomorrow's World," *Public Affairs Pamphlet Number Five* (New York: Public Affairs Committee, 1938), 3; and Robert C. Ackerson, "Some Milestones of Automotive Literature," *Michigan Historical Quarterly,* XIX (Fall 1980), 761–71; and James J. Flink, *The Car Culture* (Cambridge, 1975).

44. For a more extended discussion and summary of the auto's effect, see Avery M. Guest, "Urban History, Population Densities, and High Status Residential Location," *Economic Geography,* XLVIII (October 1972), 375–87; and Guest, "Population Suburbanization in American Metropolitan Areas," *Geographical Analysis,* VII (July 1975), 267–83.

CHAPTER 11: *Federal Subsidy and the Suburban Dream*

1. An early version of this article was presented as the first Letitia Woods Brown Memorial Lecture to the Columbia Historical Society in Washington in 1977 and published as "Federal Subsidy and the Suburban Dream, The First Quarter-Century of Government Involvement in the Housing Market," *Records of the Columbia Historical So-*

ciety of Washington, D.C., L (1980), 421–51. See also, Kenneth T. Jackson, "Race, Ethnicity, and Real Estate Appraisal: The Home Owners Loan Corporation and the Federal Housing Administration," *Journal of Urban History*, VI (August 1980), 419–52; and Kenneth T. Jackson, "The Spatial Dimensions of Social Control: Race, Ethnicity, and Government Housing Policy in the United States," in Bruce M. Stave, ed., *Modern Industrial Cities: History, Policy, and Survival* (Beverly Hills, Calif., 1981), 79–128.

2. This statement, which expresses a widely shared view, was made by political scientist Ralph A. Rossum, *Memphis Press-Scimitar*, January 6, 1977.

3. The quotation is that of Seward H. Mott, "The Case for Fringe Locations," *Planners Journal*, V (March–June 1939), 38.

4. The tax code gives older cities a built-in handicap that continues to work against them no matter how low their fortunes sink. For example, the investment-tax credit for new machinery has made a large part of the industrial plants of the Northeast and Middle West expendable. *Business Week*, December 19, 1977, p. 88.

5. In 1978 Governor Hugh L. Carey reported that New Yorkers contributed 12 percent of federal revenues and received back only 8 percent of funding ($46 billion versus $35 billion). Regina Belz Armstrong, *Regional Accounts Structure and Performance of the New York Region's Economy in the Seventies* (Bloomington, 1980); and Daniel Patrick Moynihan, "What Will They Do for New York," *New York Times Magazine*, January 27, 1980, pp. 30–40.

6. Interstate highways have provided benefits to thinly populated states as well as to suburbs. Between 1957 and 1972, Montana received $2.44 worth of highway for each dollar it paid into the highway trust fund. Wyoming received $2.71 and Nevada $1.98, while Massachusetts took back only 77 cents, New Jersey 66 cents, and New York 80 cents. *Business Week*, December 19, 1977, p. 88.

7. This is because the emphasis of federal sewer and water aid is on *new* construction. New York City's century-old water mains, in need of billions for repairs, are not eligible for the large grants.

8. For a conspiratorial view of the state, see Peter Marcuse, "The Myth of the Benevolent State: Notes Toward a Theory of Housing Conflict" (Discussion paper, Division of Urban Planning, Columbia University, 1978).

9. An example would be a New York City building regulation of 1766 creating a fire zone in which all structures had to be made of brick or stone and all roofs had to be made of tile or slate.

10. The best studies of federal housing programs are: Mark I. Gelfand, *A Nation of Cities: The Federal Government and Urban America, 1933–1965* (New York, 1975); Henry Aaron, *Shelter and Subsidies: Who Benefits from Federal Housing Policies* (Washington, 1972); William L. C. Wheaton, "The Evolution of Federal Housing Programs" (Ph.D. dissertation, University of Chicago, 1953); and Paul F. Wendt, *Housing Policy: The Search for Solutions* (Berkeley, 1962), 142–273. On the 1970s, see A. Naparstek and G. Cincotta, *Urban Disinvestment: New Implications for Community Organizations, Research, and Public Policy* (Washington, 1970); Calvin Bradford, "Financing Home Ownership: The Federal Role in Neighborhood Decline," *Urban Affairs Quarterly*, XIV (March 1979), 313–35; and Harry C. Bredemeir, "The Federal Public Housing Movement: A Case Study of Social Change" (Ph.D. dissertation, Columbia University, 1955).

11. United States Housing Corporation Records, Record Group 3, National Archives. On America's World War I government housing experience, see Miles L. Colean, *Housing for Defense: A Review of the Role of Housing in Relation to America's Defense and a Program for Action* (New York, 1940), 1–30; Roy Lubove, "Homes and a Few Well

Placed Fruit Trees: An Object Lesson in Federal Housing,'' *Social Research,* XXVII (Winter 1960), 469–86; and John F. Sutherland, ''A City of Homes: Philadelphia Slums and Reformers, 1880–1918'' (Ph.D. dissertation, Temple University, 1973), 213–50.

12. Sutherland, ''A City of Homes,'' 247–49; and Charles Abrams, *The City is the Frontier* (New York, 1965), 241–42.

13. In New York City, where about one mortgage in four was foreclosed or passed on to the Home Owners Loan Corporation between 1931 and 1934, the homeownership rate fell from 20.2 to 16.1 percent. Semer and Zimmerman, *Evolution of Federal Legislative Policy in Housing: A Report to HUD* (Consultant's Report dated June 30, 1973, available in the Department of Housing and Urban Development Library), III-1 through III-15; and Josephine Hedges Ewalt, *A Business Reborn: The Savings and Loan Story, 1930–1960* (Chicago, 1962), 21.

14. Hoover was not unusual. President Calvin Coolidge had said just a few years earlier, ''No greater contribution could be made to the stability of the Nation and the advancement of its ideals, than to make it a nation of homeowning families.'' Quoted in Glenn H. Beyer, *Housing and Society* (New York, 1965), 249.

15. Quoted in Lyle Woodyatt, ''The Origins and Evolution of New Deal Housing Programs'' (Ph.D. dissertation, Washington University, 1968), 102.

16. Amortization refers to the payment of the principle in full by the expiration date of the mortgage. Prior to 1932, many loans required only the payment of interest, with the entire original amount being due at the expiration of the loan.

17. *Final Report of the Committee on Large Scale Operations, The President's Conference on Home Building and Home Ownership* (Washington, 1932), 24.

18. David A. Bridewell, ''The Federal Home Loan Bank Board and Its Agencies'' (Manuscript in the Research Library of the Federal Home Loan Bank Board, dated May 14, 1938), 164.

19. *Emergency Relief and Construction Act,* (1932), Public Law 302, 72nd Congress, Title II, Section 201.

20. The best study of the subject is Joseph L. Arnold, *The New Deal in the Suburbs: A History of the Greenbelt Town Program* (Columbus, Oh., 1971). See also, Paul K. Conkin, *Tomorrow a New World: The New Deal Community Program* (Ithaca, 1959).

21. This refers to the fact that the price of real estate *declined* in the early 1930s.

22. Quoted in Bridewell, ''The Federal Home Loan Bank Board,'' 234.

23. The vote for HOLC was 383–4 in the House; it passed without record vote in the Senate. The best study of this agency is C. Lowell Harriss, *History and Policies of the Home Owners' Loan Corporation* (New York, 1951), 11.

24. The HOLC's active lending program ended in 1936, and it was liquidated in 1951. It was administered by and was part of the Federal Home Loan Bank Board.

25. Long-term mortgage arrangements were not common at the time because an 1864 amendment to the 1863 National Bank Act prohibited nationally chartered banks from making direct loans for real-estate transactions.

26. Harriss, *Home Owners' Loan Corporation,* 2. ''Red lining'' refers to the arbitrary decisions of government and private financial institutions not to lend in certain neighborhoods because of general characteristics of the neighborhood rather than of the particular property to be mortgaged.

27. These comments were taken from the questionnaires which ordinarily accomanied the Residential Security Maps. They are available in the HOLC City Survey Files (Record Group 195) in the National Archives. The quotations are from the New York, Newark, and St. Louis files. A few large cities, such as Washington, are missing from

the collection. Communities of fewer than 25,000 inhabitants were apparently not evaluated unless they were suburbs.

28. Even the possibility of change was sufficient to lower a rating. In suburban Mount Vernon, New York, the best neighborhoods were described as "well-maintained and evidence pride of ownership." But the security grade given was only "B" becaause of "the possible influx of less desirable elements from the Bronx."

29. Although historians may wince at the ethnic epithets contained in the evaluations, the data and the analyses undertaken by the appraisers for each neighborhood provide a matchless source of information on residential change.

30. The information in this paragraph is taken from Bradford, "Financing Home Ownership," 319–25.

31. Hoyt was a professor of real estate and Park a professor of sociology. They suggested that different groups of people "infiltrated" or "invaded" territory held by others through a process of competition.

32. Although other scholars have noted the discriminatory result of federal housing policies, this investigation is the first to make systematic use of the Residential Security Maps and the detailed reports that accompanied them. "Highly restricted" was usually a shorthand way of indicating that Jewish residents were excluded. For a recent evaluation of Ladue as "a country town with a difference," see *St. Louis Post Dispatch,* January 18, 1982.

33. Lincoln Terrace was just east of Brentwood and south of Richmond Heights along Lincoln and Stockard Streets.

34. National Resources Committee, *Regional Planning, Part II-St. Louis Region* (Washington, 1936), 53.

35. One thinks of Jane Jacobs and her love of variety and mixtures of uses as someone who might have found a great deal to admire in these older neighborhoods. Jacobs, *The Death and Life of Great American Cities* (New York, 1961), passim.

36. Those few City of St. Louis neighborhoods that were rated First or Second grade tended to be located near attractive open spaces like Forest Park, Francis Park, or Carondolet Park. By 1940 the HOLC evaluation of the city had become markedly more negative, and almost all the First and Second grade areas had been lowered in rating. See also, *Metropolitan St. Louis: Summary of an Economic, Real Estate, and Mortgage Finance Survey* (Washington, 1942), especially 4, 11, 12.

37. In this instance, as in many others, the federal appraisers were perceptive observers. For example, Paul Stellhorn has recently documented the exodus of the Newark business elite to the suburbs before World War I. Stellhorn, "Depression and Decline: Newark, N. J., 1929–1941" (Ph.D. dissertation, Rutgers University, 1983). A good description of Newark in this period is Phillip Roth's novel, *Goodbye Columbus.*

38. Ironically, many of the neighborhoods written off by the federal appraisers as hopeless in the 1930s have become chic and desirable to gentrifiers half a century later.

39. *Metropolitan St. Louis,* 16.

40. Evidence of a free interchange of ideas and information between HOLC and FHA and between government appraisers and private realtors is persuasive. As mentioned earlier, bankers and realtors helped to draw up the Residential Security Maps in the first place.

41. President Roosevelt phrased his request in somewhat the same way. He asked for legislation on May 14, 1934, "First, to return many of the unemployed to useful and gainful occupation; second, to produce tangible, useful wealth in a form for which there is great social and economic need."

42. Bradford, "Financing Home Ownership," 332; Aaron, *Shelter and Subsidies,* 76;

and Marion Clawson, *Suburban Land Conversion in the United States: An Economic and Governmental Process* (Baltimore, 1971), 80–91. For the economic significance of FHA and VA and of late public housing efforts, see Lawrence N. Bloomberg, "The Housing Problem: Long-Run Effects of Government Housing Programs," *American Economic Review,* XLI (May 1951), 589–90.

43. The Housing Act of 1964 (Section 121) authorized FHA to pay the owner of an FHA-insured home for costs incurred in correcting "substantial defects" in the structure. Clawson, *Suburban Land Coversion,* 41.

44. Although the full faith and credit of the United States government stood behind the FHA obligation, a special premium from homeowners (normally about 0.5 percent) always exceeded FHA expenses. As a result, FHA made a profit of more than $415 between 1951 and 1971 alone.

45. As noted earlier, only Iceland, Australia, and New Zealand, among all industrialized countries, exceeded the United States homeowner rate in 1972. Jim Kemeny, "Forms of Tenure and Social Structure," *British Journal of Sociology,* XXIX (March 1978), 43.

46. Carol E. Hoffecker, *Corporate Capital: Wilmington in the Twentieth Century* (Philadelphia, 1983), passim.

47. Personal Interview with Martin Winter, April 19, 1977, New York City.

48. There are important exceptions to this generalization. The notorious 608 program of the late 1940s, which Martin Winter called "the most colossal fraud of all time," offered inducements to multi-family construction that were more lucrative than the single family opportunities that FHA and VA presented. The government would lend builders as much as 30 percent more than the cost of construction, which meant that they could "mortgage out," or initiate large developments without risking any money of their own. A 1956 Senate study found that in 80 percent of the cases studied, the builder walked away with a profit of at least 25 percent. This method is recounted in Charles Abrams, *The City is the Frontier* (New York, 1965), 87–92. The National Housing Act Amendments of 1938 actually made mortgage insurance available for the first time for rental housing, but not much was written.

49. Title I of the 1934 National Housing Act was for "Housing Renovation and Modernization." It insured financial institutions against losses sustained from loans for alterations, repairs, and improvements on real property. Its ineffectiveness was admitted in 1954 by Albert M. Cole, then Administrator of the Housing and Home Finance Agency (of which FHA was one part), when he noted that Title I "is of limited assistance to families of modest income who need to finance home improvements." Committee on Banking and Currency, *Housing Act of 1954, Hearings,* 2 vols. (Washington, 1954), 52. See also, President's Advisory Committee on Government Housing Policies and Programs, *Recommendations on Government Housing Policies and Programs* (Washington, 1953), 73; Federal Housing Administration, *Remodel—Repair—Repay with FHA* (Washington, 1955), 1–6; and Mark I. Gelfand, "Cities, Suburbs and Government Policy," in Robert H. Bremner and Gary W. Reichard, eds., *Reshaping America: Society and Institutions, 1945–1960* (Columbus, 1982).

50. Sellers had a huge stake in the appraisal evaluation and decision because the FHA appraisal price often became the selling price. In the 1970s, scandals often occured when appraisers over-valued renovated homes and then guaranteed the mortgages which were secured by low down payments. When the shoddy repair work became evident and the houses started to crumble, FHA was left holding the bag. *New York Times,* March 20, 1972; June 28, 1972; and January 3, 1977.

51. *FHA Underwriting Manual* (Washington, 1947), Section 1301. Between 1934 and

1938, the *Underwriting Manual* was available only in typescript. The first edition came out in 1938, the second in 1947.

52. These comments are taken from Sections 1303 through 1316 of the 1938 *Underwriting Manual*. By 1958 the first two categories were changed to "Physical and Social Attractiveness" and "Protection Against Inharmonious Land Uses."

53. Federal Housing Administration, *Rental Housing as Investment* (Washington, 1938), 30.

54. Between 1934 and 1950 there was no FHA concern with equal opportunity in housing, and race was considered only to the extent that changing neighborhood composition could cause land values fo fall. Neither the 1938 nor the 1947 FHA *Underwriting Manual* specifically endorsed "racial" covenants, but in the context of other directives and comments, there can be little doubt but that racial covenants were deemed desirable by FHA appraisers. Such covenants, which were part of the deed itself, required that no person of African descent ever be allowed to live on the property except as a domestic servant or laborer.

55. Nathan Straus, *Two-Thirds of a Nation: A Housing Program* (New York, 1952), 222.

56. Some of these maps are filed with the Cartographic Division of the National Archives. Most of the records, however, seem to have disappeared, and senior FHA officials in the late 1970s consistently denied to me the existence not only of red-lining maps but of any information that might allow a spatial analysis of FHA mortgage insurance before 1970 other than by county. FHA did establish a statistical division to carry out a massive analysis of residential patterns in the United States, which resulted in the first national Census on Housing in 1940.

57. Quoted in Straus, *Two-Thirds of a Nation*, 221.

58. This and other aspects of discrimination in Detroit are covered in David Allan Levine, *Internal Combustion: The Races in Detroit, 1915–1926* (Westport, Conn., 1976), passim.

59. Indeed, my analysis is not of racial discrimination, but of the disparity in city and suburban aid.

60. The Single Family Computer Data Base can provide information on mortgage location by census tract, but FHA currently retains information in the data base for five years. Linda L. Royster, *HUD Operating Data Relating to FHA Mortgage Insurance Activities* (Washington, 1975).

61. Over the nation as a whole, about 40 percent of FHA mortgages were issued for new homes and 60 percent for existing dwellings between 1934 and 1972. Section 223(e) of the National Housing Act of 1968 (Part of the Housing and Urban Development Act) gave legislative sanction to relaxing FHA standards in order to increase mortgage insurance for housing in blighted sections of central cities.

62. Essex County (Newark) is not included in TABLE XII-3 because it contains affluent suburbs as well as Newark, and there was no way to disaggregate the information. In the case of St. Louis, the table actually understates the county advantage because the per-capita data are based upon 1960 population, when the suburbs were growing rapidly. During the period under review, the county normally had fewer people and the city more than in 1960. In Hudson County FHA mortgage operations increased by twenty times between 1960 and 1976, but even at the later date the county was still receiving far less than the national average.

63. The 1934 legislation creating the Federal Housing Administration authorized an 80 percent mortgage for homes costing up to $20,000. Livingston, on the western edge

of Essex County in attractive, wooded, and rolling country, was a sparsely settled area of thirteen square miles without industry or railroads in 1938. Adjacent to Newark, Irvington was about 80 percent developed in 1938.

64. Federal Housing Administration, *Washington, D.C. Housing Market Analysis* (Washington, 1939), 49.

65. Moffett was a former vice president of the Standard Oil Company.

66. The popularity of FHA, coupled with the fact that it was not a drain on the federal budget, helped to make it remarkably independent of the other agencies in Washington.

67. Jonathan Lang, "Problems Facing Urban Renewal in the Fringe City: A Study of Redevelopment Programs in Camden, New Jersey" (Seminar paper, Columbia University, 1972); Christopher Norwood, *About Paterson: The Making and Unmaking of an American City* (New York, 1974), 124; and Straus, *Two-Thirds of a Nation,* 221.

68. Mel Scott, *American City Planning Since 1890* (Berkeley, 1968), 401; and Charles Abrams, *Forbidden Neighbors: A Study of Prejudice in Housing* (New York, 1955), 229–35.

69. Bradford, "Financing Home Ownership," 314; Martin Nolan, "A Belated Effort to Save Our Cities," *The Reporter,* XXXVII (December 28, 1967), 17–20; Joseph Fried, *Housing Crisis U.S.A.* (New York, 1971); and "Ins and Outs of Home Loans," *Changing Times,* XIII (August 1959), 26–28; and Jacobs, *Death and Life,* 301. The Douglas quotation is from Paul Douglas, et al., *Building the American City: Report of the National Commission on Urban Problems to the President of the United States* (Washington, 1968), 100–101.

70. James T. Little, et al., *The Contemporary Neighborhood Succession Process: Lessons in the Dynamic of Decay from the St. Louis Experience* (St. Louis: Washington University Institute for Urban and Regional Studies, 1975), 5.

71. The literature on this subject is enormous. An excellent place to begin is the special issue on red lining of *Empire State Report,* I (March–April 1978), 5–33; and Harriett Tee Taggart and Kevin W. Smith, "Redlining: An Assessment of the Evidence of Disinvestment in Metropolitan Boston," *Urban Affairs Quarterly,* XVII (September 1981), 91–107.

72. Mark I. Gelfand, "Cities, Suburbs and Government Policy," passim; and Straus, *Two-Thirds of a Nation,* 218–21.

73. There were signs in 1984 that St. Louis was becoming a part of the nationwide trend toward restoration. Inner-city houses that a few years earlier were vacant had become occupied and attractive. An excellent analysis of the impact of federal public housing policies on St. Louis is Eugene J. Mechan, *The Quality of Federal Policymaking: Programmed Failure in Public Housing* (Columbia, Mo., 1978).

CHAPTER 12: *The Cost of Good Intentions*

1. The standard work on Veiller and his supporters is Roy Lubove, *The Progressives and the Slums: Tenement House Reform in New York City, 1890–1917* (Pittsburgh, 1963). Peter Marcuse has analyzed the way city planning in the United States ignored housing in its formative years in "Housing in Early City Planning," *Journal of Urban History,* VI (February 1980), 153–76. See also, James Ford, et al., *Slums and Housing,* I (Cambridge, 1936); Robert W. DeForest and Lawrence Veiller, eds., *The Tenement House Problem,* 2 vols. (New York, 1903); Lawrence M. Friedman, *Government and Slum*

Housing: A Century of Frustration (Chicago, 1968); and Anthony Jackson, *A Place Called Home: A History of Low-Cost Housing in Manhattan* (Cambridge, 1976).

2. In Great Britain and in the Netherlands, the national governments occasionally lowered costs for private-sector builders by buying up land and making it available at a reduced rate. In 1904 Stockholm began making municipally owned land available to workers to build their own homes. In Canada the Federal-Provincial Housing Loan Act of 1919 established the principle of government-assisted housing, but the program was ill-fated from the start and with the repeal of the Ontario Housing Accommodation Act in 1927, there was no longer any national support for low-cost shelter. L. H. Ohrbach, "Homes for Heroes: A Study in the Politics of British Social Reform" (Ph.D. dissertation, Columbia University, 1971); Shirley Spragge, "A Confluence of Interests: Housing Reform in Toronto, 1900–1920," in Alan F. J. Artibise and Gilbert A. Stelter, eds., *The Usable Urban Past: Planning and Politics in the Modern Canadian City* (Toronto, 1979), 247–67; and *New York Times,* March 30, 1980.

3. The best analysis of Wood's career is Eugenie Ladner Birch, "Edith Elmer Wood and the Genesis of Liberal Housing Thought" (Ph.D. dissertation, Division of Urban Planning, Columbia University, 1976), especially chapters 1–4. See also, Birch, "Woman-Made America: The Case of Early Public Housing Policy," *Journal of the American Institute of Planners,* XLIV (April 1978), 130–44; Edith Elmer Wood, *Recent Trends in American Housing* (New York, 1931); and John F. Sutherland, "A City of Homes: Philadelphia Slums and Reformers, 1880–1918" (Ph.D. dissertation, Temple University, 1973).

4. The public housers claimed much too much for their idea. There is actually very little evidence to suggest that improved housing is useful in anything other than a psychic or emotional sense. In terms of worker absenteeism and productivity, housing change has a negative impact. The conclusion is based on an examination of six sites in the United States, Korea, Mexico, Venezuela, and Kenya. Leland S. Burns, *Housing: Symbol and Shelter* (Los Angeles: International Housing Productivity Study, University of California, February 1970). See also, Alvin E. Coons and Bert T. Glaze, *Housing Market Analysis and the Growth of Nonfarm Home Ownership* (Columbus: Bureau of Business Research Monograph Number 115, Ohio State University, 1963), 84–86.

5. In 1926, under the leadership of Governor Alfred E. Smith, New York State pioneered in the housing field by allowing tax deductions for corporations that constructed cooperative apartments. About six thousand units were built under the program. The North Dakota program, which was tiny and which operated between 1919 and 1923, was designed to provide homes for both urban and rural residents. Friedman, *Government and Slum Housing,* 97–98.

6. 48 Statute 195 (1933), *National Industrial Recovery Act,* Title II, Section 202. See also, Federal Emergency Administration of Public Works, *Urban Housing: The Story of the PWA Housing Division, 1933–1936,* Bulletin No. 2 (Washington, 1937), 14–16.

7. An excellent discussion of the political trade-offs in federal housing policies can be found in Harold Wolman, *Politics of Federal Housing* (New York, 1971). See also, Timothy L. McDonnell, *The Wagner Housing Act* (Chicago, 1957).

8. Actually, American cities could have provided many such instances in the nineteenth century. In New York City, for example, it was common practice to tear down older homes, which had already been divided into as many as half a dozen separate apartments, and to build taller and deeper tenement structures in their place. Dumbbell tenements, so-called because of the narrow airshaft on either side of the building, also represented new private construction in fully developed lower Manhattan neighborhoods. Such housing obviously was not what Ickes had in mind, however. A fine local

study is John F. Bauman, "Safe and Sanitary Without the Costly Frills: The Evolution of Public Housing in Philadelphia," *The Pennsylvania Magazine of History and Biography,* CI (January 1977), 114–28.

9. The first public housing project in the United States—either federal, state, or local—was First Houses, a group of 120 apartments in a series of four and five-story walk-up buildings at Third Street and Avenue A on the Lower East Side of Manhattan. The 1935 project did not involve federal funds, which were just becoming available. "Public Housing in the United States," *Neighborhood: The Journal for City Preservation,* VI (Summer 1983), 2–6.

10. It was the use of eminent domain for housing, not the construction of housing, that was found unconstitutional. *United States v. Certain Lands in the City of Louisville,* 9 F. Supp. 137 (W. D., Ky., 1935). See also, William Ebenstein, *The Law of Public Housing* (Madison, 1940), 39.

11. "Federal Activities in the Housing Field," *Congressional Digest,* April 1936, p. 104. See also, Robert K. Brown, *The Development of the Public Housing Program in the United States* (Atlanta: Bureau of Business and Economic Research, 1960); and Leonard Freedman, *Public Housing: The Politics of Poverty* (New York, 1969).

12. The bill died because FDR did not pressure the conservative chairman of the committee, Representative Henry Steagall of Alabama, who would have supported it had the President so requested. The most likely reason for this is that Roosevelt preferred to avoid the political risk of endorsing the bill in the 1936 election year. He was reasonably sure that the portion of the electorate committed to the New Deal would not vote against him on this issue alone, and he did not wish to alienate powerful business interests by openly favoring the growth of the public sector at the expense of the private. McDonnell, *The Wagner Housing Act,* 210. See also, Joseph Huthmacher, *Senator Robert Wagner and the Rise of Urban Liberalism* (New York, 1958), 204–10.

13. Birch, "Woman-Made America," 139.

14. Franklin D. Roosevelt, "A Changed Moral Climate in America," *Vital Speeches,* III, February 1, 1937.

15. *New York Times,* September 3, 1937.

16. Housing Files, Franklin D. Roosevelt Library, Hyde Park, New York.

17. *Annual Report of the United States Housing Authority for the Fiscal Year 1938* (Washington, 1939), vii, 38.

18. The 1937 housing act was important in terms of jobs because the unemployment rate among construction workers averaged about 55 percent at the time. Moreover, housing construction is a key factor in economic recoveries because it uses large amounts of capital, labor, and materials. Robert M. Fisher, *Twenty Years of Public Housing: Economic Aspects of the Federal Program* (New York, 1959), 229.

19. At various times, officials of the federal government have tried to withhold certain federal grants from communities that refuse to accept public housing, but the efforts have not met with great success. For example, between 1977 and 1979, Secretary Patricia Roberts Harris of the Department of Housing and Urban Development attempted to use financial incentives to encourage suburbs to accept a fair share of public housing.

20. Certain older suburbs, such as Greenwich, Connecticut, maintain an exclusive image despite the presence of public housing, but newer suburbs tend to resist such incursions with great ferocity.

21. Federal Judge Frank J. Battisti found Parma guilty of violating the 1968 Fair Housing Act by enacting a law "motivated by a racially discriminatory and exclusionary intent and with forseeable segregative effects." *New York Times,* January 13, 1981. In 1971 the United States Supreme Court upheld the constitutionality of state laws requir-

ing approval in a public referendum before subsidized housing could be built in a community.

22. Bauman, "Safe and Sanitary," 116–25.

23. Quoted in Fisher, *Twenty Years of Public Housing,* 96.

24. For the legislative background of the Housing Act of 1949, see Mark I. Gelfand, *A Nation of Cities: The Federal Government and Urban America, 1933–1965* (New York, 1975), 105–56.

25. *Congressional Record,* April 21, 1949, 4840–52.

26. Between 1969 and 1974, President Richard M. Nixon frequently affirmed that he would not use financial leverage to compel suburbs to accept low-income housing projects against their wishes.

27. When New York's civil rights-minded Mayor John V. Lindsay attempted to force a public housing project on middle-class Forest Hills in Queens, a huge rhubarb resulted. Even though a compromise was reached, the project did result in an accelerated white flight to the suburbs. Thomas M. Gray, "Daley News: Chicago's Public Housing Fiasco," *The New Republic* CLXIV (April 3, 1971), 17. See also, *New York Times,* October 1, 1973 and April 16, 1976.

28. Interviews with black youths who defined the ghetto as the projects were conducted in conjunction with the exhibit, "Ruins and Revivals: The Architecture of Urban Devastation," by Camilo J. Vergara and Kenneth T. Jackson, which opened at the Municipal Art Society in New York in September, 1983.

29. The negative results of these huge public housing concentrations were analyzed by Oscar Newman, *Defensible Space: Crime Prevention Through Urban Design* (New York, 1973).

30. Application H-163, letter of Henry B. Immel, Mail and Files Division, Public Works Administration, Record Group 196, National Archives.

31. Chester Hartman, "The Limitations of Public Housing," *Journal of the American Institute of Planners,* XXIX (November 1963), 283–85.

CHAPTER 13: *Sewers, Services, and Schools*

1. I use the term "men" because in World War II the armed forces were overwhelmingly male.

2. The best study of this phenomenon is Barry Checkoway, "The Politics of Postwar Suburban Development" (Childhood and Government Project Working Paper #13, School of Law, University of California, Berkeley, 1977).

3. Mark I. Gelfand, "Cities, Suburbs and Government Policy," in Robert H. Bremner and Gary W. Reichard, eds., *Reshaping America: Society and Institutions, 1945–1960* (Columbus, Oh., 1982), 112–38.

4. John Keats, *The Crack in the Picture Window* (Boston, 1956). See also, Edward P. Eichler and Marshall Kaplan, *The Community Builders* (Berkeley, 1967).

5. In 1942 the National Housing Agency was set up to centralize government efforts to provide wartime shelter. Of the nine million persons who migrated to arms factories, about half were taken care of by the "Share Your Home" program. Only 1,206,000 permanent units were added to the American housing inventory during the conflict. National Housing Agency, *Fourth Annual Report* (Washington, 1946), 2–3, 26–27. See also, Barry Checkoway, "Large Builders, Federal Housing Programs, and Postwar Suburbanization," *International Journal of Urban and Regional Research,* IV (1980), 21–44.

6. The best analysis of this question is Mark Alan Willis, "The Effects of Cyclical Demand on Industry Structure and on the Rate of Technological Change: An International Comparison of the Housebuilding Sectors in the United States, Great Britain, and France" (Ph.D. dissertation, Yale University, 1979). See also, *Fortune,* CII (November 3, 1980), 15.

7. There is no biography of William Levitt or his company. The best study is John T. Liell, "Levittown: A Study in Community Development and Planning" (Ph.D. dissertation, Yale University, 1952), which was based on questionnaires sent to every sixth person, supplemented by interviews. An objective look at Levittown as a sociological phenomenon in transition is William M. Dobriner, *Class in Suburbia,* (Englewood Cliffs, N.J., 1963). See also, Joseph M. Guilfoyle and J. Howard Rutledge, "Levitt Licks the Housing Shortage," *Coronet,* September 1948, 112–16; "Levitts Deliver $6,990 House," *American Builder,* June 1947, 96–97; Harold L. Wattel, "Levittown: A Suburban Community," in William M. Dobriner, ed., *The Suburban Community* (New York, 1958), 287–313; and a work of fiction, Charles Mergendahl, *It's Only Temporary* (Garden City, N.Y., 1950). Autobiographical accounts include William J. Levitt, "More Houses and Better Values," *Journal of the American Institute of Planners,* IX (June 1948), 253–56; Alfred S. Levitt, "A Community Builder Looks at Community Planning," *Journal of the American Intitute of Planners,* XVII (Spring 1951), 80–88; and William J. Levitt, "What! Live in a Levittown," *Good Housekeeping,* July, 1958, 47, 175–76.

8. The earliest printed reference to the Levitts I have found is Boyden Sparks, "They'll Build Neighborhoods, Not Houses," *Saturday Evening Post,* October 28, 1944, 11, 43–46.

9. To the best of my knowledge, no single builder has yet surpassed the Levittown, New York, total. New communities that will ultimately be larger, such as Reston, Columbia, and Irvine, are being built by many separate housing contractors.

10. For example, lumber came from the Levitt-owned Grizzly Park Lumber Company of Blue Lake, California; appliances were purchased through the Levitt-owned North Shore Supply Company on Long Island; and parts were preassembled in a Levitt factory in Rosslyn. The development of prefabricated houses is covered in Albert F. Bemis, *The Evolving House* (Cambridge, 1936); and Burnham Kelly, *The Prefabrication of Houses* (New York, 1951). See also, Willis, "The Effects of Cyclical Demand," 2–3.

11. *Thousand Lanes,* I (November 1951), 3. See also, Eric Larrabee, "The Six Thousand Houses That Levitt Built," *Harper's Magazine,* CXCVII (September 1948), 79–88; and John T. Liell, "4000 Houses a Year," *Architectural Forum,* XC (April 1949), 84–93.

12. Levitt and one of his homes appeared on the July 3, 1950 cover of *Time.* For an explanation of why Levitt felt that the Cape Cod style was particularly efficient, see "The Cape Cod Cottage," *Architectural Forum,* March 1949, 98–106.

13. *New York Times,* April 2, 1981.

14. Howell Walker, "Long Island Outgrows the Country," *National Geographic,* March, 1951, 279–326. For an analysis of the much-touted community spirit, see Ralph G. Martin, "Life in the New Suburbia," *New York Times Magazine,* January 15, 1950, 16, 40–42. See also *New York Times,* April 18, 1972.

15. *New York Times,* April 2, 1981; and Keats, *The Crack in the Picture Window,* 55. The seminal architectural critique of the 1950s suburban house is Serge Chermayeff and Christopher Alexander, *Community and Privacy* (Garden City, 1963).

16. *New York Times,* April 18, 1972.

17. The Levittowns have fallen in status and prestige over the past three decades, especially as original residents moved out and were replaced by working class families.

18. Herbert J. Gans, *The Levittowners: Ways of Life and Politics in a New Suburban Community* (New York, second edition, 1982).

19. In 1968 William J. Levitt sold his original construction concern, Levitt and Sons, to International Telephone and Telegraph for $92 million. The consulting contract that ensued stipulated that he was not to re-enter the American housing field before 1977. In 1978 he assumed the rights to a 3,100 acre site near Orlando, Florida, and began building the first of a planned nine thousand units, about three thousand of which were to be apartments. *New York Times*, January 25, 1981.

20. The nation's dominant builder within a single city is Ray Ellison, who was accounting for about 45 percent of the San Antonio market's housing starts in 1984. *USA Today*, June 14, 1984. See also, "A New Method of Merchant Building," *Architectural Forum*, XCI (September 1949), 75–77; and "Park Forest Moves Into 1952," *House and Home*, I (March 1952), 115.

21. The strongest argument against Title VI financing, as Eric Larrabee has noted, was that it gave a builder a contingent profit for which he had not paid and that he secured a stake in a project in which he had made no personal investment. Larrabee, "The Six Thousand Houses That Levitt Built," 79–88. See also, Michael Sumichrast and Sara A. Frankel, *Profile of the Builder and His Industry* (Washington, 1970); and Alfred Steinberg, "FHA: Profits Before Housing," *The Nation*, CLXVIII (January 1, 1949), 11–13.

22. In the New York Metropolitan Region, the suburbs accounted for only 42.5 percent of new homes in 1946–1947.

23. Matthew Edel, Elliott D. Sclar, and Daniel Luria, *Shaky Palaces: Home Ownership and Social Mobility in Boston, 1870–1970* (New York, 1984); and David Halle, *America's Working Man: Work, Home and Politics Among Blue-Collar Property Owners* (Chicago, 1984); and John B. Lansing and Gary Hendricks, *Automobile Ownership and Residential Density* (Ann Arbor: Institute of Social Research, University of Michigan, June 1967).

24. My information on Swedish-American comparisons comes from David R. Goldfield, whose most recent publications on the subject include "National Urban Policy in Sweden," *APA Journal*, Winter 1982, 24–38; and "A Metropolitan Vision: Planning and Social Equity in Sweden and the United States," *Human Environment in Sweden* (New York: Swedish Information Service Paper No. 21, December, 1982), 1–8.

25. Most regional patterns in housing resulted either from climate or from the comparative cost of competing materials. In California, for example, the popularity of stucco-exterior walls initially stemmed from the fact that it was a relatively inexpensive surfacing material that was similar in appearance to early Spanish adobe construction. In the North Central and Southern regions brick homes predominated; in the heavily forested Northeast, wood was the most commonly used exterior-wall material. See, Charles Moore, Gerald Allen, and Donlyn Lyndon, *The Place of Houses* (New York, 1974), 70–74. See also *Saturday Evening Post Urban Housing Survey* (Philadelphia, 1945), 11.

26. I base this assertion on the selling price of the house and lot, which in the fifteen years after World War II was typically under $10,000 in most suburban tract developments. By contrast, the average cost of a similar type dwelling was almost as high in the 1920s, and, in real terms, was even higher in the 1890s. Similarly, the real cost of housing in the 1980s (median new home price of about $75,000 in 1983) is substantially higher than it was three decades ago.

27. The typical pattern in these turn-of-the-century elite suburbs was for the service people and minority citizens, who were essential to the functioning of the substantial homes and whose presence in no way diminished the social prestige of the town, to live

within a half mile of the railroad station, while the wealthier inhabitants lived on much larger plots farther from town. Another reason for a large servant and minority population within many communities was their status as live-in employees. The new post-World War II affluent suburbs contain reltively few full-time domestic servants and no slum districts. Howard L. Preston, *Automobile Age Atlanta: The Making of a Southern Metropolis, 1900–1935* (Athens, Ga., 1979), 96–110.

28. This pattern had changed markedly by 1980, when 38 percent of Levittown, New Jersey, was black. Gans, *Levittowners,* introduction to the second edition. See also, Ron Rosenbaum, "The House That Levitt Built," *Esquire,* December 1983, 378–90.

29. Many early Levittowners on Long Island went on to significant career successes, but at the time of their residence in the community they shared the general socioeconomic characteristics of their neighbors.

30. Gwendolyn Wright, *Building the Dream: A Social History of Housing in America* (New York, 1981), 220–39; Sidonie M. Gruenberg, "Homogenized Children of New Suburbia," *The New York Times Magazine,* September 19, 1954, 14; Lewis Mumford, *The City in History* (New York, 1961), 486; and Mumford, "The Wilderness of Suburbia," *The New Republic,* September 7, 1921, 44–45.

31. The best analysis anywhere of the process of inner-city neighborhood decline as it relates to the suburban housing market is James T. Little, Hugh O. Nourse, R. B. Read, and Charles L. Leven, *The Contemporary Neighborhood Succession Process: Lessons in the Dynamic of Decay from the St. Louis Experience* (St. Louis: Washington University Institute for Urban and Regional Studies, 1975), 79–81, 180.

CHAPTER 14: *Drive-in Culture*

1. The thesis that road-watching is a delight and that the highway might be a work of art is expressed in Donald Appleyard, Kevin Lynch, and John R. Myer, *The View From the Road* (Cambridge, 1964). See also, Paul W. Gikas, "Crashworthiness as a Cultural Ideal," *Michigan Historical Quarterly,* XIX (Fall 1980), 704. An unconvincing strident defense of the automobile and the expressway system is David Brodsly, *L.A. Freeway: An Appreciative Essay* (Berkeley, 1982).

2. Schaeffer and Sclar, *Access for All: Transportation and Urban Growth* (London, 1975), 39–41. See also, Jean-Pierre Bardou, Jean-Jacques Chanaron, Patrick Fridenson, and James M. Laux, *The Automobile Revolution: The Impact of an Industry* (Chapel Hill, trans., 1982); Joel A. Tarr, *Transportation Innovation and Changing Spatial Patterns: Pittsburgh, 1850–1934* (Pittsburgh: Transportation Research Institute, Carnegie-Mellon University, 1977); James J. Flink, *The Car Culture* (Cambridge, 1975); and James R. Dunn, Jr., *Miles to Go; European and American Transportation Policies* (Cambridge, 1981). The leading authority on the "car culture," James J. Flink, is now working on an ambitious international history of the automobile age.

3. Willard Morgan, "At Last—A Place to Park," *American Builder,* July 1929, 58–60; and *New York Times,* May 22, 1979.

4. Ed Cray, *Chrome Colossus: General Motors and Its Times* (New York, 1980), 326.

5. Cray, *Chrome Colossus,* 356–58. The best study of the origins of the Interstate Highway system is Mark H. Rose, *Interstate: Express Highway Politics, 1941–1956* (Lawrence, Ks., 1979).

6. Robert A. Caro, *The Power Broker: Robert Moses and the Fall of New York* (New York, 1974), passim.

7. James J. Flink, "The Automobile Revolution in Worldwide Comparative Perspective" (Paper presented at the Detroit Historical Society Conference on the Automobile and American Culture, Wayne State University, Detroit, October 1, 1982). Kenneth T. Jackson, "The Crabgrass Frontier: 150 Years of Suburban Growth in America," in Raymond A. Mohl and James F. Richardson, eds., *The Urban Experience: Themes in American History* (Belmont, Ca., 1973), 196–221. See also, Rose, *Interstate,* 75–79.

8. An excellent recent study of Los Angeles suburbanization is Fred W. Viehe, "Black Gold Suburbs: The Influence of the Extractive Industry on the Suburbanization of Los Angeles, 1890–1930," *Journal of Urban History,* VIII (November 1981), 3–26.

9. United States Bureau of the Census, *Census of Housing, 1950,* Volume I, chapter 1, table 32. Since 1975 downtown Los Angeles has begun to revive, especially as large corporations like Atlantic Richfield, the Bank of America, and Wells Fargo Bank have built major skyscrapers there. More importantly, recent residential complexes have added a 24-hour atmosphere to the once-desolate nighttime scene. The lively Mexican quarter along Broadway also gives diversity to the central business district. In terms of residential density, Los Angeles suburbs are typically more closely packed than the post-World War II automobile suburbs of Eastern cities like Philadelphia, New York, Washington, and Boston. Partly because of extraordinarily high land prices, partly because of the absence of rainfall, partly because of the large amount of undevelopable land, and partly because of a Spanish tradition that emphasizes enlcosed space rather than open lawns, average lot sizes even in exclusive L.A. suburbs like Palos Verdes were less than one-fifth of an acre in 1980. By contrast, equivalently located and priced homes in the eastern cities just mentioned were typically located on at least one-half of an acre and often much more.

10. The best work on the garage is Folke T. Kihlstedt, "The Automobile and the Transformation of the American Home, 1910–1935," *Michigan Historical Quarterly,* XIX (Fall 1980), 555–70. See also, Charles Moore, Gerald Allen, and Donlyn Lyndon, *The Face of Houses* (New York, 1974), 183–87; and J. B. Jackson, "The Domestication of the Garage," *Landscape,* XX (Winter 1976), 10–19.

11. The first book exclusively devoted to the problem of sheltering automobiles was Dorothy and Julian Olney, *The American Home Book of Garages* (Garden City, N.Y., 1931). See also, *New York Times,* October 11, 1984.

12. A 1922 *New York Times* advertisement for the Kindred-McAvoy homes in Long Island City in Queens promised "two 7-room apartments, with 20 windows each and a 4-car garage."

13. Robert Venturi, Denise Scott Brown, and Steven Izenour, *Learning From Las Vegas* (Cambridge, 1972).

14. Paul Lancaster, "The Great American Motel," *American Heritage,* XXXIII (June–July 1982), 100–8. In 1925 Florida alone registered 178 tourist courts.

15. An outstanding study of automotive tourism is Warren James Belasco, *Americans on the Road: From Autocamp to Motel, 1910–1945* (Cambridge, 1979). See also, David L. Lewis, "Sex and the Automobile: From Rumble Seats to Rockin' Vans," *Michigan Historical Quarterly,* XIX (Fall 1980), 518–28; and *This Fabulous Century,* 272–73.

16. Lewis, "Sex and the Automobile," passim.

17. By 1972, when there were approximately 43,500 motels in the United States, there were approximately twice as many motels as hotels in the country. *New York Times,* February 23, 1972.

18. *New York Times,* October 19, 1975; and *New York Times,* July 19, 1981.

19. According to a *New York Times* article on May 30, 1982, Hollinshead opened the drive-in in 1934 on the back wall of his machine-parts shop in Camden, New Jersey.

The number of drive-in theaters peaked at 4,063 in 1958 and was down to 3,484 in 1976, when Texas led all states with 264 and Alaska was last with only one.

20. *New York Times,* November 7, 1983.

21. The best-illustrated and most thorough study of the service station is Daniel I. Vieyra, *Fill'er Up: An Architectural History of America's Gas Stations* (New York, 1979), especially 1–14.

22. Two excellent short essays on the gas station appeared by William K. Stevens and Paul Goldberger in *New York Times,* February 7, 1982. See also, Gary Herbert Wolf, "The Gasoline Station and the Evolution of a Building Type as Illustrated Through a History of the Sun Oil Company Gasoline Stations" (Thesis, University of Virginia, 1974); K. Lonberg-Holm, "The Gasoline Filling and Service Station," *Architectural Record,* LXVII (June 1930), 561–68; *Louisville Courier-Journal,* December 11, 1983; Alexander Guth, "The Automobile Service Station," *The Architectural Forum,* XLV (July 1926), 33–56; Bruce Lohof, "The Service Station in America: The Evolution of a Vernacular Building Type," *Industrial Archeology,* XI (Spring 1974), 1–13; and Henry Ozane, "The Service Station," *Architectural Record,* XCV (February 1944), 70–82.

23. Willard Morgan, "At Last—A Place to Park," 58–60. On the deconcentration policies of Sears, Roebuck and Company, see Arthur Rubloff, "Shopping Center Development and Operation," *The Appraisal Journal,* XXX (1962), 75–77; Boris Emmet and John E. Jeuck, *Catalogues and Counters: A History of Sears, Roebuck and Company* (Chicago, 1950); and Leonard Z. Breen, "A Study of the Decentralization of Retail Trade Relative to Population in the Chicago Area, 1929–1948" (Ph.D. dissertation, University of Chicago, 1956).

24. Blaine A. Brownell, "The Automobile and Urban Structure" (Paper presented at the annual meeting of the American Studies Association, San Antonio, Tx., November 6, 1975); and Brownell, "A Symbol of Modernity: Attitudes Toward the Automobile in Southern Cities in the 1920's," *American Quarterly,* XXIV (March 1972), 20–44. See also, Howard L. Preston, *Automobile Age Atlanta: The Making of a Southern Metropolis, 1900–1935* (Athens, Ga., 1979).

25. J. C. Nichols, 'The Planning and Control of Outlying Shopping Centers," *The Journal of Land and Public Utility Economics,* II (January 1926), 17–22.

26. My list of early shopping centers is partly borrowed from John B. Rae, *The Road and the Car in American Life* (Cambridge, 1971), 230. For a quantitative analysis of the spread of this merchandising concept, see Yehoshua S. Cohen, *Diffusion of an Innovation in an Urban System* (Chicago: The University of Chicago Department of Geography Research Paper No. 140, 1972). See also, James Simmons, *The Changing Pattern of Retail Location* (Chicago: The University of Chicago Department of Geography Research Paper No. 92, 1964).

27. For example, in 1979, the percentage of metropolitan retail trade taking place outside the city was 70 percent in Boston, 67 percent in St. Louis, and 68 percent in Hartford. John C. Van Nostrand, "The Queen Elizabeth Way: Public Utility Versus Public Space," *Urban History Review,* XII (October 1983), 1–23.

28. During the 1979 gasoline shortage, the customer bases of the largest regional malls shrunk from a thirty-mile radius to a ten-mile radius, and weekend and social trips to the malls were also curtailed. On the changing image and size of Tyson's Corner, see Megan Rosenfeld, "Tyson's Corner: An Example of Suburbia's Future," *Washington Post,* February 20, 1977. See also, *New York Times,* November 10, 1981.

29. Quoted in the *New York Times,* February 5, 1971. Although the main shopping and business districts in European areas tend to be in the center of the cities, American-style malls are gaining popularity. Skarholmen Center, southwest of Stockholm, is sur-

rounded by the largest parking lot in Scandinavia and is similar to the regional shopping centers of the United States.

30. Anthony Zube-Jackson, 104–5. See also, *New York Times,* November 10, 1981; Ross J. McKeever, *Shopping Centers: Principles and Policies* (Washington: Urban Land Institute Technical Bulletin No. 20, 1953); Homer Hoyt, "The Current Trends in New Shopping Centers: Four Different Types," *Urban Land,* XII (1953), No. 4; and *New York Times,* December 31, 1982.

31. The published material on the history of mobile homes is very spare. The social aspects of the subject are covered in Donald Olen Cowell, *Mobile Homes: A Study of Trailer Life* (Philadelphia, 1941). The most comprehensive treatment is Carleton M. Edwards, *Homes for Travel and Living: The History and Development of the Recreational Vehicle and Mobile Home Industry* (East Lansing, privately printed, 1977). On the conversion of trucks into residences, see Jane Lidz, *Rolling Homes: Handmade Houses on Wheels* (New York, 1979). See also, *New York Times,* June 27, 1982.

32. Although the practice was already a decade old, beginning in 1969, Congress officially authorized the Federal Housing Administration (FHA) to issue government-insured mortgages for mobile home park sites. In April 1979, the New Jersey Supreme Court ruled that mobile homes were real property and could be taxed as such. An examination of late 1970s attitudes toward trailering and mobile homes is Michael Aaron Rockland, *Homes on Wheels* (New Brunswick, N.J., 1980).

33. A mobile home, unlike a modular structure, has a metal undercarriage that remains even when the house in place on blocks. A modular home can be set up on a site in a few days; a prefab home may take several weeks of on-site labor to put together. Partly for reasons of prejudice and partly for reasons of preference, black Americans occupy only a disproportionately small percentage (2 percent in 1960, for example) of mobile home units.

34. A campaign is presently underway to give landmark status to the earliest existing McDonald's hamburger stand, which is located in Downey, California. The Pointe Coupee Funeral Home in New Roads, Louisiana began the drive-in funeral practice in 1976 with rather substantial fanfare. James G. Huneker, *New Cosmopolis: Book of Images* (New York, 1915), 76–77, 82; Lewis, "Sex and the Automobile," 524; and Paul Hirshorn and Steven Izenour, *White Towers* (Cambridge, 1979); *New York Times,* January 15, 1984.

35. The most complete information about Schuler's drive-in church comes from *The Story of a Dream,* a 28-page booklet available at the church bookstore for one dollar. Free walking tours of the grounds are provided free of charge every half hour. See also, *Decision,* XII (March 1971), 6; Thomas Hines, "Designing for the Motor Age: Richard Neutra and the Automobile," *Oppositions: A Journal for Ideas and Criticism in Architecture,* XXI (Summer 1980), 35–51; and Judith and Neil Morgan, "Orange, a Most California County," *National Geographic,* December 1981, pp. 750–79.

36. Quoted in the *New York Times,* May 30, 1971.

37. Although Santa Clara County was the envy of civic boosters around the United States, in 1979 it instituted measures to curtail industrial development. Susan Benner, "Storm Clouds Over Silicon Valley," *Inc.,* September 1982, pp. 84–89.

38. Both the 1970 and the 1980 federal censuses contain detailed information on the journey-to-work of a selected sample of Americans. See also, Marion Clawson, *Suburban Land Conversion in the United States: An Economic and Governmental Process* (Baltimore, 1971), 232–34.

39. Leon Moses and Harold F. Williamson, "The Location of Economic Activities in Cities," *American Economic Review,* LVII (May 1967), 214–15; Kenneth T. Jack-

son, "The Effect of Suburbanization on the Cities," in Philip C. Dolce, ed., *Suburbia: The American Dream and Dilemna* (Garden City, 1976), 89–110; *New York Times,* August 15, 1981.

40. The most sensible voice on this topic is that of the Regional Plan Association, and especially its vice president for research, Boris Pushkarev. See, for example, Pushkarev, "Transportation Crawling Towards Consolidation," *New York Affairs,* V (1978), 75–90. Two fine scholarly analyses are Peter O. Muller, *The Suburbanization of Corporate Headquarters* (Washington, Conn., 1978); and Barry Bluestone et al., *Corporate Flight* (Washington, 1981).

41. *Intercorp,* February 7, 1984, p. 26; *New York Times,* February 20, 1984. A British study in 1976 found that 800 businesses had moved from London to new and expanded towns between 1966 and 1974. *New York Times,* June 30, 1977.

42. Lewis Mumford, *The City in History: Its Origins, Its Transformations, and Its Prospects* (New York, 1961), 505.

43. John Keats, *The Insolent Chariots* (New York, 1958).

44. Lewis, "Sex and the Automobile," 518–28. The argument that the automobile will continue to dominate urban transportation for the forseeable future is forcefully made by Mark S. Foster, "The Automobile in the Urban Environment: Planning for an Energy Short Future," *The Pacific Historian,* II (Fall 1981), 23–31.

CHAPTER 15: *The Loss of Community*

1. This chapter is based upon an article which appeared earlier as Kenneth T. Jackson, "The Effect of Suburbanization on the Cities," in Philip C. Dolce, ed., *Suburbia: The American Dream and Dilemna* (Garden City, 1976), 89–110.

2. Two good examples of the literature on urban imperialism are Richard C. Wade, *The Urban Frontier: The Rise of Western Cities, 1790–1830* (Cambridge, 1959); and Robert R. Dykstra, *The Cattle Towns* (New York, 1968).

3. Mel Ziegler, ed., *Amen: The Diary of Rabbi Martin Siegel* (New York, 1970), 20; or Mel Ziegler, "Diary of a Suburban Rabbi," *New York Magazine,* IV (January 18, 1971), 24–33.

4. Quoted in Kenneth T. and Barbara B. Jackson, *Two Cities: A Comparison and Analysis of White Plains and Newark* (unpublished manuscript, 1974).

5. Paul A. Stellhorn, "Depression and Decline: Newark, N.J., 1929–1941" (Ph.D. dissertation, Rutgers University, 1983).

6. The New York data are included in TABLE A-4 in the Appendix. Additional journey-to-work information may be found in TABLES A-7, A-8, A-9.

7. See, for example, Brett W. Hawkins, *Nashville Metro: The Politics of City-County Consolidation* (Nashville, 1966).

8. The argument that most problems of American urban government are due to the failure of large cities to expand their boundaries is made most effectively by William G. Colman, *Cities, Suburbs, and States: Governing and Financing Urban America* (New York, 1975).

9. The District of Columbia, which was originally one hundred square miles in area, was reduced to sixty square miles with the loss of the Virginia sections in the nineteenth century.

10. Stellhorn, "Depression and Decline," chapters 2 and 3.

11. *The Wall Street Journal,* October 15, 1982.

12. On life and leisure in an earlier society, see Lawrence Stone, *The Family, Sex, and Marriage in England, 1500–1800* (New York, 1977).

13. Robert C. Wood, *Suburbia: Its People and Their Politics* (Boston, 1958), 107–8.

14. The changing use of spare time in Westchester County is considered in a remarkable volume by George A. Lundberg, et al., *Leisure: A Suburban Study* (New York, 1934). See also, John R. Stilgoe, "The Suburbs," *American Heritage,* XXXV (February-March 1984), 20–36.

15. Philippe Aries, "The Family and the City," in Alice S. Rossi, ed., *The Family* (New York, 1978), 227–35. For an amusing comparison of nineteenth- and twentieth-century standards of comfort, see Allen J. Share, "Good Ol' Summer Time," *Louisville Courier-Journal,* July 25, 1980.

16. "The Front Porch," a late 1983 exhibition at the Craft and Folk Art Museum in Los Angeles evoked the sensation of being on a porch and documented the traditions of the porch throughout America: stoop, portico, gallery, veranda, loggia, dog trot, and arbor. See also, Hugh Stevens, "The Lost Art of Porch-Sitting," *Country Journal,* XI (July 1984), 84–85; and Charles Moore and Gere Kavanaugh, *Home Sweet Home: American Domestic Vernacular Architecture* (New York, 1983).

17. Some medical experts have recently suggested that air-conditioning, by allowing an avoidance of the natural swings in climate, may reduce the average individual's capacity to adapt to stress. See, Robert Friedman, "The Air-Conditioned Century," *American Heritage,* XXXV (August-September 1984), 20–33; and "The Great American Cooling Machine," *Time,* August 13, 1979, p. 75.

18. Home builders discovered in the 1970s that families on a limited budget would prefer eliminating the living room rather than doing without the family room. In the South, contractors finessed the problem by calling the main gathering area the "great room."

19. In the 1946 movie, "The Girl Next Door," June Haver "loved every minute" of life in Scarsdale.

20. Equally pervasive was the advertising medium. The American way of life—dishwashers, bicycles, ovens, lawnmowers, and automobiles—has long been depicted as a suburban way of life.

CHAPTER 16: *Retrospect and Prospect*

1. James Schlesinger, Secretary of Energy, *Consumer Briefing Summary No. 5,* United States Department of Energy, April 27, 1978. Because there are no incorporated municipalities in Baltimore County, it is administered by the County Executive and the seven-member County Council as if it were a city. Agnew grew up in a middle-class suburban area called Forest Park. Arthur Schlesinger, Jr., "The Amazing Success Story of Spiro Who?" *New York Times Magazine,* July 26, 1970, p. 5. See also, *New York Times,* April 29, 1970; and *New York Times,* May 2, 1971.

2. Between 1950 and 1960, 5.8 million persons moved to the suburbs; in the next decade, 4.9 million made the same decision.

3. On the new demographic and economic realities, see David R. Goldfield, "The Limits of Suburban Growth," *Urban Affairs Quarterly,* XII (September 1976), 83–102.

4. The best analysis of this process is James T. Little, Hugh O. Nourse, R. B. Read, and Charles L. Leven, *The Contemporary Neighborhood Succession Process: Lessons*

in the Dynamic of Decay from the St. Louis Experience (St. Louis: Washington University Institute for Urban and Regional Studies, 1975), 36–40.

5. The process of abandonment has been extensively photographed and documented by Camilo J. Vergara and Kenneth T. Jackson in a public exhibition, "Ruins and Revivals: The Architecture of Urban Devastation," which opened at the Urban Center in New York City in September, 1983, and which subsequently traveled to a variety of sites in the Northeast.

6. Robin H. Best, "Urban Growth and Agriculture" (Paper presented to Section MI of the British Association for the Advancement of Science, August 1983). See also, Richard Munton, *London's Green Belt: Containment in Practice* (London, 1983).

7. Using tax assessment figures, William S. Hastings has shown that some Philadelphia blocks were losing both population and value before the Civil War. Usually, however, population declined because high land values drove non-commercial uses from the area. Hastings, "Philadelphia Microcosm," *Pennsylvania Magazine of History and Biography,* XCI (April 1967), 164–80.

8. The pattern of white to black residential succession has been well documented in many American situations. For example, see Richard G. Ford, "Population Succession in Chicago," *American Journal of Sociology,* LVI (September 1950), 156–60; and Paul Frederick Cressey, "Population Succession in Chicago, 1898–1930," *American Journal of Sociology,* XLIV (July 1938), 59–69. Hispanic groups may follow a slightly different pattern according to Terry J. Rosenberg and Robert W. Lake, "Toward a Revised Model of Residential Segregation and Succession: Puerto Ricans in New York, 1960–1970," *American Journal of Sociology,* LXXXI (September 1975), 1142–50. For a British example of neighborhood change, see H. J. Dyos, *Victorian Suburb: A Study of the Growth of Camberwell* (Leicester, 1961), 50–51.

9. Dolores Hayden, *Redesigning the American Dream: The Future of Housing, Work, and Family Life* (New York, 1984), passim; and Robert A. Gross, "Transcendentalism and Urbanism: Concord, Boston, and the Wider World" (Paper presented at biennial conference of the Nordic Association for American Studies, Copenhagen, Denmark, June 27, 1982).

10. Olivier Zunz, *The Changing Face of Inequality: Urbanization, Industrial Development, and Immigrants in Detroit, 1880–1920* (Chicago, 1982) 154–55; Carolyn Tyirin Kirk and Gordon W. Kirk, Jr., "The Impact of the City on Home Ownership: A Comparison of Immigrants and Native Whites at the Turn of the Century," *Journal of Urban History,* VII (August 1981), 471–88; and Daniel D. Luria, "Wealth, Capital, and Power: The Social Meaning of Home Ownership," *Journal of Interdisciplinary History,* VII (August 1976), 277–79.

11. In 1914 average per capita income in the United States was $335, compared with $243 for Great Britain, $185 for France, and $146 for Germany. David M. Potter, *People of Plenty: Economic Abundance and the American Character* (Chicago, 1954) 79–84.

12. Homer Hoyt, "The Influence of Highways and Transportation in the Structure and Growth of Cities and Urban Land Values," in Jean Labatut and Wheaton Lane, eds., *Highways in Our National Life* (Princeton, N.J., 1950), 206; and David Harrison and John F. Kain, "An Historical Model of Urban Form," Harvard University Program on Regional and Urban Economics Discussion Paper 63, 1970.

13. Mark Alan Willis, "The Effects of Cyclical Demand on Industry Structure and on the Rate of Technological Change: An International Comparison of the Housebuilding Sectors in the United States, Great Britain, and France" (Ph.D. dissertation, Yale University, 1979), 75–79; and Kathryn Robertson Murphy, *New Housing and Its Ma-*

terials, 1940–1956 (Washington: Department of Labor Bulletin No. 1231, August 1958), especially the seventeen-page preface.

14. Henry J. Aaron, *Shelter and Subsidies: Who Benefits From Federal Housing Policies?* (Washington: The Brookings Institution, 1972); Calvin Bradford, "Financing Home Ownership: The Federal Role in Neighborhood Decline," *Urban Affairs Quarterly,* XIV (March 1979), 313–35; Anthony Downs, *Federal Housing Subsidies* (Lexington, Mass., 1973); and Cushing L. Dolbeare, "Toward a More Responsive Housing Policy," *City Limits,* VII (February 1982), 18–20.

15. In 1862 deductions for state and local income taxes were allowed; in 1864 deductions were extended to include mortgage interest, repairs, and losses from the sale of lands. Henry J. Aaron, "Income Taxes and Housing," *The American Economic Review,* LX (December 1970), 789–806; Gerald Carson, *The Golden Egg: The Personal Income Tax; Where It Came From, How it Grew* (Boston, 1977); Dan Throop Smith, *Federal Tax Reform: The Issues and a Program* (New York, 1961); and Richard E. Slitor, *The Federal Income Tax in Relation to Housing* (Washington: National Commission on Urban Problems, Research Report No. 5, 1968).

16. West Germany, Canada, Great Britain, and Sweden all offer subsidies for homeownership, but on a much smaller scale than the United States. In Canada, for example, neither mortgage interest nor property taxes are deductible. The tax inducements to homeownership in the United States are so popular that when Presidential candidate George McGovern suggested in 1972 that they be abolished in favor of a lower overall tax rate, he was immediately subjected to scorn and a loss of credibility.

17. David R. Goldfield, "National Urban Policy in Sweden," *APA Journal,* Winter 1982, 24–38; and Anthony Sutcliffe, *Towards the Planned City: Germany, Britain, the United States, and France, 1780–1914* (Oxford, England, 1981).

18. The residential construction industry usually suffers very early in a recession and the loss of jobs in the building trades subsequently becomes an important factor in the deepening of the depression.

19. Herbert J. Gans, "The White Exodus to Suburbia Steps Up," *The New York Times Magazine,* January 7, 1968, pp. 25, 85–97. Already outdated is David L. Birch, *The Economic Future of City and Suburb* (New York: Committee for Economic Development Supplementary Paper Number 30, 1970).

20. Paul Somerson, "There's No (Work) Place Like Home," *PC Magazine,* II (December 1983), 106–33. In 1974 Moshe Safde predicted that in 2024 there would be few motor vehicles, that city centers would regain their regional dominance, that many suburban houses would be demolished, and that highrise life would be typical. Safde, "Beyond the City Limits," *Saturday Review World,* August 24, 1974, pp. 54–57. In 1976 I made a similar prediction about 2076. See, Kenneth T. Jackson, "The Greater City: New York and Its Suburbs, 1876–2076," in Milton Klein, ed., *New York: The Centennial Years, 1676–2076* (New York, 1976), 169–87.

21. Herman E. Koenig, "The Impetus and Nature of Impending Change in Human Settlements," in Koenig and Lawrence M. Sommers, eds., *Energy and the Adaptation of Human Settlements: A Prototype Process in Genesee County, Michigan* (East Lansing: Center for Environmental Quality, Michigan State University, 1980), 1–15; and John Lukacs, *Outgrowing Democracy: A History of the United States in the Twentieth Century* (Garden City, 1984), 318.

22. Joel Darmstadter, Joy Dunkerly, and Jack Alterman, *How Industrial Societies Use Energy: A Comparative Analysis* (Baltimore, 1978); *New York Times,* May 28, 1977; May 28, 1979; September 16, 1979; and October 14, 1980; and *U.S. News and World Report,* July 16, 1979, pp. 37–41.

23. Because of heavy use of mechanized farm equipment, the states with the highest per capita consumption of energy are those with high proportions of agricultural workers. James S. Roberts, *Energy, Land Use and Growth Policy: Implications for Metropolitan Washington,* second edition (Washington: Real Estate Research Corporation and the Metropolitan Washington Council of Governments, 1975); George BaSalla, "The Fallacy of the Energy-Civilization Equation," *Saturday Review,* November 24, 1979, pp. 28–31; and *New York Times,* May 28, 1979.

24. Hittman Associates, Inc., *Residential Energy Consumption: Single-Family Housing, Final Report* (Washington, 1975).

25. FHA reported that from 1946 to 1960, land costs increased by 180 percent while construction costs were rising by only 77 percent. Richard F. Muth, "Urban Residential Land and Housing Markets," in Harvey S. Perloff and Lowden Wingo, Jr., eds., *Issues in Urban Economics* (Baltimore, 1968), 285–333.

26. *New York Times,* February 6, 1983. Willis, "The Effects of Cyclical Demand," 2–3. The home price figures are from the United States Department of Commerce.

27. Thus far, however, black suburbanization represents more of a resegregation of blacks in particular sectors of suburbia than a dispersal in an open housing market. Phillip L. Clay, "The Process of Black Suburbanization," *Urban Affairs Quarterly,* XIV (June 1979), 405–24. See also, Seth Reichlin, "The Aging of the Suburbs," *Fortune,* December 15, 1980, pp. 66–84; and Thomas A. Clark, *Blacks in Suburbs: A National Perspective* (Piscataway, N.J.: The Center for Urban Policy Research, 1979).

28. An excellent local study of this phenomenon is Jeffrey R. Henig, *Gentrification in Adams Morgan: Political and Commercial Consequences of Neighborhood Change* (Washington: George Washington University Center for Washington Area Studies No. 9, 1982). See also, Lukacs, *Outgrowing Democracy,* 193.

29. Thomas Falk, *Urban Sweden: Changes in the Distribution of Population in the 1960's in Focus* (Stockholm: Stockholm School of Economics, 1977); *New York Times,* October 30, 1980; Leo F. Schnore, "On the Spatial Structure of Cities in the Two Americas," in Philip M. Hauser and Leo F. Schnore, eds., *The Study of Urbanization* (New York, 1965), 347–98; and John R. Harris, "Urban and Industrial Deconcentration in Developing Economies: An Analytical Framework," *Regional and Urban Economics,* I (August 1971), 139–52.

30. Stuart W. Leslie, *Boss Kettering: Wizard of General Motors* (New York, 1983), ix–x; and Michael P. Conzen, "Analytical Approaches to the Urban Landscape," in Karl W. Butzer, ed., *Dimension in Urban Geography: Essays on Some Familiar and Neglected Topics* (Chicago: University of Chicago Department of Geography Research Paper No. 186, 1978), 146.

APPENDIX

1. In order to minimize the tendency for an area with a small population to show a higher growth rate than a densely settled areas, the entire county rather than just the urbanized area was used as the statistical base. Significantly, the zones with the smallest population and lowest density, Byberry and Moreland, were never among the growth-rate leaders between 1810 and 1870.

2. In 1850 Kensington, Northern Liberties, Southwark, and Spring Garden were all officially among the nation's largest cities, even though they would become Philadelphia neighborhoods in 1854.

3. Philadelphia Board of Trade, *Twenty-First Annual Report, February 6, 1854*

(Philadelphia, 1854). The thirty suburbs and districts sometimes competed among themselves and with Philadelphia for basic urban amenities, such as water. Nelson Manfred Blake, *Water for the Cities: A History of the Urban Water Supply Problem in the United States* (Syracuse, 1956), 87–89.

4. Between 1900 and 1950, the wave of urban expansion moved outward at the rate of about one mile per decade, and since 1950 has increased to about three miles per decade. Hans Blumenfeld, "The Tidal Wave of Metropolitan Expansion," *Journal of the American Institute of Planners,* XX (Winter 1954), 3–15.

5. "Sketchings," *The Crayon: A Journal Devoted to the Graphic Arts and the Literature Related to Them,* I (January 3, 1855), 11. See also, David Paul Schuyler, "Public Landscapes and American Urban Culture, 1800–1870: Rural Cemeteries, City Parks, and Suburbs" (Ph.D. Dissertation, Columbia University, 1979), chapter 6.

6. Although an occasional small community lost population in the nineteenth century, it was not until after 1940 that entire cities began to peak. The process, however, dates from about 1820.

7. Although this process has continued in Philadelphia in the twentieth century, the city gained population until 1950 because the huge area annexed in 1854 continued to gain on the outer edges.

8. A vapid and superficial analysis of the New York experience is Frederick P. Clark, "Concentration and Decentralization in the New York Metropolitan Region," *Journal of the American Institute of Planners,* XVI (Fall 1970), 172–78.

9. New York *American,* March 22, 1843. Quoted in Edward K. Spann, *The New Metropolis: New York City, 1840–1857* (New York, 1981), Spann, 116.

10. In Chicago the population within two miles of the business center reached its peak in 1890 and subsequently declined. Two careful studies of demography in the Windy City are Homer Hoyt, *One Hundred Years of Land Values in Chicago, 1933* (Chicago, University of Chicago Press); and Chicago Community Inventory, *Growth and Redistribution of the Resident Population in the Chicago Standard Metropolitan Area* (Chicago: Chicago Community Inventory, 1954), 18–19.

11. In Memphis the five innermost wards all lost population between 1890 and 1910 while peripheral wards were registering gains of as much as 400 percent.

12. The innermost area of London lost residence after 1860, as did eleven of its thirty districts after 1881. The administrative County of London began losing population after 1891.

13. For more detail on this point, see Kenneth T. Jackson, "Urban Deconcentration in the Nineteenth Century: A Statistical Inquiry," in Leo F. Schnore, ed., The *New Urban History: Quantitative Expectations By American Historians* (Princeton, 1975), pp. 134–40.

14. The most detailed study of Philadelphia journey-to-work patterns assumes, I think incorrectly, that industrial workers inevitably toiled at the nearest factory that might have required their skills. See Theodore Hershberg, Harold Cox, and Dale Light, Jr., "The Journey-to-Work: An Empirical Analysis of Work, Residence, and Transportation, Philadelphia, 1850–1880," in Theodore Hershberg, ed., *Toward a New Urban History: Urbanization and Industrialization in Nineteenth Century America.*

15. Quoted in Schuyler, "Public Landscapes and American Urban Culture," Chapter VI, p. 10.

16. For roughly this same early twentieth-century period, Allan R. Pred has found that Manhattan blue-collar workers who could best afford the time and money to commute tended to live farther from their workplace than the less-skilled. Pred, *The Spatial Dynamics of Urban Industrial Growth, 1800–1914* (Cambridge, Mass., 1966), 336–38.

17. Three of the better studies of the relationship between transit and residence are James W. Simmons, "Charging Residence in the City: A Review of Intra-Urban Mobility," *Geographical Review,* LVIII (October 1968), 622–51; George M. Smerk, "The Streetcar: Shaper of American Cities," *Traffic Quarterly,* XXI (1967), 569–84; and Olivier Zunz, "Technology and Society in an Urban Environment: The Case of the Third Avenue Elevated Railway," *Journal of Interdisciplinary History,* III (Summer 1972), 89–102.

18. Henry C. Binford, "The Suburban Enterprise: Jacksonian Towns and Boston Commuters," (Ph.D. dissertation, Harvard University, 1976), 120–42.

19. Data on journey-to-work were reviewed more extensively in the chapters on automobility.

Index